T0075613

ARC SCHEMES
AND
SINGULARITIES

ARC SCHEMES
AND
SINGULARITIES

Editors

David Bourqui
Université de Rennes 1, France

Johannes Nicaise
Imperial College London, UK
University of Leuven, Belgium

Julien Sebag
Université de Rennes 1, France

 **World Scientific**

NEW JERSEY · LONDON · SINGAPORE · BEIJING · SHANGHAI · HONG KONG · TAIPEI · CHENNAI · TOKYO

Published by

World Scientific Publishing Europe Ltd.

57 Shelton Street, Covent Garden, London WC2H 9HE

Head office: 5 Toh Tuck Link, Singapore 596224

USA office: 27 Warren Street, Suite 401-402, Hackensack, NJ 07601

Library of Congress Cataloging-in-Publication Data

Names: Bourqui, David, editor. | Nicaise, Johannes, editor. | Sebag, Julien, editor.
Title: Arc schemes and singularities / edited by David Bourqui (Université de Rennes 1, France),
    Johannes Nicaise (Imperial College London, UK & University of Leuven, Belgium),
    Julien Sebag (Université de Rennes 1, France).
Description: New Jersey : World Scientific, 2019. | Includes bibliographical references.
Identifiers: LCCN 2019013758 | ISBN 9781786347190 (hc)
Subjects: LCSH: Curves, Algebraic. | Geometry, Algebraic. | Algebraic spaces. |
    Geometrical constructions. | Singularities (Mathematics)
Classification: LCC QA567 .A5725 2019 | DDC 516.3/5--dc23
LC record available at https://lccn.loc.gov/2019013758

British Library Cataloguing-in-Publication Data
A catalogue record for this book is available from the British Library.

For any available supplementary material, please visit
https://www.worldscientific.com/worldscibooks/10.1142/Q0213#t=suppl

Desk Editors: Dipasri Sardar/Jennifer Brough/Shi Ying Koe

Typeset by Stallion Press
Email: enquiries@stallionpress.com

About the Editors

David Bourqui has been *Maître de conférences* at the University of Rennes 1 since 2004. He works in algebraic geometry and is specifically interested in the geometry of moduli spaces of curves and arc schemes.

Johannes Nicaise is a Professor of Mathematics at the Imperial College London and the University of Leuven. He works on the interactions among non-Archimedean geometry, birational geometry and mirror symmetry. He is the author of over 40 research articles in leading international journals and has authored two books.

Julien Sebag is Professor at the University of Rennes 1 since 2009. His current work deals with algebraic geometry in connection with arithmetic and differential aspects of algebra and geometry. He particularly focuses his mathematical activity on the study of arc schemes and related topics.

Together with their co-author Antoine Chambert-Loir, Johannes Nicaise and Julien Sebag have received the 2017 Ferran Sunyer i Balaguer Prize for their book *Motivic Integration*.

Contents

1

Introduction

David Bourqui[*,§], Johannes Nicaise[†,‡,¶] and Julien Sebag[*,‖]

[*]*Institut de recherche mathématique de Rennes*
UMR 6625 du CNRS, Université de Rennes 1
Campus de Beaulieu, 35042 Rennes cedex, France
[†]*Department of Mathematics, Imperial College,*
South Kensington Campus, London SW72AZ, UK
[‡]*Department of Mathematics, KU Leuven,*
Celestijnenlaan 200B, 3001 Heverlee, Belgium
[§]*david.bourqui@univ-rennes1.fr*
[¶]*j.nicaise@imperial.ac.uk*
[‖]*julien.sebag@univ-rennes1.fr*

This volume contains the proceedings of the conference *Arc Schemes and Singularity Theory* that was held in Rennes in November 2016. The list of participants were as follows:

1. A'CAMPO Norbert
2. AFSHARIJOO Pooneh
3. BELOTTO DA SILVA André
4. BILU Margaret
5. BITOUN Thomas
6. BOURQUI David
7. BOUTHIER Alexis
8. CARADOT Antoine
9. CASALE Guy
10. CASSOU-NOGUÈS Pierrette
11. CAUWBERGS Thomas
12. CELY Jorge
13. CHIU Christopher
14. CLUCKERS Raf
15. DACHS CADEFAU Ferran
16. FANTINI Lorenzo
17. GROS Michel
18. HAIECH Mercedes
19. HALLE Lars Halvard
20. HAUSER Herwig
21. ISHII Shihoko
22. IVORRA Florian
23. KPOGNON Kodjo
24. LANGLOIS Kevin
25. LEJEUNE-JALABERT Monique
26. LEYTON-ÁLVAREZ Maximiliano
27. LIU Wille
28. LOESER François

29. LORAY Frank
30. MAURI Mirko
31. MOUROUGANE Christophe
32. MOURTADA Hussein
33. MUKHTAR Muzammil
34. NARVÁEZ MACARRO Luis
35. NGUYEN Kien
36. NICAISE Johannes
37. PASCUAL ESCUDERO Beatriz
38. PE PEREIRA María

39. QUAREZ Ronan
40. RAIBAUT Michel
41. REGUERA Ana
42. ROCZEN Marko
43. ROND Guillaume
44. SEBAG Julien
45. SHAVIV Ary
46. SUSTRETOV Dmitry
47. TEYMURI GARAKANI Mahdi
48. VEYS Wim

The conference focused on various aspects of the geometry of arc spaces, and, in particular, the Nash problem.

Let V be an algebraic variety over a field k. The *arc scheme* $\mathscr{L}_\infty(V)$ is a k-scheme that parameterizes the formal germs of curves on V called *arcs*. John Nash is probably the first mathematician who observed that the algebraic, geometric and topological properties of the arc scheme $\mathscr{L}_\infty(V)$ are closely related to the geometry of the singularities of V. In his 1968 paper "*Arc Structure of Singularities*" (published in 1995 as [6] in a special volume of the *Duke Mathematical Journal*), Nash proved that for every subvariety W of V, the subset $\mathscr{L}_\infty(V)_W \subset \mathscr{L}_\infty(V)$ corresponding to the arcs on V with a center in W has only finitely many irreducible components, assuming that k has characteristic zero. This is a non-trivial result because the arc scheme $\mathscr{L}_\infty(V)$ is not noetherian if V has a positive dimension. When W is the singular locus of V, Nash constructed an injective map from the set of irreducible components of $\mathscr{L}_\infty(V)_W$ to the set of *essential divisors* of V; these are the divisorial valuations v on the function field of V (up to scaling) such that, for every resolution of singularities $V' \to V$, the center of v on V' is an irreducible component of the exceptional locus. This map is nowadays called the *Nash map*, and the irreducible components of $\mathscr{L}_\infty(V)_W$ are called *Nash families* (Nash used the terminology *families of arcs* instead).

The injectivity of the Nash map implies, in particular, that the number of Nash families is finite.

EXAMPLE. If V is a normal surface, then the essential components are simply the irreducible components of the exceptional locus of its minimal resolution of singularities, which are all of codimension 1. In dimension ≥ 3, on certain resolutions of singularities, the center of an essential divisor may be an irreducible component of the exceptional locus of codimension ≥ 2.

Nash observed that, on a series of two-dimensional examples, the Nash map was not only injective but also surjective. He asked the intriguing question whether this could be true in general. Precisely, he wrote in [6] the following:

> *For surfaces it seems possible that there are exactly as many families of arcs associated with a point as there are components of the image of the point in the minimal resolution of the singularities of the surface.* [...] *In higher dimensions, the arc families associated with the singular set correspond to "essential components" which must appear in the image of the singular set in all resolutions. We do not know how complete is the representation of essential components by arc families.*

The question of the surjectivity of the Nash map is now known as the *Nash problem*.

QUESTION. *Let k be an algebraically closed field of characteristic zero. Is the Nash map surjective?*

Let us note that, in dimension 1, this question has an affirmative answer: it boils down to a one-to-one correspondence between the set of branches of a curve at a singular point and the inverse image of the singular point in the normalization of the curve.

Over the course of several decades, the Nash problem has been a source of motivation and inspiration for geometers all over the world. The question was found to have a positive answer for many interesting classes of singularities (quasi-ordinary hypersurface singularities, toric singularities, rational surface singularities, etc.) and these investigations were an important stimulus for the study of the arc space in general (see, for instance, [7]). In [4], Ishii and Kollár finally proved that question 1 admits a negative answer, by giving counterexamples in dimension ≥ 4. In [1, 5], one can find more extensive lists of counterexamples, starting in dimension 3. For instance, we can mention the following example of [5, Proposition 9]. Let $m \geq 5$ be an odd integer and let X_m be the hypersurface in $\mathbf{A}^4_{\mathbf{C}}$ defined by the equation

$$x_1^2 + x_2^2 + x_3^2 + x_4^m = 0.$$

In this case, there is only one Nash family (i.e., the space of arcs centered in the singular locus is irreducible), but there are two essential divisors. It is interesting to remark that, at the end of [**6**], Nash himself had already suggested to study this hypersurface, but his idea had lain dormant for almost 50 years. The next major breakthrough came in 2011, when Fernández de Bobadilla and Pe Pereira proved that the Nash problem always has a positive answer in dimension 2; the proof was published in [**3**]. Their arguments relied on topological techniques; a more algebraic proof was later given in [**2**].

This settled the Nash problem in all dimensions (yes in dimension ≤ 2, no in dimension ≥ 3). However, it remains a challenging problem to give a suitable interpretation of the image of the Nash map: Which essential divisors are selected? Another question that is still open is: What happens in positive characteristic (assuming resolution of singularities)? Thus, the study of the Nash map is still a very relevant subject. Moreover, the arc space also plays a central role in the theory of *motivic integration*, initiated by M. Kontsevich in 1995, and some key concepts in the study of the Nash map were directly inspired by Denef and Loeser's proof of the change of variables formula for motivic integrals.

Contents of this volume: This book contains the contributions on various topics related to arc spaces, including the following:

- arc spaces and differential algebra;
- the solution of Nash problem in dimension 2;
- the local structure of arc spaces and the theorem of Drinfeld–Grinberg–Kazhdan;
- recent interactions between the theory of motivic zeta functions and the theory of motives;
- dual complexes of resolutions of singularities.

They are meant to give an overview of current research on the geometry of arc spaces and applications in different domains of algebraic geometry.

Bibliography

[1] T. de Fernex, Three-dimensional counter-examples to the Nash problem. *Compos. Math.*, 149(9):1519–1534, 2013.
[2] T. de Fernex and R. Docampo, Terminal valuations and the Nash problem. *Invent. Math.*, 203(1):303–331, 2016.
[3] J. Fernández de Bobadilla and María Pe Pereira, The Nash problem for surfaces. *Ann. of Math.* (2), 176(3):2003–2029, 2012.
[4] S. Ishii and J. Kollár, The Nash problem on arc families of singularities. *Duke Math. J.*, 120(3):601–620, 2003.

[5] J. M. Johnson and J. Kollár, Arc spaces of cA-type singularities. *J. Singul.*, 7:238–252, 2013.

[6] J. F. Nash, Jr., Arc structure of singularities. *Duke Math. J.*, 81(1):31–38 (1996), 1995. A celebration of John F. Nash, Jr.

[7] A. J. Reguera, A curve selection lemma in spaces of arcs and the image of the Nash map. *Compos. Math.*, 142(1):119–130, 2006.

Arc Schemes in Geometry
and Differential Algebra

David Bourqui[*,§], Johannes Nicaise[†,‡,¶] and Julien Sebag[*,‖]

[*]*Institut de recherche mathématique de Rennes*
UMR 6625 du CNRS, Université de Rennes 1
Campus de Beaulieu, 35042 Rennes cedex, France
[†]*Department of Mathematics, Imperial College, South Kensington Campus, London SW72AZ, UK*
[‡]*Department of Mathematics, KU Leuven, Celestijnenlaan 200B, 3001 Heverlee, Belgium*
[§]*david.bourqui@univ-rennes1.fr*
[¶]*j.nicaise@imperial.ac.uk*
[‖]*julien.sebag@univ-rennes1.fr*

For every scheme X over a field k, the *arc scheme* of X is a k-scheme that parameterizes formal arcs on X: for every field extension k' of k, the k'-points of the arc scheme correspond canonically to points on X with values in the ring of formal power series $k'[[t]]$. This construction can be generalized to a relative setting, replacing X by a morphism of schemes. In this chapter, we present three constructions of the arc scheme: one in terms of Weil restrictions, and the two others in terms of differential algebra. Each construction highlights some particular features of the arc scheme. Further results on arc schemes can be found in [A. Chambert-Loir, J. Nicaise and J. Sebag, *Motivic Integration*, Progress in Mathematics, Vol. 325, Birkhäuser, Basel, 2018].

1. Weil Restrictions

Let S be a scheme, and let S' be an S-scheme. In this section, we construct the *Weil restriction with respect to $S' \to S$*. The presentation here is broadly inspired by [2, Chapter 7, §6]. For every scheme T, we denote by $\mathbf{Sch}_T$ the category of schemes over T. We will identify every T-scheme Y with its associated presheaf of points on $\mathbf{Sch}_T$ by means of the Yoneda lemma.

1.1. Let X be an S'-scheme. For every S-scheme Y, we set

$$(1.1.1) \qquad \mathcal{R}_{S'/S}(X)(Y) = \mathrm{Hom}_{\mathbf{Sch}_{S'}}(Y \times_S S', X).$$

For every morphism of S-schemes $f\colon Y' \to Y$, we denote by $\mathcal{R}_{S'/S}(X)(f)$ the map

$$(1.1.2) \qquad \mathcal{R}_{S'/S}(X)(f)\colon \mathcal{R}_{S'/S}(X)(Y) \to \mathcal{R}_{S'/S}(X)(Y')\colon \alpha \mapsto \alpha \circ f_{S'},$$

where $f_{S'}$ is the base change of f to S'. Finally, for every morphism of S'-schemes $g\colon X' \to X$ and every S-scheme Y, we denote by $\mathcal{R}_{S'/S}(g)(Y)$ the map

$$(1.1.3) \qquad \mathcal{R}_{S'/S}(g)(Y)\colon \mathcal{R}_{S'/S}(X')(Y) \to \mathcal{R}_{S'/S}(X)(Y)\colon \beta \mapsto g \circ \beta.$$

These data define a bifunctor

$$\mathcal{R}_{S'/S}\colon \mathbf{Sch}_{S'} \times \mathbf{Sch}_S^{\mathrm{op}} \to \mathbf{Sets}.$$

DEFINITION 1.1.4. Let $h\colon S' \to S$ be a morphism of schemes. Let X be an S'-scheme. The presheaf $\mathcal{R}_{S'/S}(X)$ on $\mathbf{Sch}_S$ is called the *Weil restriction of X with respect to the morphism h*. If $\mathcal{R}_{S'/S}(X)$ is representable by an S-scheme, then this scheme is also called the Weil restriction of X with respect to h.

EXAMPLE 1.1.5. The presheaf $\mathcal{R}_{S'/S}(S')$ is canonically isomorphic to S, and $\mathcal{R}_{S'/S'}(X)$ is canonically isomorphic to X.

1.2. Let T be a scheme, and let $F \to G$ be a morphism of presheaves on $\mathbf{Sch}_T$. We say that $F \to G$ is an open immersion (resp. closed immersion, immersion) if, for every T-scheme Y and every morphism of presheaves $Y \to G$, the fibered product $F \times_G Y$ is representable by a scheme X, and the natural morphism $X \to Y$ is an open immersion (resp. closed immersion, immersion).

PROPOSITION 1.2.1. Let T be a scheme, and let $\alpha\colon F \to G$ be an immersion of presheaves on $\mathbf{Sch}_T$. Then α is a monomorphism. If α is an open immersion, then it is an isomorphism if and only if $\alpha(k)\colon F(\mathrm{Spec}(k)) \to G(\mathrm{Spec}(k))$ is a bijection for every field k and every morphism $\mathrm{Spec}(k) \to T$.

PROOF. Let T' be any scheme over T. By Yoneda's lemma, we can canonically identify $F(T')$ with the set of morphism of functors $T' \to F$, and the analogous statement holds for G. Thus, to prove that $\alpha\colon F(T') \to G(T')$ is injective (resp. bijective) it suffices to show that the map $\mathrm{Hom}(T', F) \to \mathrm{Hom}(T', G)$ defined by composition with α is injective (resp. bijective). For every morphism $\gamma\colon T' \to G$, we can identify the fiber of $\mathrm{Hom}(T', F) \to \mathrm{Hom}(T', G)$ over γ with the set of sections of the morphism $F \times_G T' \to T'$, the base change of α over γ. If α is an immersion, then $F \times_G T' \to T'$ is an immersion of schemes, and thus has at most one section. If α is an open immersion, then $F \times_G T' \to T'$ is an open immersion; thus, it is an isomorphism if and only if it is bijective on the underlying sets of points. $\qquad\square$

PROPOSITION 1.2.2. *Let* $S' \to S$ *be a morphism of schemes, and let* X *be an* S'-*scheme.*

(1) *The functor* $\mathcal{R}_{S'/S}(X)$ *is a sheaf for the Zariski topology on* $\mathbf{Sch}_S$.
(2) *Assume that* $h\colon S' \to S$ *is proper, flat, and of finite presentation. If* $X' \to X$ *is an open immersion (resp. closed immersion, immersion), then so is the morphism of functors* $\mathcal{R}_{S'/S}(X') \to \mathcal{R}_{S'/S}(X)$.

PROOF. Point (1) follows immediately from the definitions. Point (2) is proven in [**2**, Chapter 7, §6, Proposition 2]. $\qquad\square$

From Proposition 1.2.2, we formally deduce the following corollary.

COROLLARY 1.2.3. *Assume that* $h\colon S' \to S$ *is proper, flat, and of finite presentation. Let* S *be a scheme, and let* X *be an* S'-*scheme. Let* $u\colon X' \to X$ *be an immersion (resp. an open immersion, a closed immersion). Assume that the functor* $\mathcal{R}_{S'/S}(X)$ *is representable. Then, the functor* $\mathcal{R}_{S'/S}(X')$ *is representable, and* $\mathcal{R}_{S'/S}(u)$ *is an immersion (resp. an open immersion, a closed immersion).*

The key question is that of the representability of the Weil restrictions. Even in simple cases, the functor $\mathcal{R}_{S'/S}(X)$ may not be representable. The following example is provided in [**3**, Chapter 3, Remark 1.2.4].

EXAMPLE 1.2.4. Let k be an infinite field. We denote by R the localization of $k[T]$ at the ideal (T), and by K is fraction field $k(T)$. Set $S = \mathsf{Spec}(k[T])$ and $S' = \mathsf{Spec}(k)$, and let $S' \to S$ be the zero section. Then, the functor $\mathcal{R}_{S'/S}(\mathbf{A}^1_{S'})$ is not representable. If it were representable by an S-scheme X, then we would have $X(R) = \mathbf{A}^1_k(k) = k$, and $X(K) = \mathbf{A}^1_k(K \otimes_{k[T]} k)$ would be a singleton because $K \otimes_{k[T]} k$ is the trivial ring. This already shows that X cannot be separated. A refinement of this argument, producing an affine open subscheme U of X such that $U(R)$ is still infinite, shows that X does not exist.

We will now establish some sufficient conditions for the representability of $\mathcal{R}_{S'/S}(X)$.

LEMMA 1.2.5. *Let* S, S' *be two affine schemes. Let* $h\colon S' \to S$ *be a finite and free morphism of schemes. Let* X *be an affine* S'-*scheme. Then, the functor* $\mathcal{R}_{S'/S}(X)$ *is representable by an affine* S-*scheme. Moreover, if* the S'-scheme X is of finite type (resp. of finite presentation), then so is the S-scheme $\mathcal{R}_{S'/S}(X)$.

PROOF. We write $S = \mathsf{Spec}(A)$ and $S' = \mathsf{Spec}(B)$. We fix a basis $b_1, \ldots, b_n$ of the A-module B. We write $X = \mathsf{Spec}(B[(x_e)_{e \in E}]/I)$ for some family of indeterminates $(x_e)_{e \in E}$ and some ideal I in the polynomial algebra $B[(x_e)_e]$. For every ring R, we write $\mathbf{A}_R^E$ for the affine space $\mathsf{Spec}(R[(x_e)_{e \in E}])$.

Let $(x_{e,i})_{e,i}$ be a family of indeterminates indexed by $E \times \{1, \ldots, n\}$. For every polynomial $f \in B[(x_e)_{e \in E}]$, let $f_1, \ldots, f_n$ be the polynomials in $A[(x_{e,i})_{e,i}]$ defined by the relation

$$f((x_{e,1}b_1 + x_{e,2}b_2 + \cdots + x_{e,n}b_n)_e) = \sum_{i=1}^{n} f_i b_i.$$

Let J be the ideal in $A[(x_{e,i})_{e,i}]$ generated by the polynomials $f_1, \ldots, f_n$, where f runs over a generating set I_0 of the ideal I. We will prove that $\mathcal{R}_{S'/S}(X)$ is represented by the S-scheme $\mathsf{Spec}(A[(x_{e,i})_{e,i}]/J)$.

Let C be an A-algebra. We use the direct sum decomposition of C-modules

(1.2.6) $$C \otimes_A B \cong \bigoplus_{i=1}^{n} C b_i.$$

Giving a morphism of B-schemes $\mathsf{Spec}(C \otimes_A B) \to X$ is the same thing as choosing a tuple $\gamma = (\gamma_e)_{e \in E}$ of elements in $C \otimes_A B$ such that $f(\gamma) = 0$ for every $f \in I_0$. Following the decomposition (1.2.6), we write $\gamma_e = \sum_{i=1}^{n} \gamma_{e,i} b_i$, with $\gamma_{e,i} \in C$ for every $e \in E$. Then, the condition $f(\gamma) = 0$ can be rewritten as

$$f_1((\gamma_{e,i})_{e,i}) = \cdots = f_n((\gamma_{e,i})_{e,i}) = 0$$

for every $f \in I_0$. This establishes a bijection between the set

$$\mathcal{R}_{S'/S}(X)(C) = X(C \otimes_A B)$$

and the set of C-points on $\mathsf{Spec}(A[(x_{e,i})_{e,i}]/J)$, functorial in C.   □

EXAMPLE 1.2.7. Let $S' \to S$ be a free morphism of rank n between affine schemes, and let E be any set. The proof of Lemma 1.2.5 shows, in particular, that $\mathcal{R}_{S'/S}(\mathbf{A}_{S'}^E)$ is represented by the S-scheme $(\mathbf{A}_S^E)^n$.

THEOREM 1.2.8. *Let $S' \to S$ be a finite and locally free morphism of schemes, and let X be an S'-scheme. Assume that one of the following conditions is satisfied:*

    (1) *For every $s \in S$ and every finite set of points $P \subset X \times_S s$, there is an affine open subscheme U of X containing P.*

    (2) *The morphism $S' \to S$ is a universal homeomorphism.*

Then, the functor $\mathcal{R}_{S'/S}(X)$ is representable by an S-scheme $\mathcal{R}_{S'/S}(X)$.

PROOF. Case (1) is Theorem 4 in [2, Chapter 7, §6]. We will sketch the proof given there, and explain why it also applies to case (2).

One can easily reduce to the case where S and S' are affine schemes, and S' is free of rank n over S. Let $(X_i)_{i \in I}$ be the system of all affine open subschemes of X. From Lemma 1.2.5, we deduce that each functor $\mathcal{R}_{S'/S}(X_i)$ is representable by an affine S-scheme. By Proposition 1.2.2, we know then that the natural morphisms $\mathcal{R}_{S'/S}(X_i) \to \mathcal{R}_{S'/S}(X)$ are open immersions of functors. The gluing data for the open subschemes X_i give rise to gluing data for the S-schemes $\mathcal{R}_{S'/S}(X_i)$. Let us denote by Y the S-scheme obtained by gluing the schemes $\mathcal{R}_{S'/S}(X_i)$. Then, the natural morphism $Y \to \mathcal{R}_{S'/S}(X)$ is an open immersion.

To prove that $Y \to \mathcal{R}_{S'/S}(X)$ is an isomorphism, it is enough to show that $Y(k) \to \mathcal{R}_{S'/S}(X)(k)$ is surjective for every field k and every morphism $\mathsf{Spec}(k) \to S$, by Proposition 1.2.1. This amounts to showing that the image of any S'-morphism $S' \otimes_S k \to X$ is contained in one of the affine open subschemes X_i of X. Since the underlying set of $S' \otimes_S k$ is a finite collection of points, this is guaranteed by the assumption in (1); in case (2), the underlying set of $S' \otimes_S k$ consists of a single point, and the result is obvious. □

2. Jet Schemes

2.1. Let S be a scheme and let X be an S-scheme. For every $n \in \mathbf{N}$, we set $S_n = S \otimes_{\mathbf{Z}} (\mathbf{Z}[t]/(t^{n+1}))$. The morphism $S_n \to S$ is a finite free morphism of schemes, and it is a universal homeomorphism. Thus, it follows from Theorem 1.2.8 that the Weil restriction

$$\mathcal{R}_{S_n/S}(X \times_S S_n) = \mathcal{R}_{S_n/S}(X \otimes_{\mathbf{Z}} (\mathbf{Z}[t]/(t^{n+1})))$$

is representable by an S-scheme.

DEFINITION 2.1.1. The S-scheme

$$\mathscr{L}_n(X/S) = \mathcal{R}_{S_n/S}(X \otimes_{\mathbf{Z}} (\mathbf{Z}[t]/(t^{n+1})))$$

is called the *jet scheme of X over S of level n*.

EXAMPLE 2.1.2. The S-scheme $\mathscr{L}_0(X/S)$ is canonically isomorphic to X, and $\mathscr{L}_1(X/S)$ is canonically isomorphic to the tangent scheme $\mathsf{Spec}(\mathrm{Sym}(\Omega^1_{X/S}))$.

2.2. By the functorial properties of the Weil restriction, $\mathscr{L}_n(\cdot/S)$ defines an endofunctor on the category $\mathbf{Sch}_S$ of S-schemes. For all S-schemes X and Y, we have a natural bijection

$$\mathrm{Hom}_{\mathbf{Sch}_S}(Y, \mathscr{L}_n(X/S)) \to \mathrm{Hom}_{\mathbf{Sch}_S}(Y \times_S S_n, X).$$

Thus, $\mathscr{L}_n(\cdot/S)$ is right adjoint to the functor

$$\mathbf{Sch}_S \to \mathbf{Sch}_S \colon Y \mapsto Y \times_S S_n.$$

2.3. Let $m, n \in \mathbf{N}$ be non-negative integers, with $n \geq m$. The closed immersion of schemes $S_m \to S_n$ defined by reduction modulo t^{m+1} induces a morphism of functors

$$\mathcal{R}_{S_n/S}(X \otimes_{\mathbf{Z}} \mathbf{Z}[t]/(t^{n+1})) \to \mathcal{R}_{S_m/S}(X \otimes_{\mathbf{Z}} \mathbf{Z}[t]/(t^{m+1})).$$

This map corresponds to a morphism of S-schemes

$$\theta_{m,X}^n : \mathscr{L}_n(X/S) \to \mathscr{L}_m(X/S),$$

called the *truncation morphism from level n to level m*. It is functorial in X. In particular, for $m = 0$, we get a natural truncation morphism

$$\theta_{0,X}^n : \mathscr{L}_n(X/S) \to \mathscr{L}_0(X/S) = X$$

endowing $\mathscr{L}_n(X/S)$ with the structure of an X-scheme. For all non-negative integers $p \geq n \geq m$, we have

$$\theta_{m,X}^p = \theta_{m,X}^n \circ \theta_{n,X}^p.$$

2.4. In a similar way, for every n in $\mathbf{N}$, the projection $S_n \to S$ induces a morphism

$$s_{n,X} : X \to \mathscr{L}_n(X/S)$$

which is a section of the truncation morphism $\theta_{0,X}^n : \mathscr{L}_n(X/S) \to X$. If follows that the morphism $\theta_{0,X}^n$ is surjective for every $n \in \mathbf{N}$. Let us also note that

$$s_{m,X} = \theta_{m,X}^n \circ s_{n,X}$$

for all non-negative integers m, n with $n \geq m$.

EXAMPLE 2.4.1. The jet schemes and truncation morphisms are easy to describe explicitly when X and S are affine. Set $S = \mathsf{Spec}(R)$ and $X = \mathsf{Spec}(A)$ for some R-algebra A. We write A as $R[(x_e)_{e \in E}]/I$ for some set E and some ideal I. Let I_0 be a set of generators for I. Let n be a non-negative integer. For every element f of I_0, we develop the expression $f((x_{e,0}+x_{e,1}t+ \cdots + x_{e,n}t^n)_{e \in E})$ as a polynomial in t with coefficients in $R[(x_{e,i})_{e,i}]$, and we reduce the result modulo t^{n+1}:

$$f((x_{e,0} + x_{e,1}t + \cdots + x_{e,n}t^n)_{e \in E})$$
$$\equiv f_0((x_{e,0})_e) + f_1((x_{e,0}, x_{e,1})_e)t + \cdots + f_n((x_{e,i})_{e,i})t^n$$

modulo t^{n+1}. Then, $\mathscr{L}_n(X/S)$ is the spectrum of the R-algebra

$$R[(x_{e,i})_{e \in E, i=0,\ldots,n}]/(f_0, \ldots, f_n)_{f \in I_0}$$

and the truncation morphism $\theta_{m,X}^n$ is given by projection onto the coordinates $x_{e,i}$ with $i \leq m$.

2.5. Let $T \to S$ be a morphism of schemes, and let Y be a T-scheme. For every non-negative integer n and every T-scheme Z, the natural morphism $Z \times_T T_n \to Z \times_S S_n$ is an isomorphism. This yields a canonical forgetful map

$$\mathrm{Hom}_{\mathbf{Sch}_T}(Z \times_T T_n, Y) \to \mathrm{Hom}_{\mathbf{Sch}_S}(Z \times_S S_n, Y)$$

and thus a canonical morphism of S-schemes

$$\mathscr{L}_n(Y/T) \to \mathscr{L}_n(Y/S).$$

Setting $Y = X \times_S T$, we obtain a canonical morphism of T-schemes

$$\mathscr{L}_n(X \times_S T/T) \to \mathscr{L}_n(X/S) \times_S T$$

and this is an isomorphism because, for every T-scheme Z, the map

$$\mathrm{Hom}_{\mathbf{Sch}_T}(Z \times_T T_n, X \times_S T) \to \mathrm{Hom}_{\mathbf{Sch}_S}(Z \times_S S_n, X)$$

is a bijection. Thus, the formation of jet schemes commutes with base change.

LEMMA 2.5.1. *Let $X \to S$ be a morphism of schemes. Let T be an open subscheme of S, and let $Y \to X \times_S T$ be an étale morphism. Then, the natural morphism*

$$\mathscr{L}_n(Y/T) \to \mathscr{L}_n(X/S) \times_X Y$$

is an isomorphism for every $n \geq 0$.

PROOF. Since the ideal sheaf of the closed immersion $S \to S_n$ is nilpotent, this follows from the infinitesimal lifting criterion for étale morphisms. $\qquad\square$

THEOREM 2.5.2. *Let $X \to S$ be a morphism of schemes, and let $n \geq m$ be non-negative integers.*

(1) *If the schemes X and S are affine, then so is the S-scheme $\mathscr{L}_n(X/S)$.*

(2) *The truncation morphism $\theta^n_{m,X} \colon \mathscr{L}_n(X/S) \to \mathscr{L}_m(X/S)$ is affine.*

(3) *If X is locally of finite type (resp. locally of finite presentation) over S, then $\mathscr{L}_n(X/S)$ is of finite type (resp. of finite presentation) over X.*

(4) *If $f \colon X' \to X$ is an open immersion (resp. a closed immersion, resp. an immersion) of S-schemes, then so is the morphism $\mathscr{L}_n(f)$.*

(5) *Assume that $X \to S$ is smooth of pure relative dimension d. Then, locally over X, the truncation morphism $\mathscr{L}_n(X/S) \to \mathscr{L}_m(X/S)$ is a trivial fibration with fiber $\mathbf{A}_S^{d(n-m)}$.*

PROOF. Point (1) follows from Lemma 1.2.5. For points (2) and (3), we may assume that X and S are affine by Lemma 2.5.1; then the result follows again from Lemma 1.2.5. Point (4) is a direct consequence of Proposition 1.2.2. Finally, for point (5), Lemma 2.5.1 allows us to reduce to the case where S is affine and $X = \mathbf{A}_S^d$; then $\mathscr{L}_n(X/S) = \mathbf{A}_X^{d(n+1)}$ for every

$n \geq 0$ by Example 1.2.7, and the truncation morphisms $\theta_{m,X}^n$ are linear projections. $\qquad\square$

COROLLARY 2.5.3. *Let $X \to S$ be a morphism of schemes. Then, the morphism $s_{n,X}\colon X \to \mathscr{L}_n(X/S)$ is a closed immersion, for every $n \geq 0$.*

PROOF. Since the morphism $\theta_{0,X}^n$ is affine, and thus separated, any section of this morphism is a closed immersion, by [**5**, 5.4.6]. $\qquad\square$

Theorem 2.5.2(5) explains why jet schemes are useful in the study of singularities: we can detect and describe singularities by studying how $\mathscr{L}_n(X)$ deviates from being isomorphic to $\mathbf{A}_X^{d(n+1)}$, locally over X.

3. Arc Schemes

3.1. Let $X \to S$ be a morphism of schemes. The family of jet schemes $\mathscr{L}_n(X/S)$, endowed with the transition morphisms $(\theta_{m,X}^n)_{n \geq m}$, forms a projective system of S-schemes. We will construct the arc scheme $\mathscr{L}_\infty(X/S)$ as a limit of these jet schemes.

PROPOSITION 3.1.1. *The projective system of S-schemes $(\mathscr{L}_n(X/S))_{n \in \mathbf{N}}$ admits a limit $\varprojlim_{n \in \mathbf{N}} \mathscr{L}_n(X/S)$ in the category $\mathbf{Sch}_S$.*

PROOF. We know by Theorem 2.5.2(2) that the transition morphisms $(\theta_{m,X}^n)_{n \geq m}$ are affine; thus, the existence from the limit follows from [**6**, 8.2.3]. $\qquad\square$

We denote the S-scheme $\varprojlim_{n \in \mathbf{N}} \mathscr{L}_n(X/S)$ by $\mathscr{L}_\infty(X/S)$.

DEFINITION 3.1.2. *The S-scheme $\mathscr{L}_\infty(X/S)$ is called the arc scheme of X over S. A point of $\mathscr{L}_\infty(X/S)$ is called an arc of X over S.*

EXAMPLE 3.1.3. By Example 1.1.5, we know that the S-scheme $\mathscr{L}_n(S/S)$ is canonically isomorphic to S. It follows that $\mathscr{L}_\infty(S/S) \cong S$, as well.

3.2. By the functorial properties of the jet schemes $\mathscr{L}_n(X/S)$, the assignment $(X \mapsto \mathscr{L}_\infty(X/S))$ defines an endofunctor on the category $\mathbf{Sch}_S$. Since the formation of jet schemes commutes with base change, the same is true for the arc scheme. By its construction as a projective limit of schemes with affine transition morphisms, the arc scheme is endowed with an affine morphism of S-schemes

$$\theta_{n,X}^\infty \colon \mathscr{L}_\infty(X/S) \to \mathscr{L}_n(X/S),$$

for every $n \geq 0$, called the *truncation morphism of level n*. For all non-negative integers $n \geq m$, the following diagram of morphisms of S-schemes commutes:

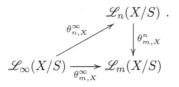

The truncation morphism $\theta^\infty_{n,X}$ is often simply denoted by $\theta_{n,X}$.

3.3. By passing to the limit, the closed immersions $s_{n,X}\colon X \to \mathscr{L}_n(X)$ induce a closed immersion

$$s_{\infty,X}\colon X \to \mathscr{L}_\infty(X/S)$$

such that $\theta^\infty_{n,X} \circ s_{\infty,X} = s_{n,X}$ for every $n \geq 0$. In particular, $s_{\infty,X}$ is a section of $\theta^\infty_{0,X}$. The subset $s_{\infty,X}(X)$ of $\mathscr{L}_\infty(X/S)$ is called the set of the *constant arcs* on X.

EXAMPLE 3.3.1. The arc scheme and truncation morphisms are again easy to describe explicitly when X and S are affine. Set $S = \mathsf{Spec}(R)$ and $X = \mathsf{Spec}(A)$ for some R-algebra A. We write A as $R[(x_e)_{e \in E}]/I$ for some set E and some ideal I. Let I_0 be a set of generators for I. For every element f of I_0, we develop the expression $f((x_{e,0} + x_{e,1}t + \cdots + x_{e,n}t^n + \cdots)_{e \in E})$ as a power series in t with coefficients in $R[(x_{e,i})_{e,i}]$:

$$f\left(\left(\sum_{i \geq 0} x_{e,i}t^i\right)_{e \in E}\right) = \sum_{j \geq 0} f_j((x_{e,i})_{e \in E, i \leq j})t^j.$$

Then, $\mathscr{L}_\infty(X/S)$ is the spectrum of the R-algebra

$$R[x_{e,i} \,|\, e \in E,\, i \in \mathbf{N}]/(f_j)_{f \in I_0,\, j \geq 0}$$

and the truncation morphism $\theta^\infty_{n,X}$ is given by projection onto the coordinates $x_{e,i}$ with $i \leq n$.

3.4. So how can we interpret the points on $\mathscr{L}_\infty(X/S)$ with values in B, for any ring B and any morphism of schemes $\mathsf{Spec}(B) \to S$? By the definition of the arc scheme, we have

$$\mathscr{L}_\infty(X/S)(B) = \varinjlim_{n \in \mathbf{N}} \mathsf{Hom}_S(\mathsf{Spec}(B[t]/(t^{n+1})), X)$$

and this is precisely the set of morphisms of *formal S-schemes* $\mathsf{Spf}\, B[[t]] \to X$, where $B[[t]]$ carries the t-adic topology, and $\mathcal{O}_X$ the discrete topology. When X is affine, then any such morphism is the t-adic completion of a morphism of *algebraic S-schemes* $\mathsf{Spec}(A[[t]]) \to X$. A deep theorem of Bhatt, generalizing algebraization results by Grothendieck to the non-noetherian setting, guarantees that this is still true when $X \to S$ is only assumed to be quasi-compact and quasi-separated. This leads to the following theorem.

THEOREM 3.4.1 (Bhatt [1]). *Let $X \to S$ be a quasi-separated and quasi-compact morphism of schemes. Then, the arc scheme $\mathscr{L}_\infty(X/S)$ represents the presheaf*

$$\mathbf{Aff}_S^{op} \to \mathbf{Sets}\colon \mathrm{Spec}(B) \to \mathrm{Hom}_S(\mathrm{Spec}(B[[t]]), X)$$

on the category affine S-schemes.

To conclude this section, we summarize various useful properties of arc schemes, which can be deduced from the corresponding properties of jet schemes.

THEOREM 3.4.2. *Let $X \to S$ be a morphism of schemes. We have the following properties*:

(1) *If the schemes X and S are affine, then so is the S-scheme $\mathscr{L}_\infty(X/S)$.*

(2) *The truncation morphism $\theta_{n,X}\colon \mathscr{L}_\infty(X/S) \to \mathscr{L}_n(X/S)$ is affine, for every $n \geq 0$.*

(3) *If $f\colon X' \to X$ is an open immersion (resp. a closed immersion, resp. an immersion) of S-schemes, then so is the morphism of S-schemes $\mathscr{L}_\infty(f)$.*

(4) *Assume that $X \to S$ is smooth of pure relative dimension d. Then, locally over X, the truncation morphism $\mathscr{L}_\infty(X/S) \to \mathscr{L}_n(X/S)$ is a trivial fibration with fiber $\mathbf{A}_S^{\mathbf{N}}$, for every $n \in \mathbf{N}$.*

Beware that $\mathscr{L}_\infty(X/S)$ is rarely of finite type over S; for instance, if $X = \mathbf{A}_S^1$, then $\mathscr{L}_\infty(X/S) = \mathbf{A}_S^{\mathbf{N}}$ (see Example 3.3.1).

4. Some Brief Reminders on Differential Algebra

In this section, motivated by the applications to the descriptions of arc/jet schemes, we present some of the basics of differential algebra. The key notion is that of *derivations* of R-algebras.

4.1. Let R be a ring. Let A be an R-algebra. An *R-derivation* of A with values in an A-module M is an R-linear map $D\colon A \to M$ that satisfies the Leibniz rule

$$(4.1.1) \qquad D(aa') = a \cdot D(a') + a' \cdot D(a)$$

for all elements a, a' in A. Then, $D(1^2) = 2D(1)$ so that $D(1) = 0$, and, since D is R-linear, $D(r \cdot 1) = 0$ for every $r \in R$. It follows immediately from the definition that a derivation D is completely determined by its values on a set of generators for the R-algebra A.

4.2. The set of the R-derivations of A with values in M is denoted by $\mathrm{Der}_R(A, M)$. It is a sub-A-module of the A-module M^A of maps from A to M. Its neutral element is the map defined by $a \mapsto 0$ and is called the *trivial R-derivation*. We denote by $\mathrm{Der}_R(A)$ the A-module of the R-derivations of A to itself. It can be identified with the dual of the A-module $\Omega^1_{A/R}$ of

Kähler differentials of A over R. More generally, for every A-module M, the canonical A-linear map

$$\operatorname{Hom}_{\mathbf{Mod}_A}(\Omega^1_{A/R}, M) \to \operatorname{Der}_R(A, M),$$

defined by $\varphi \mapsto \varphi \circ d_{A/R}$, is an isomorphism, where $d_{A/R} \colon A \to \Omega^1_{A/R}$ is the exterior differential; thus, $d_{A/R}$ is the universal derivation of A over R.

4.3. For every $D \in \operatorname{Der}_R(A)$ and every $n \in \mathbf{N}$, one denotes by $D^{(n)}$ the nth power of the map D with respect to the composition of maps. Repeatedly applying the Leibniz rule, one sees that

$$(4.3.1) \qquad D^{(n)}(aa') = \sum_{i=0}^{n} \binom{n}{i} D^{(i)}(a) D^{(n-i)}(a')$$

for all elements a, a' in A.

DEFINITION 4.3.2. A *differential ring* (A, D_A) is a ring A endowed with a derivation $D \in \operatorname{Der}_{\mathbf{Z}}(A)$.

EXAMPLE 4.3.3. Every ring can be viewed as a differential ring endowed with the trivial derivation.

4.4. A ring morphism $f \colon A \to B$ between two differential rings (A, D_A), (B, D_B) is called a *differential morphism* if it satisfies the formula

$$f(D_A(a)) = D_B(f(a))$$

for every element $a \in A$. A differential ring (A, D_A), endowed with a differential morphism from a differential ring (R, D_R) to (A, D_A), is called a *differential* (R, D_R)*-algebra* (or more simply a *differential R-algebra* if the derivation D_R is clear from the context). The differential morphism $(R, D_R) \to (A, D_A)$ is then called the *structural morphism* of the differential (R, D_R)-algebra (A, D_A). If A is an R-algebra and D_A is an element of $\operatorname{Der}_{\mathbf{Z}}(A)$, then (A, D_A) is a differential algebra over the ring R endowed with the trivial derivation if and only if D_A is a R-derivation.

4.5. If (A, D_A), (B, D_B) are two differential (R, D_R)-algebras, then a *differential morphism of differential* (R, D_R)*-algebras* f from (A, D_A) to (B, D_B) is a commutative diagram of differential morphisms:

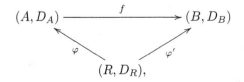

$$(A, D_A) \xrightarrow{\;\;f\;\;} (B, D_B)$$
$$\varphi \searrow \qquad \swarrow \varphi'$$
$$(R, D_R),$$

where the differential morphisms φ, φ' are the structural morphisms of the differential (R, D_R)-algebras (A, D_A) and (B, D_B).

EXAMPLE 4.5.1. Let (R, D_R) be a differential ring and let $(x_e)_{e \in E}$ be a family of indeterminates. Then, the polynomial R-algebra $R[(x_{e,n})_{e \in E, n \in \mathbf{N}}]$ has a unique structure of a differential (R, D_R)-algebra such that $D(x_{e,n}) = x_{e,n+1}$ for every $e \in E$ and every non-negative integer $n \in \mathbf{N}$. This differential (R, D_R)-algebra is denoted by $R\{(x_e)_{e \in E}\}$ and is called the *differential R-algebra obtained by adjunction of the differential indeterminates* $(x_e)_{e \in E}$. Note that $R\{(x_e)_{e \in E}\}$ has a natural $R[(x_e)_{e \in E}]$-algebra structure, given by $x_e \mapsto x_{e,0}$.

DEFINITION 4.5.2. Let (A, D_A) be a differential ring. An ideal I of the ring A is called a *differential ideal* of A if $D_A(i)$ lies in I for every $i \in I$.

4.6. Let (A, D_A) be a differential ring. A subset I in A is a differential ideal if and only if there exists a surjective differential morphism of differential rings $f \colon (A, D_A) \to (B, D_B)$ whose kernel is I. In this case, the differential ring (B, D_B) can be identified with the ring A/I endowed with the derivation defined by

$$D_B(\pi(a)) = \pi(D_A(a))$$

for every element $a \in A$, where $\pi \colon A \to A/I$ is the projection morphism.

4.7. Let (A, D_A) be a differential ring. Let S be a subset of A. The intersection of all the differential ideals of A that contain S is still a differential ideal; it is the smallest differential ideal of (A, D_A) containing S. We call it the *differential ideal generated by* S and denote it by $[S]$.

4.8. The notion of derivation can be extended to that of *higher derivations* or *Hasse–Schmidt derivations*. This generalization in particular provides a suitable framework to treat rings of positive characteristic. To uniformize notation, if m is a non-negative integer then we write $\{m, \dots, \infty\}$ for the set $\{n \in \mathbf{N}, n \geq m\}$.

Let R be a ring and let A, B be R-algebras. Let $n \in \mathbf{N} \cup \{\infty\}$. A *higher R-derivation of order* n from A to B over R is a sequence $H = (H_i)_{i \in \{0, \dots, n\}}$ of morphisms of R-modules $H_i \colon A \to B$ such that $H_0(1) = 1$ and

$$(4.8.1) \qquad\qquad H_i(aa') = \sum_{j=0}^{i} H_{i-j}(a) H_j(a')$$

for all a, a' in A and every i in $\{0, \dots, n\}$. These axioms imply that $H_0 \colon A \to B$ is an morphism of R-algebras, H_1 is an R-derivation from A over R to the A-algebra B, and $H_i(r \cdot 1_A) = 0$ for every $r \in R$ and every i in $\{1, \dots, n\}$. Note that, for every $m \in \{0, \dots, n\}$, the truncated family $(H_j)_{j \in \{0, \dots, m\}}$ is still a higher R-derivation (of order m).

We denote the set of higher R-derivations of order n from A to B by $\mathrm{Der}_R^n(A, B)$. We also write $\mathrm{Der}_R^n(A)$ for the set of higher R-derivations of order n from A to itself satisfying $H_0 = \operatorname{Id}_A$. When we speak about a higher R-derivation H on A, it will always be tacitly assumed that $H_0 = \operatorname{Id}_A$.

4.9. The relation between jets (or arcs) and higher derivations is rooted in the following simple observation. Let R be a ring and denote by $\mathbf{Alg}_R$ the category of R-algebras. Let $n \in \mathbf{N}$, and let A, B be R-algebras. With each element $(H_i)_{i \in \{0,\dots,n\}} \in \operatorname{Der}_R^n(A, B)$, one can associate the map

$$A \to B[t]/(t^{n+1})\colon a \mapsto \sum_{i=0}^{n} H_i(a)t^i.$$

The definition of a higher derivation readily implies that this is a morphism of R-algebras, and that this construction defines a bijection

$$\operatorname{Der}_R^n(A, B) \xrightarrow{\sim} \operatorname{Hom}_{\mathbf{Alg}_R}(A, B[t]/(t^{n+1})).$$

Note that with our conventions, the set $\operatorname{Der}_R^n(A)$ then identifies with the set of those $f \in \operatorname{Hom}_{\mathbf{Alg}_R}(A, A[t]/(t^{n+1}))$ such that the composition with the quotient by (t) is Id_A.

Likewise, for $n = \infty$, we obtain bijections

$$\operatorname{Der}_R^\infty(A, B) \xrightarrow{\sim} \operatorname{Hom}_{\mathbf{Alg}_R}(A, B[[t]])$$

which identify $\operatorname{Der}_R^\infty(A)$ with the set of morphisms in $\operatorname{Hom}_{\mathbf{Alg}_R}(A, A[[t]])$ whose composition with the reduction modulo (t) is Id_A.

In the sequel, we shall frequently make use of the above identifications without further notification.

DEFINITION 4.9.1. Let R be a ring and let A be an R-algebra. A higher derivation $(H_i)_{i \in \mathbf{N}}$ in $\operatorname{Der}_R^\infty(A)$ is called *iterative* if for all $i, j \in \mathbf{N}$, one has

$$H_i \circ H_j = \binom{i+j}{i} H_{i+j}.$$

4.10. We can reformulate Definition 4.9.1 in the following way: an element $H = (H_i)_{i \in \mathbf{N}}$ of

$$\operatorname{Der}_R^\infty(A) \subset \operatorname{Hom}_{\mathbf{Alg}_R}(A, A[[t]])$$

is iterative if and only if the diagram

$$
\begin{array}{ccc}
A & \xrightarrow{\ \ H\ \ } & A[[t]] \\[2pt]
{\scriptstyle H}\big\downarrow & & \big\downarrow{\scriptstyle t \mapsto t+u} \\[2pt]
A[[t]] & \xrightarrow{\ H[[t]]\ } & A[[t, u]]
\end{array}
$$

commutes, where we denote by $H[[t]]$ the morphism

$$A[[t]] \to A[[t, u]], \quad \sum_{i \in \mathbf{N}} a_i t^i \mapsto \sum_{i \in \mathbf{N}} \left(\sum_{j \in \mathbf{N}} H_j(a_i) u^j \right) t^i.$$

EXAMPLE 4.10.1. Let R be a ring that contains $\mathbf{Q}$, and let A be an R-algebra. Let D be a derivation in $\mathrm{Der}_R(A)$. For every $i \geq 0$, we set

$$(4.10.2) \qquad\qquad H_i = \frac{D^{(i)}}{i!}.$$

Then, the family $(H_i)_{i \geq 0}$ is an iterative higher derivation of order ∞. Indeed, for all elements a, a' in A and every $m \in \mathbf{N}$, we have by (4.8.1) that

$$(m!)H_m(aa') = \sum_{i=0}^{m} \binom{m}{i} D^{(i)}(a) D^{(m-i)}(a')$$

$$= (m!) \sum_{i=0}^{m} \left(D^{(i)}(a)/i! \right) \left(D^{(m-i)}(a')/(m-i)! \right)$$

$$= (m!) \sum_{i=0}^{m} H_i(a) H_{m-i}(a').$$

Thus, $(H_i)_{i \geq 0}$ is an element of $\mathrm{Der}_R^{\infty}(A)$, and it is obvious from (4.10.2) that it is iterative. The higher derivation $(H_i)_{i \geq 0}$ is called the *formal flow* of the derivation D and denoted by $\exp(D)$. The map $D \mapsto \exp(D)$ defines a bijection from the set of derivations $\mathrm{Der}_R(A)$ to the set of iterative higher derivations H in $\mathrm{Der}_R^{\infty}(A)$. The inverse of this map is given by $H \mapsto H_1$. Thus, if R contains $\mathbf{Q}$, giving an iterative higher derivation on A over R is the same thing as giving an ordinary derivation on A over R. This may be expressed as an equivalence of categories (see Section 5.8). In general, the two notions are different: there exist derivations that cannot be extended to a higher derivation (iterative or not) or that can be extended to more than one iterative higher derivation.

4.11. Let R be a ring, let A and B be R-algebras, and let

$$H \in \mathrm{Hom}_{\mathbf{Alg}_R}(A, B[[t]])$$

be a higher R-derivation from A to B of order ∞. Let I be an ideal of A. We denote by $[I]_H$ the ideal of B generated by the elements $H_i(a)$ for $i \in \mathbf{N}$ and $a \in I$. Then, the composition of H with the projection morphism

$$B[[t]] \to (B/[I]_H)[[t]]$$

factors through a morphism

$$\widetilde{H} \in \mathrm{Hom}_{\mathbf{Alg}_R}(A/I, (B/[I]_H)[[t]]),$$

i.e., a higher R-derivation from A to $B/[I]_H$ of order ∞.

Assume now that $H \in \mathrm{Der}_R^{\infty}(A)$ and $[I]_H = I$, or equivalently $H(I) \subset I[[t]]$. Then, $\widetilde{H}$ is an element of $\mathrm{Der}_R^{\infty}(A/I)$. Now, consider the following diagram:

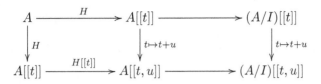

where the left-hand square is defined in Section 4.10 and the horizontal morphisms reduce the coefficients of power series modulo I. The right-hand square is obviously commutative. If H is iterative, the left-hand square is also commutative, and we conclude that $\widetilde{H}$ is iterative.

5. Adjunction Formulas in Differential Algebra

In this section, we prove various adjunction formulas in the context of differential algebra. These will be used in Section 6 to give local descriptions of arc/jet schemes, providing an alternative viewpoint on the local equations of arc/jet schemes given in Examples 2.4.1 and 3.3.1. This paves the way for the use of techniques and results from differential algebra in the study of the geometry of arc schemes.

5.1. Let (R, D_R) be a differential ring. We denote by $\mathbf{DAlg}_{(R, D_R)}$ (or $\mathbf{DAlg}_R$ when D_R is the trivial derivation) the category of differential (R, D_R)-algebras, whose objects are the differential R-algebras and whose morphisms are the differential morphisms of differential R-algebras. We have a forgetful functor

$$\text{For}\colon \mathbf{DAlg}_{(R, D_R)} \to \mathbf{Alg}_R$$

to the category of R-algebras. We will now prove that this functor has a left adjoint.

THEOREM 5.1.1. *Let (R, D_R) be a differential ring. Let A be an R-algebra. Then, there exists an A-algebra A_∞, endowed with a derivation δ_∞ making A_∞ a differential (R, D_R)-algebra, which satisfies the following universal property: given an (R, D_R)-differential algebra (B, D_B), and a morphism of R-algebras $f\colon A \to B$, there exists a unique differential morphism of differential (R, D_R)-algebras*

$$f_\infty\colon (A_\infty, \delta_\infty) \to (B, D_B)$$

that makes the following diagram of morphisms of R-algebras commute:

$$\begin{CD} A @>f>> B \\ @VVV \nearrow \\ A_\infty \end{CD}$$

PROOF (sketch, see also [**4**]). When A is a polynomial R-algebra $R[(x_e)_{e \in E}]$, then one can take for $(A_\infty, \delta_\infty)$ the differential R-algebra $R\{(x_e)_{e \in E}\}$ obtained by adjunction of differential indeterminates (see Example 4.5.1); in particular, for every e and i, one has $\delta_\infty(x_{e,i}) = x_{e,i+1}$. If (B, D_B) is an object of $\mathbf{DAlg}_{(R,D_R)}$ and $f \in \mathrm{Hom}_{\mathbf{Alg}_R}(A, B)$, for every $e \in E$ and $n \in \mathbf{N}$, set $f_\infty(x_{e,i}) = D_B^{(i)}(f(x_e))$. In general, A can be written as a quotient $R[(x_e)_{e \in E}]/I$, and one can set $A_\infty = R\{(x_e)_{e \in E}\}/[I]$, where $[I]$ is the differential ideal of $R\{(x_e)_{e \in E}\}$ generated by the image of I in $R\{(x_e)_{e \in E}\}$ (or equivalently by the image of a set of generators of I in $R\{(x_e)_{e \in E}\}$). $\square$

5.2. By means of Theorem 5.1.1, we can define a functor

$$(\cdot)_\infty \colon \mathbf{Alg}_R \to \mathbf{DAlg}_{(R,D_R)} \colon A \mapsto (A_\infty, \delta_\infty)$$

from the category of R-algebras to the category of differential (R, D_R)-algebras such that for every R-algebra A, and every differential (R, D_R)-algebra (B, D_B), the composition with the structural morphism $A \to A_\infty$ defines a bijection

$$\mathrm{Hom}_{\mathbf{DAlg}_{(R,D_R)}}((A_\infty, \delta_\infty), (B, D_B)) \to \mathrm{Hom}_{\mathbf{Alg}_R}(A, \mathrm{For}((B, D_B))).$$

In other words, the functor $(\cdot)_\infty$ is left adjoint to the forgetful functor

$$\mathrm{For} \colon \mathbf{DAlg}_{(R,D_R)} \to \mathbf{Alg}_R.$$

5.3. Let R be a ring, and let A be an R-algebra. Let ∂_t be the unique A-derivation on $A[[t]]$ such that $\partial_t t = 1$. Then, $(A[[t]], \partial_t)$ is a differential R-algebra, where we endow R with the trivial derivation. Moreover, for every morphism of R-algebras $f \colon A \to B$, $f[[t]] \colon A[[t]] \to B[[t]]$ is a morphism of differential R-algebras, and the assignments

(5.3.1) $A \mapsto (A[[t]], \partial_t), \quad f \in \mathrm{Hom}_{\mathbf{Alg}_R}(A, B) \mapsto f[[t]]$

define a functor $\mathbf{Alg}_R \to \mathbf{DAlg}_R$.

PROPOSITION 5.3.2 (Proposition 3.12 in [**7**]). *Let R be a $\mathbf{Q}$-algebra endowed with the trivial derivation. Then, the functor (5.3.1) is right adjoint to the forgetful functor* For$\colon \mathbf{DAlg}_R \to \mathbf{Alg}_R$.

If R is not assumed to contain $\mathbf{Q}$, we will prove an analogous statement in Proposition 5.9.1 using the theory of higher derivations.

PROOF. Let (A, D_A) be a differential R-algebra and B be an R-algebra. Let $\varphi \colon A \to B$ be a morphism of R-algebras. For every $a \in A$, we set

$$\psi(a) = \sum_{i \geq 0} \frac{\varphi(D_A^{(i)}(a))}{i!} t^i \in B[[t]].$$

By (4.3.1), this defines a morphism of R-algebras $\psi\colon A \to B[[t]]$. Moreover,

$$\partial_t \left(\sum_{i\geq 0} \frac{\varphi(D_A^{(i)}(a))}{i!} t^i \right) = \sum_{i\geq 0} \frac{\varphi(D_A^{(i+1)}(a))}{i!} t^i = \psi(D_A(a))$$

so that ψ is a differential morphism of differential R-algebras.

Now, one checks by direct computation that the mapping $\varphi \mapsto \psi$ defines a bijection

$$\mathrm{Hom}_{\mathbf{Alg}_R}(A, B) \to \mathrm{Hom}_{\mathbf{DAlg}_R}((A, D_A), (B[[t]], \partial_t))$$

with inverse given by reduction modulo t. $\qquad\square$

5.4. We will now extend the constructions of the left and right adjoints of the functor For to the setting of higher derivations. Let R be a ring. We denote by $\mathbf{HDAlg}_R^\infty$ the category of R-algebras with higher derivations of order ∞, whose objects are the pairs (A, H_A), where A is an R-algebra and H_A is a higher R-derivation on A of order ∞, and whose morphisms from (A, H_A) to (B, H_B) are the R-algebra morphisms $f\colon A \to B$ such that $H_{B,i}(f(a)) = f(H_{A,i}(a))$ for every $a \in A$ and every $i \in \mathbf{N}$. This condition on morphisms is equivalent to the commutativity of the following diagram:

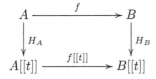

5.5. Let R be a ring, let A be an R-algebra, let H be a higher derivation in $\mathrm{Der}_R^\infty(A)$, and let I be an ideal of A be such that $H(I) \subset I[[t]]$. We denote by $\pi\colon A \to A/I$ the quotient morphism, and by $\widetilde{H} \in \mathrm{Der}_R^\infty(A/I)$ the higher derivation constructed in Section 4.11. By construction, the following diagram is commutative:

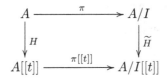

This shows that π defines a morphism in $\mathrm{Hom}_{\mathbf{HDAlg}_R^\infty}((A, H), (A/I, \widetilde{H}))$ and that for every object $(B, H_B) \in \mathbf{HDAlg}_R^\infty$, one has an identification

$$\mathrm{Hom}_{\mathbf{HDAlg}_R^\infty}((A/I, \widetilde{H}), (B, H_B)) = \{f \in \mathrm{Hom}_{\mathbf{HDAlg}_R^\infty}((A, H), (B, H_B)),$$
$$I \subset \mathrm{Ker}(f)\}$$

given by $g \mapsto g \circ \pi$.

5.6. We denote by $\mathbf{IHDAlg}_R^\infty$ the full subcategory of $\mathbf{HDAlg}_R^\infty$ whose objects are the pairs (A, H_A) such that H_A is iterative (see Definition 4.9.1).

The following theorem is the analog of Theorem 5.1.1 for higher derivations.

THEOREM 5.6.1. *Let R be a ring and let A be an R-algebra. Then, there exists an A-algebra $A_{(\infty)}$, endowed with an iterative higher R-derivation $H_{(\infty)}$ of order ∞, that satisfies the following universal property: given an object (B, H_B) of $\mathbf{IHDAlg}_R^\infty$ and a morphism of R-algebras $f \colon A \to B$, there exists a unique element $f_{(\infty)} \in \mathrm{Hom}_{\mathbf{IHDAlg}_R^\infty}((A_{(\infty)}, H_{(\infty)}), (B, H_B))$ that makes the diagram*

commute.

PROOF. First assume that A is a polynomial R-algebra $R[(x_e)_{e \in E}]$. Let $A_{(\infty)}$ be the R-algebra

$$R[x_{e,i} \,|\, e \in E, i \in \mathbf{N}].$$

We endow $A_{(\infty)}$ with the unique higher R-derivation $H_{(\infty)}$ of order ∞ that satisfies

$$(5.6.2) \qquad H_{(\infty),j}(x_{e,i}) = \binom{i+j}{j} x_{e,i+j}$$

for all $i, j \in \mathbf{N}$ and all $e \in E$. It is easily seen that $H_{(\infty)}$ is iterative (see also Remark 5.6.3). The morphism

$$\iota \colon A \to A_{(\infty)}, \ x_e \mapsto x_{e,0}$$

endows $A_{(\infty)}$ with the structure of an A-algebra.

Now, let (B, H_B) be an object of $\mathbf{IHDAlg}_R^\infty$, and let $f \colon A \to B$ be a morphism of R-algebras, we seek

$$f_{(\infty)} \in \mathrm{Hom}_{\mathbf{IHDAlg}_R^\infty}((A_{(\infty)}, H_{(\infty)}), (B, H_B))$$

such that $f_{(\infty)} \circ \iota = f$. Necessarily, for every $e \in E$ and every $i \in \mathbf{N}$, one must have $f_{(\infty)}(x_{e,i}) = H_{B,i}(f(x_e))$. Using the iterativity of H_B, one checks that this formula indeed defines a morphism in the category $\mathbf{IHDAlg}_R^\infty$. Thus, $f \mapsto f_{(\infty)}$ induces a bijection

$$\mathrm{Hom}_{\mathbf{Alg}_R}(A, B) \xrightarrow{\sim} \mathrm{Hom}_{\mathbf{IHDAlg}_R^\infty}((A_{(\infty)}, H_{(\infty)}), (B, H_B))$$

which is easily seen to be functorial in B.

In general, A can be written as a quotient $R[(x_e)_{e \in E}]/I$. One sets

$$A_{(\infty)} = R[x_{e,i} \,|\, e \in E, i \in \mathbf{N}]/[I]_{(\infty)},$$

where $[I]_{(\infty)}$ is the ideal generated by the set

$$\{H_{(\infty),i}(f) \mid f \in I, \, i \in \mathbf{N}\}.$$

Note that by the Leibniz rule, if I_0 is a set of generators for I, we could replace the condition $f \in I$ by $f \in I_0$ in the definition of $[I]_{(\infty)}$.

By iterativity of H_∞, the image $H_{(\infty)}([I]_{(\infty)})$ of $I_{(\infty)}$ in

$$R[x_{e,i} \mid e \in E, \, i \in \mathbf{N}][[t]]$$

is contained in the ideal consisting of series with coefficients in $[I]_{(\infty)}$. Thus, $H_{(\infty)}$ induces an iterative higher derivation of order ∞ on $A_{(\infty)}$, by Section 4.11. It follows from Section 5.5 that the pair $(A_{(\infty)}, H_{(\infty)})$ has the required universal property. $\qquad\square$

REMARK 5.6.3. Let $n \in \mathbf{N} \cup \{\infty\}$ and let A be the polynomial R-algebra $R[(x_e)_{e \in E}]$. For every $e \in E$, one has $H_{(\infty)}(x_e) = \sum_{i \geq 0} x_{e,i} t^i$. Retaining the notation of Examples 2.4.1 and 3.3.1, this shows that for every f in A and every $i \in \mathbf{N}$, one has $H_{(\infty),i}(f) = f_i$. Note in particular that for every $i \in \mathbf{N}$, one has $H_{(\infty),i}(f) \in R[x_{e,i} \mid e \in E, j \in \{0, \ldots, i\}]$.

Now assume that R is a $\mathbf{Q}$-algebra. On $A_\infty = R[(x_{e,i})_{e \in E, i \in \mathbf{N}}]$ we consider the derivation $\widetilde{\delta}_\infty$ defined by $\widetilde{\delta}_\infty(x_{e,i}) = (i+1)x_{e,i+1}$ for $e \in E$ and $i \in \mathbf{N}$ (compare with δ_∞ in the proof of Theorem 5.1.1). Then, $H_{(\infty)}$ is nothing but the formal flow $\exp(\widetilde{\delta}_\infty)$ associated with $\widetilde{\delta}_\infty$. We shall see that this observation can be useful also for computations over an arbitrary base ring R (see Section 6.4). In some sense, this is because the formal flow of $\widetilde{\delta}_\infty$ is "defined over $\mathbf{Z}$", which is not the case for δ_∞. On the other hand, the map $x_{e,i} \mapsto \frac{x_{e,i}}{i!}$ defines an automorphism of the $\mathbf{Q}$-algebra A_∞ transforming δ_∞ into $\widetilde{\delta}_\infty$.

5.7. By means of Theorem 5.6.1, we can define a functor

$$(\cdot)_{(\infty)} \colon \mathbf{Alg}_R \to \mathbf{IHDAlg}_R^\infty \colon A \mapsto (A_{(\infty)}, H_{(\infty)})$$

from the category of R-algebras to the category of R-algebras with iterative higher derivations of order ∞ This functor is left adjoint to the forgetful functor

$$\mathrm{For} \colon \mathbf{IHDAlg}_R^\infty \to \mathbf{Alg}_R.$$

In Section 7.6, we shall see another construction of the left adjoint, based on the notion of universal higher derivation.

5.8. Assume that R is a $\mathbf{Q}$-algebra. Then, the construction of the formal flow $\exp(\delta)$ in Example 4.10.1 gives rise to a functor

$$\mathbf{DAlg}_R \to \mathbf{IHDAlg}_R^\infty, \, (A, \delta) \mapsto (A, \exp(\delta)).$$

If (A, δ_A) and (B, δ_B) are objects of $\mathbf{DAlg}_R$, and $\phi \colon A \to B$ is a morphism of R-algebras, then the conditions

$$f \circ \frac{\delta_A^{(i)}}{i!} = \frac{\delta_B^{(i)}}{i!} \circ f \quad \text{for all } i \geq 1$$

are easily seen to be equivalent to the single condition $f \circ \delta_A = \delta_B \circ f$. Moreover, if (C, H) is an object of $\mathbf{IHDAlg}_R^\infty$, then it follows directly from the definitions of an iterative derivation and of the formal flow that $H = \exp(H_1)$. We conclude that the formal flow functor

$$\mathbf{DAlg}_R \to \mathbf{IHDAlg}_R^\infty$$

defines an equivalence of categories, and the forgetful functor

$$\mathbf{IHDAlg}_R^\infty \to \mathbf{DAlg}_R, (A, H) \mapsto (A, H_1)$$

is a quasi-inverse. Thus, if A is an R-algebra, then $(A_{(\infty)}, H_{(\infty)})$ and $(A_\infty, \exp(\delta_\infty))$ are naturally isomorphic. When A is a polynomial R-algebra $R[(x_e)_{e \in E}]$, the canonical isomorphism is given by

$$x_{e,i} \mapsto \frac{x_{e,i}}{i!}$$

(see Remark 5.6.3).

5.9. Finally, we construct a right adjoint to For: $\mathbf{IHDAlg}_R^\infty \to \mathbf{Alg}_R$. Let A be an R-algebra. Recall from Section 4.9 that we may identify a higher R-derivation H_A on A of order ∞ with an element of $\mathrm{Hom}_{\mathbf{Alg}_R}(A, A[[t]])$ such that the composition with the reduction modulo t is the identity on A. We define a higher derivation $H_{A,u}$ on $A[[u]]$ by means of the unique morphism of A-algebras

$$H_{A,u} \colon A[[u]] \to A[[t, u]]$$

that maps u to $t + u$. By Section 4.10, the commutativity of the diagram

$$
\begin{array}{ccc}
A[[u]] & \xrightarrow{\ t \mapsto t+u\ } & A[[t, u]] \\
{\scriptstyle t \mapsto t+u}\downarrow & & \downarrow{\scriptstyle t \mapsto t+v} \\
A[[t, u]] & \xrightarrow{\ u \mapsto u+v\ } & A[[t, u, v]]
\end{array}
$$

implies that $H_{A,u}$ is iterative.

For every morphism of R-algebras $f \colon A \to B$, the morphism

$$f[[u]] \colon A[[u]] \to B[[u]], \sum_{i \in \mathbf{N}} a_i u^i \mapsto \sum_{i \in \mathbf{N}} f(a_i) u^i$$

induces a morphism

$$(A[[u]], H_{A,u}) \to (B[[u]], H_{B,u}))$$

in the category $\mathbf{IHDAlg}_R^\infty$. This defines a functor

$$\mathrm{HD}^\infty \colon \mathbf{Alg}_R \to \mathbf{IHDAlg}_R^\infty \colon A \mapsto (A[[u]], H_{A,u}).$$

The following result is the analog of Proposition 5.3.2 for higher derivations.

PROPOSITION 5.9.1. Let R be a ring. Let (A, H_A) be an object in $\mathbf{IHDAlg}_R^\infty$. Let B be an R-algebra. For every morphism of R-algebras $f \colon A \to B$, there exists a unique morphism of R-algebras $\tilde{f} \colon A \to B[[u]]$ whose composition with the reduction modulo u equals f, and such that the following diagram commutes:

(5.9.2)
$$
\begin{array}{ccc}
A & \xrightarrow{\ \tilde{f}\ } & B[[u]] \\
{\scriptstyle H_A}\downarrow & & \downarrow{\scriptstyle H_{B,u}} \\
A[[t]] & \xrightarrow[\tilde{f}[[t]]]{} & B[[t,u]]
\end{array}
$$

Here, the morphism $\tilde{f}[[t]]$ is defined by applying $\tilde{f}$ to the coefficients of power series in $A[[t]]$.

Thus, the functor HD^∞ is right adjoint to the forgetful functor For: $\mathbf{IHDAlg}_R^\infty \to \mathbf{Alg}_R$.

PROOF. We first prove that $\tilde{f}$ is unique if it exists. Let a be an element of A, and set

$$\varphi_a(t,u) = \sum_{i \in \mathbf{N}} \tilde{f}(H_{A,i}(a))t^i.$$

Since the reduction of $\tilde{f}$ modulo u is equal to f, we know that

$$\varphi_a(t,0) = \sum_{i \in \mathbf{N}} f(H_{A,i}(a))t^i.$$

On the other hand, the commutativity of diagram (5.9.2) implies that

$$\tilde{f}(a) = \varphi_a(t,0)|_{t \mapsto u}.$$

Thus, if $\tilde{f}$ exists, then it is given by

(5.9.3)
$$\tilde{f} \colon A \mapsto B[[u]], \quad a \mapsto \sum_{i \in \mathbf{N}} f(H_{A,i}(a))u^i.$$

To prove the existence of $\tilde{f}$, we take the expression (5.9.3) as a definition, and we check that it satisfies the required properties. It is clear that $\tilde{f}$ is a morphism of R-algebras, and that its reduction modulo u equals f. Now, consider the diagram

$$
\begin{array}{ccccc}
A & \xrightarrow{\ H_A\ } & A[[t]] & \xrightarrow{\hspace{2cm}} & B[[u]] \\
{\scriptstyle H_A}\downarrow & & {\scriptstyle t \mapsto t+u}\downarrow & & \downarrow{\scriptstyle H_{B,u} \,:\, u \mapsto t+u} \\
A[[t]] & \xrightarrow[H_A[[t]]]{} & A[[t,u]] & \xrightarrow{\hspace{2cm}} & B[[t,u]]
\end{array}
$$

The left-hand square is the square from Section 4.10; thus, it is commutative because H_A is iterative. The horizontal morphisms in the right-hand square are defined by applying f to the coefficients of power series, and substituting t by u in the top row; then the right-hand square is obviously commutative, as well. The composition of both squares is precisely the diagram (5.9.2), and we conclude that this diagram is commutative. $\qquad\square$

6. Algebro-differential Description of Jet/Arc Schemes

In this section, we show that the tools of differential algebra introduced in the previous sections can be used to provide local descriptions of jet schemes and arc schemes.

6.1. We first consider the case of arc schemes.

PROPOSITION 6.1.1. *Let R be a ring, and let A be an R-algebra. We set $S = \mathsf{Spec}(R)$ and $X = \mathsf{Spec}(A)$. Then, the arc scheme $\mathscr{L}_\infty(X/S)$ is isomorphic to $\mathsf{Spec}(A_{(\infty)})$ over X, where $A_{(\infty)}$ is the A-algebra constructed in Theorem 5.6.1.*

PROOF. Let us show that $A_{(\infty)}$ represents the covariant functor of arcs

$$\mathbf{Alg}_R \to \mathbf{Sets}\colon B \mapsto \mathrm{Hom}_{\mathbf{Alg}_R}(A, B[[t]]).$$

Let B be an R-algebra. We can write $B[[t]]$ as $(\mathrm{For} \circ \mathrm{HD}^\infty)(B)$, where

$$\mathrm{For}\colon \mathbf{IHDAlg}_R^\infty \to \mathbf{Alg}_R$$

is the forgetful functor and HD^∞ is the functor defined in Section 5.9. Since $(\cdot)_{(\infty)}$ is left adjoint to For by Theorem 5.6.1, and For is left adjoint to HD^∞ by Proposition 5.9.1, we have canonical bijections

$$\mathrm{Hom}_{\mathbf{Alg}_R}(A, B[[t]]) \xrightarrow{\sim} \mathrm{Hom}_{\mathbf{IHDAlg}_R^\infty}((A_{(\infty)}, H_{(\infty)}), \mathrm{HD}^\infty(B))$$

$$\xrightarrow{\sim} \mathrm{Hom}_{\mathbf{Alg}_R}(A_{(\infty)}, B).$$

This shows that $A_{(\infty)}$ is R-isomorphic to the algebra of functions of $\mathscr{L}_\infty(X/S)$. But since the composition of the above bijections commutes with the maps to $\mathrm{Hom}_{\mathbf{Alg}_R}(A, B)$ induced by reduction modulo t (resp. the structural morphism $A \to A_{(\infty)}$), the algebra $A_{(\infty)}$ is even A-isomorphic to the algebra of functions of $\mathscr{L}_\infty(X/S)$. $\qquad\square$

COROLLARY 6.1.2. *Assume that R contains $\mathbf{Q}$. Then, the arc scheme $\mathscr{L}_\infty(X/S)$ is isomorphic to $\mathsf{Spec}(A_\infty)$ over X, where A_∞ is the A-algebra constructed in Theorem 5.1.1.*

PROOF. If R contains $\mathbf{Q}$, then the A-algebras A_∞ and $A_{(\infty)}$ are canonically isomorphic (see Section 5.8). Alternatively, one can mimic the proof of Proposition 6.1.1, using the adjunction statements in Theorem 5.1.1 and Proposition 5.3.2. $\qquad\square$

6.2. Proposition 6.1.1 and Corollary 6.1.2 provide a very convenient way to write down equations for arc schemes of affine schemes. Let us make this explicit. First, assume that R is a ring that contains $\mathbf{Q}$. Let $A = R[(x_e)_{e \in E}]/I$ be an R-algebra, and let I_0 be a set of generators for the ideal I. Then, $\mathcal{L}_\infty(X/S)$ is the closed subscheme of

$$\mathbf{A}_R^{E \times \mathbf{N}} = \mathsf{Spec}(R\{(x_e)_{e \in E}\})$$

defined by the equations $\delta_\infty^{(i)}(f) = 0$, where i runs through the set $\mathbf{N}$ and f runs through the set of generators I_0. Thus, we find equations for the arc space by repeatedly applying the derivation δ_∞ to the equations for $\mathsf{Spec}(A)$.

EXAMPLE 6.2.1. Let R be a ring that contains $\mathbf{Q}$, let $S = \mathsf{Spec}(R)$, and let X be the S-scheme $\mathsf{Spec}(R[x,y]/(f))$ with $f = x^3 - y^2$. Then, $\mathcal{L}_\infty(X/S)$ is the closed subscheme of

$$\mathsf{Spec}(R[x_i, y_i \mid i \in \mathbf{N}])$$

defined by the equations

$$\begin{cases} f = x_0^3 - y_0^2 = 0, \\ \delta_\infty(f) = 3x_0^2 x_1 - 2y_0 y_1 = 0, \\ \delta_\infty^{(2)}(f) = 6x_0 x_1^2 + 3x_0^2 x_2 - 2y_1^2 - 2y_0 y_2 = 0, \\ \delta_\infty^{(3)}(f) = 3x_0^2 x_3 + 18x_0 x_1 x_2 + 6x_1^3 - 2y_0 y_3 - 6y_1 y_2 = 0, \\ \vdots \end{cases}$$

6.3. Now, let R be an arbitrary ring. Let $A = R[(x_e)_{e \in E}]/I$ be an R-algebra, and let I_0 be a set of generators for the ideal I. Then, by Proposition 6.1.1, the arc scheme $\mathcal{L}_\infty(X/S)$ is the closed subscheme of

$$\mathbf{A}_R^{E \times \mathbf{N}} = \mathsf{Spec}(R[(x_{e,i})_{e \in E, i \in \mathbf{N}}])$$

defined by the equations $H_{(\infty),i}(f) = 0$, where i runs through the set $\mathbf{N}$ and f runs through the set of generators I_0. In other words, we find equations for the arc space by applying the higher derivation $H_{(\infty)}$ to the equations for $\mathsf{Spec}(A)$.

EXAMPLE 6.3.1. Let R be a ring of characteristic three, let $S = \mathsf{Spec}(R)$, and let X be the S-scheme $\mathsf{Spec}(R[x,y]/(f))$ with $f = x^3 - y^2$. Then, $\mathcal{L}_\infty(X/S)$ is the closed subscheme of

$$\mathsf{Spec}(R[x_i, y_i \mid i \in \mathbf{N}])$$

defined by the equations given by the coefficients of $H_{(\infty)}(x^3 - y^2)$. Note that $H_{(\infty)}(x^3 - y^2) = H_{(\infty)}(x)^3 - H_\infty(y)^2$ and

$$H_{(\infty)}(x)^3 = \left(\sum_{i=0}^{\infty} x_i t^i \right)^3 = \sum_{i=0}^{\infty} x_i^3 t^{3i}$$

$$H_{(\infty)}(y)^2 = \left(\sum_{i=0}^{\infty} y_i t^i \right)^2 = \sum_{i=0}^{\infty} \left(\sum_{j+k=i} y_j y_k \right) t^i.$$

Thus, the defining ideal of $\mathscr{L}_\infty(X/S)$ is generated by the set

$$\left\{ x_i^3 - \sum_{j+k=3i} y_j y_k \right\}_{i \in \mathbf{N}} \cup \left\{ \sum_{j+k=i} y_j y_k \right\}_{i \in \mathbf{N} \backslash 3\mathbf{N}}$$

$$= \{ x_0^3 - y_0^2, 2y_0 y_1, 2y_0 y_2 + y_1^2, x_1^3 - 2y_0 y_3 - 2y_1 y_2, \dots \}$$

6.4. Of course, in practice, there is no difference between the procedure described in Example 3.3.1 for computing the local equations of arc schemes and the procedure described in Example 6.3.1: this is explained by Remark 5.6.3.

Note, however, that in general the coefficients of $H_{(\infty)}$ of an element of $R[(x_e)_{e \in E}]$ can be computed by a procedure akin to the repeated application of δ_∞ when R is a $\mathbf{Q}$-algebra. Let us explain this procedure. First note that since $H_{(\infty),i}$ is a morphism of R-modules, it is enough to compute $H_{(\infty),i}(f)$ when $f = \prod_{e \in E} x_e^{\nu_e}$ is a monomial. Denote by φ the unique ring morphism $\mathbf{Z} \to R$ as well as its natural extension $\mathbf{Z}[(x_{e,i})_{e \in E, i \in \mathbf{N}}] \to R[(x_{e,i})_{e \in E, i \in \mathbf{N}}]$.

It follows from (4.8.1) and (5.6.2) that

$$H_{(\infty),i}(f) = \varphi(H_{(\infty),i}(f))$$

where the latter occurrence of $H_{(\infty),i}(f)$ is computed in $\mathbf{Z}[(x_{e,i})]$.

Now, on $\mathbf{Q}[(x_{e,i})]$, one considers the $\mathbf{Q}$-derivation $\widetilde{\delta}_\infty$ defined by $\widetilde{\delta}_\infty(x_{e,i}) = (i+1)x_{e,i+1}$ for $e \in E$ and $i \in \mathbf{N}$. By Remark 5.6.3, one knows that $H_{(\infty),i}(f) = \frac{\widetilde{\delta}_\infty^{(i)}(f)}{i!}$. This gives a practical way of computing $H_{(\infty),i}(f)$ by repeatedly applying $\widetilde{\delta}_\infty^{(i)}$, dividing by $i!$ and mapping the result to the ring R.

EXAMPLE 6.4.1. Retain the notation in Example 6.3.1, setting for simplicity $\delta := \widetilde{\delta}_{(\infty)}$ and $H := H_{(\infty)}$. Then (working in $\mathbf{Z}[x, y]$), one has

$$\delta(x_0^3) = 3x_0^2 x_1, \quad \delta^{(2)}(x_0^3) = 6x_0^2 x_2 + 6x_0 x_1^2,$$

$$\delta^{(3)}(x_0^3) = 18x_0^2 x_3 + 36x_0 x_1 x_2 + 6x_1^3.$$

Thus

$$H_1(x_0^3) = 3x_0^2 x_1, \quad H_2(x_0^3) = 3x_0^2 x_2 + 3x_0 x_1^2, \quad H_3(x_0^3) = 3x_0^2 x_3 + 6x_0 x_1 x_2 + x_1^3$$

and

$$\delta(y_0^2) = 2y_0 y_1, \quad \delta^{(2)}(y_0^2) = 4y_0 y_2 + 2y_1^2, \quad \delta^{(3)}(y_0^2) = 12y_0 y_3 + 12y_1 y_2;$$

thus

$$H_1(y_0^2) = 2y_0y_1, \quad H_2(y_0^2) = 2y_0y_2 + y_1^2, \quad H_3(y_0^2) = 2y_0y_3 + 2y_1y_2.$$

In particular, in $R[x, y]$, where R is a ring of characteristic three, one has

$$H_1(x_0^3) = H_2(x_0^3) = 0, \quad H_3(x_0^3) = x_1^3$$

and we recover the first coefficients computed in Example 6.3.1.

6.5. We will now extend Proposition 6.1.1 to jet schemes. Let R be an arbitrary ring, and let $A = R[(x_e)_{e \in E}]/I$ be an R-algebra. For every integer $n \geq 0$, we set

$$A_{(n)} := \frac{R[x_{e,i} \mid e \in E, i \in \{0, \ldots, n\}]}{(H_{(\infty),i}(f))_{f \in I, i \in \{0, \ldots, n\}}}.$$

Note that by the Leibniz rule, if I_0 is a set of generators for I, we could replace the condition $f \in I$ by $f \in I_0$ in the above definition. We endow $A_{(n)}$ with the A-algebra structure induced by $x_e \mapsto x_{e,0}$.

PROPOSITION 6.5.1. *Let R be a ring, and let A be an R-algebra. We set $S = \mathsf{Spec}(R)$ and $X = \mathsf{Spec}(A)$. Let $n \in \mathbf{N}$. Then, the jet scheme $\mathscr{L}_n(X/S)$ is isomorphic to $\mathsf{Spec}(A_{(n)})$ over X.*

Thus, we find equations for the jet space of level n associated with $\mathsf{Spec}(A)$ by computing the $n + 1$ first coefficients of the higher derivation $H_{(\infty)}$ applied to equations for $\mathsf{Spec}(A)$. In particular, the procedure of Section 6.4 still applies in this context.

PROOF. Retaining the notation of Example 2.4.1, for every f in A and every $i \in \{0, \ldots, n\}$, one has $H_{(\infty),i}(f) = f_i$. Now, the result is an immediate consequence of Example 2.4.1 $\qquad \square$

From Proposition 6.5.1 and Remark 5.6.3, one obtains the following corollary.

COROLLARY 6.5.2. *Let R be a $\mathbf{Q}$-algebra, I an ideal of $R[(x_e)_{e \in E}]$, I_0 a set of generators for the ideal I and let A be $R[(x_e)_{e \in E}]/I$. We set $S = \mathsf{Spec}(R)$ and $X = \mathsf{Spec}(A)$. Let $n \in \mathbf{N}$ and*

$$\widetilde{A}_n := R[(x_{e,i})_{e \in E, i \in \{0, \ldots, n\}}]/(\widetilde{\delta}_\infty^{(i)}(f))_{f \in I_0, i \in \{0, \ldots, n\}},$$

$$A_n := R[(x_{e,i})_{e \in E, i \in \{0, \ldots, n\}}]/(\delta_\infty^{(i)}(f))_{f \in I_0, i \in \{0, \ldots, n\}}.$$

We endow $\widetilde{A}_n$ (resp. A_n) with the A-algebra structure induced by $x_e \mapsto x_{e,0}$.

Then, the A-algebras A_n and $\widetilde{A}_n$ are canonically isomorphic and we have natural isomorphisms of X-schemes

$$\mathscr{L}_n(X/S) \cong \mathsf{Spec}(\widetilde{A}_n) \cong \mathsf{Spec}(A_n).$$

7. The Universal Algebra of Higher Derivations

In this section, we present another construction of arc/jet schemes with an algebro-differential flavor based on the notion of universal higher derivation. A connection with the adjunction results in Section 5 is to be found in Section 7.6.

7.1. Let $n \in \mathbf{N} \cup \{\infty\}$, and let R be a ring. If $f \colon A \to A'$ and $g \colon B \to B'$ are two morphisms of R-algebras, then we can consider the maps

$$\mathrm{Der}^n_R(A', B) \to \mathrm{Der}^n_R(A, B) \; : \; (H_i)_{i \in \{0,\ldots,n\}} \mapsto (H_i \circ f)_{i \in \{0,\ldots,n\}},$$

$$\mathrm{Der}^n_R(A, B) \to \mathrm{Der}^n_R(A, B') \; : \; (H_i)_{i \in \{0,\ldots,n\}} \mapsto (g \circ H_i)_{i \in \{0,\ldots,n\}}.$$

This defines a bifunctor

$$(7.1.1) \qquad\qquad \mathrm{Der}^n_R(\cdot, \cdot) \colon \mathbf{Alg}^{\mathrm{op}}_R \times \mathbf{Alg}_R \to \mathbf{Sets}.$$

For all $n \geq m$ in $\mathbf{N} \cup \{\infty\}$, we have a *truncation map*

$$\psi^n_{m,A,B} \colon \mathrm{Der}^n_R(A, B) \to \mathrm{Der}^m_R(A, B) \colon (H_i)_{i \in \{0,\ldots,n\}} \mapsto (H_i)_{i \in \{0,\ldots,m\}},$$

functorial in A and B. These truncation maps satisfy

$$(7.1.2) \qquad\qquad \psi^n_{m,A,B} \circ \psi^p_{n,A,B} = \psi^p_{m,A,B}$$

for all $p \geq n \geq m$ in $\mathbf{N} \cup \{\infty\}$.

7.2. Let R be an ring. Let A be an R-algebra with structural morphism $\varphi \colon R \to A$. Let $n \in \mathbf{N} \cup \{\infty\}$. We define the A-algebra $\mathbf{Hs}^n_{A/R}$ to be the quotient of the polynomial A-algebra

$$A[T^{(i)}_a \mid a \in A, \, i \in \{0, \ldots, n\}]$$

by the ideal generated by the following elements, for all $a, b \in A$, all $r \in R$, and all $i \in \{0, \ldots, n\}$:

$$\begin{cases} T^{(0)}_a - a \\ T^{(i)}_{\varphi(r)} & (i \neq 0). \\ (T_{a+b})^{(i)} - T^{(i)}_a - T^{(i)}_b \\ (T_{ab})^{(i)} - \sum_{j=0}^{i} T^{(i-j)}_a T^{(j)}_b \end{cases}$$

For every $\alpha \in A[T^{(i)}_a \mid a \in A, i \in \{0, \ldots, n\}]$, we denote by $[\alpha]$ its class in the A-algebra $\mathbf{Hs}^n_{A/R}$. For every $i \in \{0, \ldots, n\}$, we consider the map

$$d_{A,i} \colon A \to \mathbf{Hs}^n_{A/R} \colon a \mapsto [T^{(i)}_a].$$

In particular, since $[T^{(0)}_a] = [a]$, the map $d_{A.0}$ is the structural morphism of the A-algebra $\mathbf{Hs}^n_{A/R}$. We call the family $(d_{A,i})_{i \in \{0,\ldots,n\}}$ the *exterior higher derivation of order n* associated with the R-algebra A. It follows immediately from the relations in the definition of $\mathbf{Hs}^n_{A/R}$ that the family $(d_{A,i})$ defines a higher derivation of order n, i.e., an element of $\mathrm{Der}^n_R(A, \mathbf{Hs}^n_{A/R})$.

7.3. For $a \in A$ and $i \in \{0, \ldots, n\}$, set $\mathrm{wt}(a) = 0$ and $\mathrm{wt}(T_a^{(i)}) = i$. This defines a graduation on $A[T_a^{(i)} \mid a \in A, i \in \{0, \ldots, n\}]$ which in turn induces a graduation on the A-algebra $\mathbf{Hs}_{A/R}^n$, since the defining relations are homogeneous. We denote by $\mathbf{Hs}_{A/R}^{n,(i)}$ the sub-A-module of $\mathbf{Hs}_{A/R}^n$ constituted by the elements of weight i. Note that by construction, $\mathbf{Hs}_{A/R}^{n,(0)} \overset{\sim}{\to} A$ and $\mathbf{Hs}_{A/R}^{n,(1)}$ is isomorphic to the A-module $\Omega_{A/R}^1$ of Kähler differentials of A over R. Still by construction, for every $i \in \{0, \ldots, n\}$, one has $d_{A,i}(A) \subset \mathbf{Hs}_{A/R}^{n,(i)}$.

7.4. The higher derivation $(d_{A,i})_{i \in \{0, \ldots, n\}}$ is *universal* in the following sense. For every R-algebra B, and every higher derivation $H \colon A \to B[[t]]_n$ of order n, there exists a unique morphism of R-algebras $f \colon \mathbf{Hs}_{A/R}^n \to B$ such that $H_i = f \circ d_{A,i}$ for every i in $\{0, \ldots, n\}$. This property follows immediately from the definition of $\mathbf{Hs}_{A/R}^n$: the relations in the definition encode precisely the axioms in the definition of a higher derivation. Thus, the R-algebra $\mathbf{Hs}_{A/R}^n$ represents the covariant functor

$$\mathbf{Alg}_R \to \mathbf{Sets} \colon B \mapsto \mathrm{Der}_R^n(A, B).$$

Since $\mathrm{Der}_R^n(A, B)$ is functorial also in A, the R-algebra $\mathbf{Hs}_{A/R}^n$ is functorial in A.

7.5. Let $m, n \in \mathbf{N} \cup \{\infty\}$ with $n \geq m$. The truncation morphism of functors

$$\psi_{m,A}^n \colon \mathrm{Der}^n(A, \cdot) \to \mathrm{Der}^m(A, \cdot)$$

corresponds to a morphism of R-algebras

$$\theta_{m,A}^n \colon \mathbf{Hs}_{A/R}^m \to \mathbf{Hs}_{A/R}^n.$$

For every $a \in A$ and $i \in \{0, \ldots, m\}$, one has $\theta_{m,A}^n([T_a^{(i)}]) = [T_a^{(i)}]$. For all $p \geq n \geq m$, these morphisms satisfy the formula

$$\theta_{m,A}^p = \theta_{n,A}^p \circ \theta_{m,A}^n.$$

In this way, we obtain an inductive system of R-algebras $(\mathbf{Hs}_{A/R}^n)_{n \in \mathbf{N}}$. The morphisms $(\theta_{n,A}^\infty)_{n \in \mathbf{N}}$ induce a morphism of A-algebras

$$\varinjlim_{n \in \mathbf{N}} \mathbf{Hs}_{A/R}^n \to \mathbf{Hs}_{A/R}^\infty,$$

which is easily seen to be an isomorphism.

PROPOSITION 7.5.1. *Let R be a ring, and let A be an R-algebra. We set $S = \mathsf{Spec}(R)$ and $X = \mathsf{Spec}(A)$. Then, for every $n \in \mathbf{N} \cup \{\infty\}$, there exists an isomorphism of X-schemes*

$$\mathscr{L}_n(X/S) \to \mathsf{Spec}(\mathbf{Hs}_R^n(A)),$$

functorial in A, that commutes with the truncation morphisms. In particular, $\mathbf{Hs}_R^\infty(A)$ and $A_{(\infty)}$ are isomorphic as A-algebras.

PROOF. The bijections

$$\mathrm{Der}_R^n(A, B) \xrightarrow{\sim} \mathrm{Hom}_{\mathbf{Alg}_R}(A, B[[t]]_n) \colon (H_i)_{i=0,\dots,n} \mapsto \left(a \mapsto \sum_{i=0}^{n} H_i(a) t^i \right)$$

(see Section 4.9) are functorial in the R-algebra B and commute with the truncation morphisms. Thus, by the universal property of the R-algebra $\mathbf{Hs}_{A/R}^n$, the scheme $\mathrm{Spec}(\mathbf{Hs}_{A/R}^n)$ represents the functor of jets (resp. arcs) on X over S, via an isomorphism that is compatible with the truncation maps. It then follows from Proposition 6.1.1 that $\mathbf{Hs}_R^\infty(A)$ and $A_{(\infty)}$ are isomorphic as A-algebras. $\qquad\square$

7.6. In [4], it is explained how the left adjoint of For: $\mathbf{DAlg}_R \to \mathbf{Alg}_R$ can be constructed using the universal properties of the module of differentials.

Here, we explain how the universal algebra of higher derivation $\mathbf{Hs}_{A/R}^\infty$ can be used to give an alternative construction of the left adjoint of the forgetful functor For: $\mathbf{IHDAlg}_R^\infty \to \mathbf{Alg}_R$ (Theorem 5.6.1).

Composing the universal higher derivation $d_A \colon A \to \mathbf{Hs}_{A/R}^\infty[[t]]$ with $t \mapsto t + u$ gives an element of $\mathrm{Hom}_{\mathbf{Alg}_R}(A, \mathbf{Hs}_{A/R}^\infty[[t, u]])$, i.e., a higher R-derivation of order ∞ from A to $\mathbf{Hs}_{A/R}^\infty[[u]]$. By the universal property of $\mathbf{Hs}_{A/R}^\infty$, it corresponds to an element d_A^{ad} of $\mathrm{Hom}_{\mathbf{Alg}_R}(\mathbf{Hs}_{A/R}^\infty, \mathbf{Hs}_{A/R}^\infty[[u]])$, i.e., to a higher R-derivation of order ∞ from $\mathbf{Hs}_{A/R}^\infty$ to itself. Let us show that $d_A^{\mathrm{ad}} \in \mathrm{Der}_R^\infty(\mathbf{Hs}_{A/R}^\infty)$, or, equivalently, that $d_{A,0}^{\mathrm{ad}} = \mathrm{Id}_{\mathbf{Hs}_{A/R}^\infty}$. By construction, the following diagram is commutative:

(7.6.1)
$$
\begin{array}{ccc}
A & \xrightarrow{\ d_A\ } & \mathbf{Hs}_{A/R}^\infty[[t]] \\
\downarrow{\scriptstyle d_A} & & \downarrow{\scriptstyle t \mapsto t+u} \\
\mathbf{Hs}_{A/R}^\infty[[t]] & \xrightarrow{\ d_A^{\mathrm{ad}}[[t]]\ } & \mathbf{Hs}_{A/R}^\infty[[t, u]]
\end{array}
$$

Composing with the morphism $\mathbf{Hs}_{A/R}^\infty[[t, u]] \to \mathbf{Hs}_{A/R}^\infty[[t]]$ of reduction modulo u, one finds that $d_{A,0}^{ad}[[t]] \circ d_A = d_A$. By the universal property of $\mathbf{Hs}_{A/R}^\infty$, it implies that $d_{A,0}^{ad} = \mathrm{Id}_{\mathbf{Hs}_{A/R}^\infty}$.

Let us now show that $\mathbf{Hs}_{A/R}^\infty$ is iterative. We consider the diagram

$$
\begin{array}{ccc}
\mathbf{Hs}_{A/R}^\infty & \xrightarrow{\ d_A^{\mathrm{ad}}\ } & \mathbf{Hs}_{A/R}^\infty[[t]] \\
\downarrow{\scriptstyle d_A^{\mathrm{ad}}} & & \downarrow{\scriptstyle t \mapsto t+u} \\
\mathbf{Hs}_{A/R}^\infty[[t]] & \xrightarrow{\ d_A^{\mathrm{ad}}[[t]]\ } & \mathbf{Hs}_{A/R}^\infty[[t, u]]
\end{array}
$$

To show that it is commutative, by the universal property of $\mathbf{Hs}^\infty_{A/R}$, it suffices to show that the higher derivations

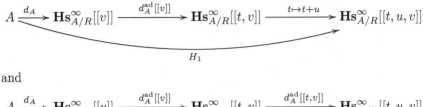

and

$$A \xrightarrow{d_A} \mathbf{Hs}^\infty_{A/R}[[v]] \xrightarrow{d_A^{\mathrm{ad}}[[v]]} \mathbf{Hs}^\infty_{A/R}[[t,v]] \xrightarrow{d_A^{\mathrm{ad}}[[t,v]]} \mathbf{Hs}^\infty_{A/R}[[t,u,v]]$$

$$H_2$$

coincide. But using diagram (7.6.1) (once for H_1 and twice for H_2), one sees that $H_1 = H_2 = (v \mapsto t + u + v) \circ d_A$.

One can now check that $A \mapsto (\mathbf{Hs}^\infty_{A/R}, d_A^{\mathrm{ad}})$ defines a functor from $\mathbf{Alg}_R$ to $\mathbf{IHDAlg}^\infty_R$ which is left adjoint to the forgetful functor For: $\mathbf{IHDAlg}^\infty_R \to \mathbf{Alg}_R$.

Bibliography

[1] B. Bhatt, Algebraization and Tannaka duality. *Camb. J. Math.*, 4(4):403–461, 2016.

[2] S. Bosch, W. Lütkebohmert and M. Raynaud, *Néron Models*, Ergebnisse der Mathematik und ihrer Grenzgebiete (3) [Results in Mathematics and Related Areas (3)], Vol. 21, Springer-Verlag, Berlin, 1990.

[3] A. Chambert-Loir, J. Nicaise and J. Sebag, *Motivic Integration*, Progress in Mathematics, Vol. 325, Birkhäuser, Basel, 2018.

[4] H. Gillet, Differential algebra — a scheme theory approach. In *Differential Algebra and Related Topics*, pp. 95–123. World Scientific Publishing, River Edge, NJ, 2002.

[5] A. Grothendieck, Éléments de géométrie algébrique. I. Le langage des schémas. *Publ. Math. Inst. Hautes Études Sci.*, 4:228, 1960.

[6] A. Grothendieck, Éléments de géométrie algébrique. IV. Étude locale des schémas et des morphismes de schémas, *Troisième partie*. *Publ. Math. Inst. Hautes Études Sci.*, 28:255, 1966.

[7] J. Nicaise and J. Sebag, Greenberg approximation and the geometry of arc spaces. *Comm. Algebra*, 38(11):4077–4096, 2010.

3

The Grinberg–Kazhdan Formal Arc Theorem and the Newton Groupoids

Vladimir Drinfeld

*Department of Mathematics, University of Chicago,
5734 University Ave., Chicago, IL 60637, USA*

We first prove the Grinberg–Kazhdan formal arc theorem without any
assumptions on the characteristic. This part of the article is equivalent
to arXiv:math-AG/0203263. Then, we try to clarify the geometric ideas
behind the proof by introducing the notion of Newton groupoid (which
is related to Newton's method for finding roots). Newton groupoids are
certain groupoids in the category of schemes associated to any generically
étale morphism from a locally complete intersection to a smooth variety.

1. Introduction

1.1. Subject of the article

Section 2 (which is equivalent to the preprint [2]) formulates and proves
the Grinberg–Kazhdan theorem [4] without any assumptions on the characteristic of the field. The proof is based on the ideas that go back to the
17th century (i.e., the implicit function theorem, or equivalently, Newton's
method for finding roots) and the 19th century (the Weierstrass division
theorem).

The goal of the remaining chapter is to clarify the geometric ideas behind the proof from Section 2. In particular, we introduce *Newton groupoids*.
These are certain groupoids in the category of schemes, which are related to
Newton's method for finding roots. They are associated to any generically
étale morphism from a locally complete intersection to a smooth variety
(both schemes are assumed to be separated). Let us note that Newton
groupoids are already used in Section 2 behind the scenes.

The main message of Sections 4 and 5 is that the notion of groupoid in the category of schemes is useful "in real life" and that Lie algebroids provide intuition that helps us understand groupoids. We also demonstrate that it is easy to construct smooth groupoids acting on a given variety in a generically transitive way (this is in contrast with the situation for algebraic group actions).

1.2. Structure of the article

In Section 2, we formulate and prove the Grinberg–Kazhdan theorem. In Section 3, we slightly rephrase the proof. In Section 4, we first recall basic facts about groupoids and stacks. Then, we formulate the main properties of the Newton groupoids (see Sections 4.3.3, 4.5.4, 4.6 and 4.7). In Section 5, we define the Newton groupoids and verify their properties. The key formula is (5.10).

2. The Grinberg–Kazhdan Theorem

2.1. Formulation of the theorem

Let X be a scheme of finite type over a field k and $X^\circ \subset X$ the smooth part of X. Consider the scheme $\mathcal{L}(X)$ of formal arcs in X. The k-points of $\mathcal{L}(X)$ are just maps $\operatorname{Spec} k[[t]] \to X$. Let $\mathcal{L}^\circ(X)$ be the open subscheme of arcs whose image is not contained in $X \setminus X^\circ$. Fix an arc $\gamma_0 : \operatorname{Spec} k[[t]] \to X$ in $\mathcal{L}^\circ(X)$, and let $\mathcal{L}(X)_{\gamma_0}$ be the formal neighborhood of γ_0 in $\mathcal{L}(X)$. We will give a simple proof of the following theorem, which was proved by Grinberg and Kazhdan [4] for fields k of characteristic zero.

THEOREM 2.1.1. *Suppose that $\gamma_0(0)$ is not an isolated point[1] of X. Then, there exists a scheme $Y = Y(\gamma_0)$ of finite type over k, and a point $y \in Y(k)$, such that $\mathcal{L}(X)_{\gamma_0}$ is isomorphic to $D^\infty \times Y_y$ where Y_y is the formal neighborhood of y in Y and D^∞ is the product of countably many copies of the formal disk $D := \operatorname{Spf} k[[t]]$.*

The proof will be given in Section 2.4; its key idea is formulated at the end of the second paragraph of Section 2.4.

2.2. Convention

Throughout this article, a *test-ring* A is a local commutative unital k-algebra with residue field k whose maximal ideal m is nilpotent. If S is a scheme over k and $s \in S(k)$ is a k-point, we think of the formal neighborhood S_s in terms of its functor of points $A \mapsto S_s(A)$ from test rings to sets. For instance, A-points of $\mathcal{L}(X)_{\gamma_0}$ are $A[[t]]$-points of X whose reduction modulo m equals γ_0.

---

[1]If $\gamma_0(0)$ is an isolated point of X then, γ_0 is an isolated point of $\mathcal{L}^\circ(X)$, so $\mathcal{L}(X)_{\gamma_0} = \operatorname{Spec} k$.

2.3. Warning

REMARK 2.3.1. $\mathcal{L}(X)_{\gamma_0} = \mathcal{L}(X_1)_{\gamma_0}$ where $X_1 \subset X$ is the closure of the connected component of X° containing $\gamma_0(\operatorname{Spec} k((t)))$. So, we can assume that X is reduced and irreducible. But Y is, in general, neither reduced nor irreducible (see the following example).

EXAMPLE 2.3.2. Let X be the hypersurface $yx_{n+1} + g(x_1, \ldots, x_n) = 0$, where g is a polynomial vanishing at 0. Let $\gamma_0(t)$ be defined by

$$x_{n+1}^0(t) = t, y^0(t) = x_1^0(t) = \cdots = x_n^0(t) = 0.$$

Then, one can define Y to be the hypersurface $g(x_1, \ldots, x_n) = 0$ and y to be the point $0 \in Y$. Indeed, by the Weierstrass division theorem (a.k.a preparatory lemma) for any test ring A, every A-deformation of $x_{n+1}^0(t) = t$ can be uniquely written as

$$x_{n+1}(t) = (t - \alpha)u(t),$$

where α belongs to the maximal ideal $m \subset A$ and $u \in 1 + m[[t]]$. Given α, u, and $x_1(t), \ldots, x_n(t) \in m[[t]]$, there is at most one $y(t) \in m[[t]]$, such that

$$y(t)x_{n+1}(t) + g(x_1(t), \ldots, x_n(t)) = 0,$$

and $y(t)$ exists if and only if $g(x_1(\alpha), \ldots, x_n(\alpha)) = 0$. Writing $x_1(t), \ldots, x_n(t)$ as

$$x_i(t) = \xi_i + (t - \alpha)\tilde{x}_i(t), \quad \xi_i \in m, \tilde{x}_i \in m[[t]],$$

we see that the set of A-points of $\mathcal{L}(X)_{\gamma_0}$ identifies with the set of collections

$$(2.1) \qquad (\alpha, u(t), \tilde{x}_1(t), \ldots, \tilde{x}_n(t), \xi_1, \ldots, \xi_n),$$

where $\xi_1, \ldots, \xi_n \in m$ satisfy the equation

$$(2.2) \qquad g(\xi_1, \ldots, \xi_n) = 0,$$

and $\alpha \in m$, $u \in 1 + m[[t]]$, $\tilde{x}_i \in m[[t]]$ are "free variables".

2.4. Proof of Theorem 2.1.1

We can assume that X is a closed subscheme of an affine space. Then, there is a closed subscheme X' of the affine space, such that $X' \supset X$, X' is a complete intersection, and the image of our arc γ_0 is not contained in the closure of $X' \setminus X$. Clearly, $\mathcal{L}(X)_{\gamma_0} = \mathcal{L}(X')_{\gamma_0}$, so we can assume that $X = X'$ is the subscheme of $\operatorname{Spec} k[x_1, \ldots, x_n, y_1, \ldots, y_l]$ defined by equations $f_1 = \ldots = f_l = 0$, such that the arc $\gamma_0(t) = (x^0(t), y^0(t)) = (x_1^0(t), \ldots, x_n^0(t), y_1^0(t), \ldots, y_l^0(t))$ is not contained in the subscheme of X defined by $\det \frac{\partial f}{\partial y} = 0$. Here, $\frac{\partial f}{\partial y}$ is the matrix of partial derivatives $\frac{\partial f_i}{\partial y_j}$.

Let γ be an A-deformation of γ_0 for some test ring A, so $\gamma(t) = (x(t), y(t))$, where $x(t) \in A[[t]]^n$, $y(t) \in A[[t]]^l$. Then, by the Weierstrass division theorem, $\det \frac{\partial f}{\partial y}(x(t), y(t))$ has a unique representation as $q(t)u(t)$, where $u \in A[[t]]$ is invertible and q is a monic polynomial whose reduction modulo the maximal ideal $m \subset A$, is a power of t. Let d denote the degree of q; it depends only on γ_0, not on its deformation γ. We assume that $d > 0$ (otherwise, we can eliminate y). *The idea of what follows is to consider q as one of the unknowns.*

More precisely, A-deformations of γ_0 are identified with solutions of the following system of equations. The unknowns are $q(t) \in A[t]$, $x(t) \in A[[t]]^n$, and $y(t) \in A[[t]]^l$, such that q is monic of degree d, $q(t)$ is congruent to t^d modulo m, and the reduction of $(x(t), y(t))$ modulo m equals $\gamma_0(t) = (x^0(t), y^0(t))$. The equations are as follows:

$$(2.3) \qquad \det \frac{\partial f}{\partial y}(x(t), y(t)) \equiv 0 \bmod q,$$

$$(2.4) \qquad f(x(t), y(t)) = 0,$$

where $f := (f_1, \ldots, f_l)$. (Note that if (2.3) is satisfied, then $q(t)^{-1} \det \frac{\partial f}{\partial y}(x(t), y(t))$ is automatically invertible because it is invertible modulo m.)

Now, fix $r \geq 2$ and consider the following system of equations. The unknowns are $q(t) \in A[t]$, $x(t) \in A[[t]]^n$, and $\bar{y} \in A[t]^l/(q^{r-1})$, such that q is monic of degree d, $q(t)$ is congruent to t^d modulo m, the reduction of $x(t)$ modulo m equals $x^0(t)$, and the reduction of $\bar{y}$ modulo m equals the reduction of y^0 modulo t^{r-1}. The equations are as follows:

$$(2.5) \qquad \det \frac{\partial f}{\partial y}(x(t), \bar{y}) \equiv 0 \bmod q,$$

$$f(x(t), \bar{y}) \in \mathrm{Im}\left(q^{r-1} \frac{\partial f}{\partial y}(x(t), \bar{y}) : A[t]^l/qA[t]^l \to q^{r-1}A[t]^l/q^r A[t]^l \right).$$
$$(2.6)$$

Condition (2.6) makes sense because $f(x(t), \bar{y})$ is well-defined modulo the image of $q^{r-1} \frac{\partial f}{\partial y}(x(t), \bar{y})$. Note that (2.6) is indeed an equation because it is equivalent to the condition $\hat{C}f(x(t), y(t)) \equiv 0 \bmod q^r$, where $y(t) \in A[t]^l$ is a preimage of $\bar{y}$ and $\hat{C}$ is the matrix adjugate to $C := \frac{\partial f}{\partial y}(x(t), y(t))$ (so, $C\hat{C} = \hat{C}C = \det C$). This condition is equivalent to the following equations, which do not involve a choice of $y(t) \in A[t]^l$, such that $y(t) \mapsto \bar{y}$:

$$(2.7) \qquad f(x(t), \bar{y}) \equiv 0 \bmod q^{r-1},$$

$$(2.8) \qquad \hat{B}f(x(t), \bar{y}) \equiv 0 \bmod q^r,$$

where $B := \frac{\partial f}{\partial y}(x(t), \bar{y})$; note that (2.8) makes sense as soon as (2.7) holds.

LEMMA 2.4.1. *The natural map from the set of solutions of (2.3)–(2.4) to the set of solutions of (2.5)–(2.6) is bijective.*

Proof. Let a be the minimal number such that $m^a = 0$. We proceed by induction on a, so we can assume that $a \geq 2$ and the lemma is proved for A/m^{a-1}. Then, there exists $\tilde{y}(t) \in A[t]^l$, such that $\tilde{y} \bmod q^{r-1} = \bar{y}$ and $f(x(t), \tilde{y}(t)) \in m^{a-1}[t]^l$; such $\tilde{y}$ is unique modulo $q^{r-1}A[t]^l \cap m^{a-1}[t]^l$. We have to find $z(t) \in q^{r-1}A[t]^l \cap m^{a-1}[t]^l$, such that $f(x(t), \tilde{y}(t) - z(t)) = 0$, i.e., $Cz(t) = f(x(t), \tilde{y}(t))$, where $C := \frac{\partial f}{\partial y}(x(t), \tilde{y}(t))$. Equation (2.5) implies that $\det C = q(t)u(t)$ for some invertible $u \in A[t]$. So, $z(t)$ is unique. By (2.6), $f(x(t), \tilde{y}(t)) \in q^{r-1}CA[t]^l + q^r A[t]^l$. But $CA[t]^l \supset (\det C)A[t]^l = qA[t]^l$, so $f(x(t), \tilde{y}(t)) = Cz(t)$ for some $z(t) \in q^{r-1}A[t]^l$. We have $Cz(t) = f(x(t), \tilde{y}(t)) \equiv 0 \bmod m^{a-1}$, so $q(t)z(t) \equiv 0 \bmod m^{a-1}$ and finally $q(t) \equiv 0 \bmod m^{a-1}$. $\qquad\square$

So, the set of A-deformations of γ_0 can be identified with the set of solutions of the system (2.5)–(2.6). This system is essentially finite because $x(t)$ is relevant only modulo q^r. For example, if $r = 2$, we can write $x(t)$ as $q^2(t)\xi(t) + \bar{x}$, $\xi \in A[[t]]^n$, $\bar{x} \in A[t]^n$, $\deg \bar{x} < 2d$, and consider $\xi(t)$, $\bar{x}$, $q(t)$, and $\bar{y}$ to be the unknowns (rather than $x(t)$, $q(t)$, $\bar{y}$); then (2.5)–(2.6) becomes a finite system of equations for q, $\bar{x}$, $\bar{y}$ (and ξ is not involved in these equations). So, $\mathcal{L}(X)_{\gamma_0}$ is isomorphic to $D^\infty \times Y_y$, where the k-scheme Y of finite type and the point $y \in Y(k)$ are defined as follows: for every k-algebra R, the set $Y(R)$ consists of triples $(q, \bar{x}, \bar{y})$ where $q \in R[t]$ is monic of degree d, $\bar{x} \in R[t]^n/(q^2)$, $\bar{y} \in R[t]^l/(q)$, $\det B \equiv 0 \bmod q$, $B := \frac{\partial f}{\partial y}(\bar{x}, \bar{y})$, $f(\bar{x}, \bar{y}) \equiv 0 \bmod q$, and $\hat{B}f(\bar{x}, \bar{y}) \equiv 0 \bmod q^2$; $y \in Y(k)$ corresponds to $q = t^d$, $\bar{x} = x^0(t) \bmod t^{2d}$, $\bar{y} = y^0(t) \bmod t^d$.

3. Rephrasing the Proof from Section 2

3.1. Deducing Theorem 2.1.1 from Proposition 3.1.2

3.1.1. *The setting*

Let $\mathbb{A}^n$ denote the n-dimensional affine space over the field k. Let f be a morphism $\mathbb{A}^{n+l} \to \mathbb{A}^l$, i.e., $f = (f_1, \ldots, f_l)$ and every f_i is a polynomial $f_i(x, y)$, where $x = (x_1, \ldots, x_n) \in \mathbb{A}^n$ and $y = (y_1, \ldots, y_l) \in \mathbb{A}^l$. Set $Q := \det(\frac{\partial f}{\partial y})$.

Set $X := f^{-1}(0) \subset \mathbb{A}^{n+l}$; for a k-algebra A, we let $X(A)$ denote the set of A-points of X. Let $\Delta_X \subset X$ be the subscheme of zeros of Q. Let N be a non-negative integer.

In Section 3.2, we will prove the following proposition.

PROPOSITION 3.1.2. (i) *There exists a k-scheme Z representing the following functor: for any k-algebra R, an R-point of Z is a pair consisting of a monic polynomial $q \in R[t]$ of degree N and an element of the set $\varprojlim_{r} X(R[t]/(q^r))$, such that the scheme-theoretic preimage of Δ_X in* $\operatorname{Spec} R[t]/(q^2)$ *equals* $\operatorname{Spec} R[t]/(q)$.

(ii) *Z is a product of a k-scheme of finite type and a (typically infinite-dimensional) affine space.*

REMARK 3.1.3. The property from statement (i) clearly implies that for *every* $r \geq 2$ the scheme-theoretic preimage of Δ_X in $\operatorname{Spec} R[t]/(q^r)$ equals $\operatorname{Spec} R[t]/(q)$.

3.1.4. *Deducing Theorem 2.1.1 from Proposition 3.1.2*

As explained at the beginning of Section 2.4, we can assume that the scheme X from Theorem 2.1.1 equals $f^{-1}(0)$ for some morphism $f : \mathbb{A}^{n+l} \to \mathbb{A}^l$ and the image of the formal arc $\gamma_0 : \operatorname{Spec} k[[t]] \to X$ is not contained in Δ_X. Then, $\gamma_0^{-1}(\Delta_X) = \operatorname{Spec} k[t]/(t^N)$ for some $N \in \mathbb{N}$. Let Z be the scheme from Proposition 3.1.2(i) corresponding to this N. Let $q = t^N$, then the pair (q, γ_0) defines a k-point of Z. The Weierstrass division theorem implies that the formal neighborhood of (q, γ_0) in Z is equal to the formal neighborhood of γ_0 in the scheme of formal arcs in X. So, Theorem 2.1.1 follows from Proposition 3.1.2(ii).

3.2. Proof of Proposition 3.1.2

3.2.1. *Representing Z as a limit*

We will represent Z as a projective limit of certain k-schemes Z_r of finite type, $r \geq 2$, so that for $r \geq 3$, the scheme Z_{r+1} is isomorphic to a product of Z_r and an affine space. Let us note that the schemes Z_r defined below were secretly used in the proof of Theorem 2.1.1.

We define Z_r to represent the functor that associates to a k-algebra R the set of triples $(q, \bar{x}, \bar{y})$ satisfying the following conditions:

(1) $q \in R[t]$ is a monic polynomial of degree N;
(2) $\bar{x} \in (R[t]/(q^r))^n$, $\bar{y} \in (R[t]/(q^{r-1}))^l$;
(3) $f(\bar{x}, \bar{y}) \equiv 0 \bmod q^{r-1}$;
(4) $Q(\bar{x}, \bar{y}) \equiv 0 \bmod q$;
(5) let $C := \frac{\partial f}{\partial y}(\bar{x}, \bar{y})$ and let $\hat{C}$ be the matrix adjugate to C, then

$$\hat{C} f(\bar{x}, \bar{y}) \equiv 0 \bmod q^r;$$

using (3), (4), and the equality $\hat{C}C = Q(\bar{x}, \bar{y})$, one easily checks that the congruence modulo q^r makes sense (even though $\bar{y}$ is defined only modulo q^{r-1});
(6) if $r \geq 3$, then the element $q^{-1}Q(\bar{x}, \bar{y}) \in R[t]/(q^{r-2})$ is invertible.

It is easy to check that this functor is indeed representable by an affine scheme of finite type over k. Now, the following lemma remains to be proved.

LEMMA 3.2.2. (i) *The canonical morphism* $Z_{r+1} \to Z_r$ *is smooth. Its fibers have dimension* nN.

(ii) *If* $r \geq 3$, *then* Z_{r+1} *is isomorphic (as a scheme over* Z_r*) to a product of* Z_r *and an affine space.*

PROOF. Let V_r be the k-scheme whose R-points are pairs $(q, \bar{x})$, where $q \in R[t]$ is a monic polynomial of degree N and $\bar{x} \in (R[t]/(q^r))^n$; of course, this scheme is isomorphic to an affine space of dimension $N(1+nr)$. For each r, we have a canonical morphism $Z_r \to V_r$ (forgetting $\bar{y}$). These morphisms are compatible with each other, so we get a morphism $\varphi_r : Z_{r+1} \to Z_r \underset{V_r}{\times} V_{r+1}$. To prove the lemma, one checks straightforwardly that φ_r is étale, and for $r \geq 3$, the map φ_r is an isomorphism.[2]

Let us only describe φ_r^{-1} assuming that $r \geq 3$. An R-point of $Z_r \underset{V_r}{\times} V_{r+1}$ is a triple

$$(q, \bar{x}, \bar{y}), \quad \bar{x} \in (R[t]/(q^{r+1}))^n, \quad \bar{y} \in (R[t]/(q^{r-1}))^l,$$

satisfying properties (3)–(6) from Section 3.2.1 (as usual, $q \in R[t]$ is monic of degree N). It is easy to check that φ_r^{-1} is given by Newton's formula

$$\varphi_r^{-1}(q, \bar{x}, \bar{y}) = (q, \bar{x}, \tilde{y} - h), \quad h := C^{-1} f(\bar{x}, \tilde{y}),$$

where $\tilde{y} \in (R[t]/(q^r))^l$ is any preimage of $\bar{y} \in (R[t]/(q^{r-1}))^l$. Note that h is a well-defined element of $(q^{r-1}R[t]^l)/(q^r R[t]^l)$: indeed, $h = (q^{-1}Q(\bar{x}, \bar{y}))^{-1} q^{-1}\hat{C}f(\bar{x}, \tilde{y})$. $\square$

The rest of the article is devoted to the geometric interpretation of the schemes Z_r in terms of the *Newton groupoids*. In some sense, the idea goes back to Finkelberg and Mirković (see Section 4.4).

4. Introduction to the Newton Groupoids

4.1. The language of groupoids

4.1.1. *Abstract groupoids*

Recall that an (abstract) groupoid is just a category in which all morphisms are invertible. So the data defining a groupoid are as follows: the set of objects X, the set of morphisms Γ, the "source" map $p_1 : \Gamma \to X$, the "target" map $p_2 : \Gamma \to X$, and the composition map $c : \Gamma \times_X \Gamma \to \Gamma$. These data should have certain properties; in particular, one should have the "unit" map $e : X \to \Gamma$, $x \mapsto \mathrm{id}_x$; one should also have the inversion map $i : \Gamma \xrightarrow{\sim} \Gamma$. (More details can be found in [**8**, Exposé V] or [**1**]).

---

[2]The map φ_2 is neither surjective nor injective in general; both phenomena occur already if $n = l = N = 1$ and $f(x, y) = y(y - P(x))$, where $P(x)$ is a polynomial.

4.1.2. Groupoids in the category of k-schemes

In the situation of Section 4.1.1 one can consider e and i as a part of the data; then all the properties become *identities* (i.e., they do not involve existence quantifiers). After this, the notion of groupoid in any category with fiber products becomes clear. In particular, one has the notion of groupoid in the category of k-schemes (more details can be found in [**8**, Exposé V] or [**1**]).

One can also define the notion of groupoid in any category using the language of S-points, see [**8**, Exposé V].

4.1.3. Conventions

In the situation of Section 4.1.1 or Section 4.1.2, one says that Γ is a groupoid on X or that Γ is a groupoid acting on X.

We usually write a groupoid as $\Gamma \overset{p_1}{\underset{p_2}{\rightrightarrows}} X$ or as $\Gamma \overset{(p_1,p_2)}{\longrightarrow} X \times X$ (without mentioning the composition map explicitly). This will not lead to confusion because we are mostly interested in the situation, where the map $\Gamma \overset{(p_1,p_2)}{\longrightarrow} X \times X$ is a birational isomorphism.

From now on, we consider only groupoids in the category of k-schemes (unless stated otherwise).

4.1.4. Smooth groupoids and quotient stacks

A groupoid $\Gamma \overset{p_1}{\underset{p_2}{\rightrightarrows}} X$ is said to be *smooth* if p_1 is smooth. This is equivalent to p_2 being smooth (indeed, the inversion map $i : \Gamma \overset{\sim}{\longrightarrow} \Gamma$ interchanges p_1 and p_2).

Let $\Gamma \overset{p_1}{\underset{p_2}{\rightrightarrows}} X$ be a smooth groupoid, such that the morphism $\Gamma \overset{(p_1,p_2)}{\longrightarrow} X \times X$ is quasi-compact and quasi-separated (this is a very mild assumption). Then, one defines the *quotient stack* X/Γ, which is an *algebraic stack*, see [**6, 7, 9**]. "Almost all" algebraic stacks can be obtained this way (to get all of them, one has to allow Γ to be an algebraic space rather than a scheme).

4.2. Pointy stacks and Maps°

4.2.1. Pointy stacks

By a *pointy stack*, we mean an algebraic k-stack locally of finite type which has a dense open substack isomorphic to the point $\operatorname{Spec} k$. Note that such an open substack is clearly unique.

If an action of an algebraic group G on a k-scheme X locally of finite type has a dense open orbit on which the action is free then the stack X/G is pointy. For example $\mathbb{A}^1/\mathbb{G}_m$ is a pointy stack.

More generally, let X be a k-scheme locally of finite type and Γ a smooth groupoid acting on X so that the corresponding morphism $\Gamma \to X \times X$ is an isomorphism over $U \times U$ for some dense open $U \subset X$. Then, the stack X/Γ is pointy.

4.2.2. *Maps from a curve to a pointy stack*

Let $\mathcal{Y} \supset \operatorname{Spec} k$ be a pointy stack; let $\mathcal{Y}' \subset \mathcal{Y}$ be the reduced closed substack, such that $\mathcal{Y} \setminus \mathcal{Y}' = \operatorname{Spec} k$. We assume that the diagonal morphism $\mathcal{Y} \to \mathcal{Y} \times \mathcal{Y}$ is separated (The assumption is mild because usually the diagonal morphism is affine).

On the other hand, let C be a smooth curve over k (e.g., $\mathbb{A}^1$).

In this situation, we define $\operatorname{Maps}^\circ(C, \mathcal{Y})$ to be the functor that associates to a k-scheme S the set[3] of morphisms $f : C \times S \to \mathcal{Y}$, such that $f^{-1}(\mathcal{Y}')$ is finite over S.

CONJECTURE 4.2.3. *In this situation, the functor* $\operatorname{Maps}^\circ(C, \mathcal{Y})$ *is representable by an algebraic space locally of finite type over* k.

4.2.4. *Easy example*

Let $\mathcal{Y} = \mathbb{A}^1/\mathbb{G}_m$. Then, a morphism $C \times S \to \mathcal{Y}$ is just a line bundle on $C \times S$ equipped with a section. So, $\operatorname{Maps}^\circ(C, \mathcal{Y})$ is the scheme parameterizing effective divisors on C; in other words, $\operatorname{Maps}^\circ(C, \mathcal{Y})$ is the disjoint union of the symmetric powers $\operatorname{Sym}^N C$, $N \geq 0$.

4.2.5. *Important example*

Let G be a reductive group, $B \subset G$ a Borel subgroup, and U its unipotent radical. Let $\mathcal{Y} := \overline{U \setminus G}/B$, where $\overline{U \setminus G}$ is the affine closure of $U \setminus G$ (i.e., the spectrum of the ring of regular functions on $U \setminus G$). Then, $\operatorname{Maps}^\circ(C, \mathcal{Y})$ is known to be representable by a scheme locally of finite type over k, which is called the *open Zastava scheme*, see [3] ("Zastava" is the Croatian for "flag").

4.2.6. *Remark*

Let $\mathcal{Y} \supset \operatorname{Spec} k$ be a pointy substack with the following property: there exists an effective Cartier divisor $\Delta \subset \mathcal{Y}$, such that $\mathcal{Y} \setminus \Delta = \operatorname{Spec} k$. Moreover, let us fix such Δ. Then, for any morphism $f : C \times S \to \mathcal{Y}$ as in Section 4.2.2, the subscheme $f^{-1}(\Delta) \subset C \times S$ is an S-family of effective divisors on C, so we get a morphism from $\operatorname{Maps}^\circ(C, \mathcal{Y})$ to the disjoint union of the symmetric powers $\operatorname{Sym}^N C$, $N \geq 0$. The preimage of $\operatorname{Sym}^N C$ in $\operatorname{Maps}^\circ(C, \mathcal{Y})$ will be denoted by $\operatorname{Maps}^\circ_N(C, \mathcal{Y})$.

---

[3]*A priori*, such morphisms form a groupoid rather than a set. But separateness of the diagonal morphism $\mathcal{Y} \to \mathcal{Y} \times \mathcal{Y}$ easily implies that this groupoid is a set.

4.3. The goal

4.3.1. *The setting*

Let X and Y be separated k-schemes locally of finite type and $\varphi : X \to Y$ be a k-morphism. Let $U \subset X$ be the locus where φ is étale. We assume that Y is smooth, X is a locally complete intersection,[4] and U is dense in X.

In this situation, one defines the *different*; this is a canonical effective Cartier divisor $\Delta_X \subset X$, such that $X \setminus \Delta_X = U$, namely, Δ_X is the divisor associated by Knudsen–Mumford [5] to the relative cotangent sheaf $\Omega^1_{X/Y}$ (note that in our situation $\Omega^1_{X/Y}$ has homological dimension 1 and vanishes on U, so the construction from [5] is applicable).

The reader may prefer to focus on the following particular case.

4.3.2. *Particular case*

Let $X \subset \mathbb{A}^{n+l}$ be as in Section 3.1.1, $Y = \mathbb{A}^n$ and $\varphi : X \to \mathbb{A}^n$ the projection. Assume that $\varphi : X \to \mathbb{A}^n$ is étale on a dense open subset of X. This assumption means that the subscheme $\Delta_X \subset X$ from Section 3.1.1 is a Cartier divisor; this is the different.

4.3.3. *The goal*

In the situation of Section 4.3.1, we will construct in Section 5 for each $r \geq 2$ a smooth groupoid Γ_r acting on X, which is called the rth *Newton groupoid* of $\varphi : X \to Y$. It has the following properties:

  (i) the morphism $\Gamma_r \to X \times X$ is an isomorphism over $(X \setminus \Delta_X) \times (X \setminus \Delta_X)$, so the stack X/Γ_r is point in the sense of Section 4.2.1;
  (ii) the action of Γ_r becomes the identity[5] when restricted to Δ_X, so the stack X/Γ_r has the property from Section 4.2.6 with $\Delta := \Delta_X/\Gamma_r$;
  (iii) in the situation of Section 4.3.2, the functor $\mathrm{Maps}^\circ_N(\mathbb{A}^1, X/\Gamma_r)$ (see Section 4.2.6) is representable by an open subscheme of the scheme Z_r from Section 3.2.1; if $r \geq 3$, the open subscheme equals Z_r, and if $r = 2$, it equals $\mathrm{Im}\,(Z_3 \to Z_2)$.

More properties of Γ_r will be formulated in Section 4.5.4 and Sections 4.6–4.7.

4.4. Relation to Finkelberg–Mirković

Finkelberg and Mirković [3] proved Theorem 2.1.1 in the particular case that $X = \overline{U \setminus G}$, where G and U are as in Section 4.2.5. They did it

---

[4]We do *not* assume that X is a relative locally complete intersection with respect to φ.

[5]By definition, this means that the morphisms $p_1, p_2 : \Gamma_r \to X$ have equal restrictions to $p_1^{-1}(\Delta_X)$.

by considering $\mathrm{Maps}^\circ(\mathbb{A}^1, \overline{U \setminus G}/B)$. The proof of Theorem 2.1.1 given in Section 2 or Section 3 secretly uses a similar strategy, with the groupoid Γ_r playing the role of B; this is clear from Section 4.3.3(iii).

4.5. The Lie algebroid of Γ_r

4.5.1. *The notion of Lie algebroid*

Let X be a scheme locally of finite type over k and Θ_X its tangent sheaf.

Recall that a *Lie algebroid* on X is a sheaf $\mathfrak{a}$ on X equipped with an $\mathcal{O}_X$-module structure, a Lie ring structure, and an *anchor map* $\tau : \mathfrak{a} \to \Theta_X$, which is supposed to be both an $\mathcal{O}_X$-module morphism and a Lie morphism; moreover, if f is a regular function on an open subset $U \subset X$ and $v_1, v_2 \in H^0(U, \mathfrak{a})$, then one should have

$$[v_1, fv_2] = f[v_1, v_2] + ((\tau(v_1))(f)) \cdot v_2.$$

For instance, Θ_X is a Lie algebroid with the anchor map being the identity.

We say that a Lie algebroid $\mathfrak{a}$ on X is *locally free* if $\mathfrak{a}$ is a locally free coherent $\mathcal{O}_X$-module.

4.5.2. *The Lie algebroid of a smooth groupoid*

Let Γ be a smooth groupoid on X and let $e : X \to \Gamma$ be its unit. Define $\mathrm{Lie}(\Gamma)$ to be the normal bundle of $X = e(X) \subset \Gamma$. It is well known that $\mathrm{Lie}(\Gamma)$ has a natural structure of locally free Lie algebroid on X (e.g., see [1]). In particular, the anchor map $\tau : \mathrm{Lie}(\Gamma) \to \Theta_X$ is just the map from the normal bundle of $e(X) \subset \Gamma$ to the normal bundle of $X_{\mathrm{diag}} \subset X \times X$ induced by the morphism $\Gamma \xrightarrow{(p_1, p_2)} X \times X$.

4.5.3. *An example of Lie algebroid*

In the situation of Section 4.3.1, set

$$\mathfrak{a}_r := (\varphi^* \Theta_Y)(-r\Delta_X).$$

It is easy to see that if $r \geq 1$, then $\mathfrak{a}_r \subset \Theta_X$ and moreover, $\mathfrak{a}_r$ is a Lie subalgebroid of Θ_X. It is clear that this Lie algebroid is locally free, and the restriction of its anchor map to $X \setminus \Delta_X$ is an isomorphism (this is parallel to Section 4.3.3(i)). Moreover, if $r \geq 2$, then the image of the anchor map of $\mathfrak{a}_r$ is contained in $\Theta_X(-\Delta_X)$ (this is parallel to Section 4.3.3(ii)).

4.5.4. *A key property of Γ_r*

The groupoid Γ_r that we will construct has the following property: the anchor map induces an isomorphism $\mathrm{Lie}(\Gamma_r) \xrightarrow{\sim} \mathfrak{a}_r$.

4.6. Relation between Γ_r and Γ_{r+1}

By Section 4.5.4, we have

$$\mathrm{Lie}(\Gamma_{r+1}) = (\mathrm{Lie}(\Gamma_r))(-\Delta_X).$$

It turns out that in a certain sense,

(4.1) $\Gamma_{r+1} = \Gamma_r(-\Delta_X).$

The precise meaning of (4.1) is as follows. First, one has a morphism of groupoids $\Gamma_{r+1} \to \Gamma_r$ inducing the identity on X (which is the scheme of objects for both groupoids). Second, for any scheme S flat over X, the map

$$\mathrm{Mor}_X(S, \Gamma_{r+1}) \to \mathrm{Mor}_X(S, \Gamma_r)$$

is injective, and an X-morphism $f : S \to \Gamma_r$ belongs to its image if and only if the restriction of f to $S \times_X \Delta_X$ is equal to the composition

$$S \times_X \Delta_X \to X \xrightarrow{e} \Gamma_r,$$

where $e : X \to \Gamma_r$ is the unit.

4.7. The restriction of Γ_r to Δ_X

By Section 4.3.3(ii), $\Gamma_r \underset{X}{\times} \Delta_X$ is a smooth group scheme over Δ_X. Let us describe its fiber $(\Gamma_r)_z$ over a point $z \in \Delta_X$.

Let $n := \dim_z X$. Let m be the multiplicity of z in $X \times_Y z$ (i.e., in the fiber of φ corresponding to z); then $m \in \mathbb{N} \cup \{\infty\}$, and since $z \in \Delta_X$, we have $m \geq 2$. In Section 5.5, we will show the following:

  (i) if either $r \geq 3$ or $m \geq 3$, then $(\Gamma_r)_z \simeq \mathbb{G}_a^n$;
  (ii) if $m = r = 2$, z is a non-singular point of X, and the characteristic of k is not 2, then $(\Gamma_r)_z \simeq \mathbb{G}_m \ltimes \mathbb{G}_a^{n-1}$, where $\lambda \in \mathbb{G}_m$ acts on $\mathbb{G}_a^{n-1}$ as multiplication by λ^2;
  (iii) if $m = r = 2$, z is a non-singular point of X, and k has characteristic two, then $(\Gamma_r)_z \simeq \mathbb{G}_a^n$;
  (iv) if $m = r = 2$ and z is a singular point of X, then $(\Gamma_r)_z \simeq (\mathbb{Z}/2\mathbb{Z}) \times \mathbb{G}_a^n$.

5. Newton Groupoids (Details)

Let $\varphi : X \to Y$ be as in Section 4.3.1. Just as in Section 4.3.1, let $\Delta_X \subset X$ be the different and let $U = X \setminus \Delta_X$. We are going to define the groupoids Γ_r on X, $r \geq 2$, which were promised in Section 4.3.3.

5.1. Γ_r as a scheme over $X \times X$

This scheme will be obtained from $X \times X$ by a kind of "affine blow-up".

Note that $X \times_Y X$ and the diagonal X_{diag} are closed subschemes of $X \times X$ (because X and Y are separated). Let $I_1 \subset \mathcal{O}_{X \times X}$ be the sheaf of ideals of X_{diag}. Let $I_2 \subset \mathcal{O}_{X \times X}$ be the sheaf of ideals of $X \times_Y X$. Then $\mathcal{O}_{X \times X} \supset I_1 \supset I_2$.

Let $D := \Delta_X \times X$, $\tilde{D} := X \times \Delta_X$; then D and $\tilde{D}$ are effective Cartier divisors on $X \times X$ and $(X \times X) \setminus (D \cup \tilde{D}) = U \times U$. Let $j : U \times U \hookrightarrow X \times X$ be the open immersion.

We define $\mathcal{A}_r \subset j_* \mathcal{O}_{U \times U}$ to be the $\mathcal{O}_{X \times X}$-subalgebra generated by $I_1((r-1)D)$, $I_2(rD)$, $I_1((r-1)\tilde{D})$, $I_2(r\tilde{D})$.

Finally, we set $\Gamma_r^\varphi = \operatorname{Spec} \mathcal{A}_r$; this is a scheme affine over $X \times X$. Usually, we write Γ_r instead of Γ_r^φ.

LEMMA 5.1.1. (i) *The morphism $\Gamma_r \to X \times X$ is an isomorphism over $U \times U$.*

(ii) *The morphisms $p_1, p_2 : \Gamma_r \to X$ have equal restrictions to $p_1^{-1}(\Delta_X)$. Similarly, $p_1|_{p_2^{-1}(\Delta_X)} = p_2|_{p_2^{-1}(\Delta_X)}$.*

(iii) *Let $X' \subset X$ and $Y' \subset Y$ be open subschemes such that $\varphi(X') \subset Y'$. Let $\varphi' : X' \to Y'$ be induced by $\varphi : X \to Y$. Then, $\Gamma_r^{\varphi'}$ is obtained from Γ_r^φ by base change $X' \times X' \to X \times X$.*

(iv) *Γ_r^φ is not changed if $\varphi : X \to Y$ is composed with an étale morphism $Y \to \tilde{Y}$.*

PROOF. Checking (i) and (iii) is straightforward. Statement (ii) follows from the inclusions $I_1 \subset \mathcal{A}_r(-D)$ and $I_1 \subset \mathcal{A}_r(-\tilde{D})$. To prove (iv), use (ii) and the fact that $X \times_Y X$ and $X \times_{\tilde{Y}} X$ are equal in a neighborhood of X_{diag} (because the morphism $Y \to \tilde{Y}$ is étale). $\qquad\square$

5.2. Γ_r as a groupoid

In Section 5.4, we will prove that the two morphisms $\Gamma_r \to X$ are flat. Assuming this fact, we prove

PROPOSITION 5.2.1. *There is a unique way to make $\Gamma_r \to X \times X$ into a groupoid on X.*

$\Gamma_r = \Gamma_r^\varphi$ is called the rth *Newton groupoid* of $\varphi : X \to Y$.

PROOF. Because of Lemma 5.1.1(ii), we can assume that X and Y are affine.

By Lemma 5.1.1(i), we have an open embedding $U \times U \hookrightarrow \Gamma_r$. Its image is schematically dense in Γ_r (this is clear from the definition of Γ_r). Moreover, using flatness of the two morphisms $\Gamma_r \to X$, we see that the open embeddings

$$U \times U \times U \hookrightarrow \Gamma_r \times_X \Gamma_r, \quad U \times U \times U \times U \hookrightarrow \Gamma_r \times_X \Gamma_r \times_X \Gamma_r$$

also have schematically dense images. So, there is at most one morphism

$$(5.1) \qquad\qquad \Gamma_r \times_X \Gamma_r \to \Gamma_r$$

over $X \times X$, and if it exists, it automatically has the associativity property.

Let us check that (5.1) exists. Recall that $\Gamma_r := \operatorname{Spec} \mathcal{A}_r$. The ideals $I_1, I_2 \subset \mathcal{O}_{X \times X}$ that were used in the definition of $\mathcal{A}_r$ from Section 5.1 have the following properties:

$$p_{13}^* I_1 \subset p_{12}^* I_1 + p_{23}^* I_1, \quad p_{13}^* I_2 \subset p_{12}^* I_2 + p_{23}^* I_2,$$

where p_{12}, p_{13}, p_{23} are the three projections $X^3 \to X^2$. This implies that if f is a regular function on Γ_r, then its pullback with respect to the composed morphism

$$U \times U \times U \xrightarrow{p_{13}} U \times U \hookrightarrow \Gamma_r$$

extends to $\Gamma_r \times_X \Gamma_r$. This proves the existence of (5.1).

We also need the unit and the inversion map for Γ_r. The automorphism of $X \times X$ that takes (x_1, x_2) to (x_2, x_1) clearly has a unique lift to an isomorphism $i : \Gamma_r \xrightarrow{\sim} \Gamma_r$. The diagonal embedding $X \to X \times X$ has a unique lift to a morphism $e : X \to \Gamma_r$ (to prove its existence, one checks that if f is any regular function on Γ_X, then the restriction of f to $U_{\text{diag}} \subset U \times U \subset \Gamma_r$ extends to X_{diag}). Then, e is the unit and i is the inversion map for Γ_r (it suffices to check the required identities on schematically dense open subschemes).   □

5.3. Relation between Γ_r and Γ_{r+1}

In this section (which is not used in the rest of Section 5), we verify the claim of Section 4.6.

PROPOSITION 5.3.1. (i) *There is a unique morphism of $(X \times X)$-schemes $\Gamma_{r+1} \to \Gamma_r$. Moreover, it is a morphism of groupoids.*

(ii) *The corresponding morphism $\Gamma_{r+1} \underset{X \times X}{\times} (\Delta_X \times \Delta_X) \to \Gamma_r \underset{X \times X}{\times}$ $(\Delta_X \times \Delta_X)$ is trivial, i.e., it is equal to the composition*

$$\Gamma_{r+1} \underset{X \times X}{\times} (\Delta_X \times \Delta_X) \xrightarrow{p_i} \Delta_X \xrightarrow{e} \Gamma_r \underset{X \times X}{\times} (\Delta_X \times \Delta_X),$$

where e is the unit section and i equals 1 or 2.

(iii) *The affine morphism $\Gamma_{r+1} \to \Gamma_r$ can be described as follows.[6] Define $\Delta_r \subset \Gamma_r$ to be the preimage of the divisor Δ_X with respect to $p_1 : \Gamma_r \to X$ (or equivalently, with respect to $p_2 : \Gamma_r \to X$). Let $\nu : U \times U \to \Gamma_r$ be the open immersion. Let $\mathcal{B} \subset \nu_* \mathcal{O}_{U \times U}$ be the quasi-coherent unital $\mathcal{O}_{\Gamma_r}$-algebra generated by $I(\Delta_r)$, where $I \subset \mathcal{O}_{\Gamma_r}$ is the ideal of $e(\Delta_X)$. Then $\Gamma_{r+1} = \operatorname{Spec} \mathcal{B}$.*

(iv) *For any scheme S flat over X, the map*

$$(5.2) \qquad\qquad \operatorname{Mor}_X(S, \Gamma_{r+1}) \to \operatorname{Mor}_X(S, \Gamma_r)$$

---

[6]This description means that Γ_{r+1} is obtained from Γ_r by blowing up $e(\Delta_X)$ and then removing the strict transform of Δ_r.

is injective, and an X-morphism $f : S \to \Gamma_r$ belongs to its image if and only if the restriction of f to $S \times_X \Delta_X$ is equal to the composition

$$S \times_X \Delta_X \to \Delta_X \xrightarrow{\ e\ } \Gamma_r,$$

where e is the unit.

PROOF. Statement (i) is clear: the morphism $\Gamma_{r+1} \to \Gamma_r$ comes from the obvious inclusion $\mathcal{A}_r \subset \mathcal{A}_{r+1}$.

Statement (ii) essentially says that the composed map

$$\mathcal{A}_r \hookrightarrow \mathcal{A}_{r+1} \twoheadrightarrow \mathcal{A}_{r+1} \underset{\mathcal{O}_{X \times X}}{\otimes} \mathcal{O}_D$$

is equal to the composed map $\mathcal{A}_r \xrightarrow{e^*} \mathcal{O}_{X_{\mathrm{diag}}} \to \mathcal{A}_{r+1} \underset{\mathcal{O}_{X \times X}}{\otimes} \mathcal{O}_D$. This is checked straightforwardly.

Let us prove (iii). Let $J \subset \mathcal{A}_r$ be the ideal corresponding to the closed subscheme $e(\Delta_X) \subset \Gamma_r$. By Lemma 5.1.1(ii), $\mathcal{A}_r(-D) = \mathcal{A}_r(-\tilde{D})$, so $\mathcal{A}_r(D) = \mathcal{A}_r(\tilde{D})$ and $J(D) = J(\tilde{D})$. Statement (iii) essentially says that

(5.3) $$\mathcal{A}_{r+1} = \mathcal{C},$$

where $\mathcal{C} \subset j_* \mathcal{O}_{U \times U}$ is the $\mathcal{A}_r$-subalgebra generated by $J(D) = J(\tilde{D})$. Let us prove (5.3). Since J contains $I_1((r-1)D)$, $I_2(rD)$, $I_1((r-1)\tilde{D})$, and $I_2(r\tilde{D})$, we see that $\mathcal{A}_{r+1} \subset \mathcal{C}$. On the other hand, statement (ii) means that $J\mathcal{A}_{r+1} = \mathcal{A}_{r+1}(-D)$, so $J(D) \subset \mathcal{A}_{r+1}$ and therefore, $\mathcal{C} \subset \mathcal{A}_{r+1}$. This proves (5.3) and statement (iii).

Statement (iv) follows from (iii). □

5.4. Flatness of Γ_r

Let us prove that the two morphisms $\Gamma_r \to X$ are flat (later, we will show that they are smooth, see Proposition 5.5.1). By Lemma 5.1.1, it is enough to consider the situation of Section 4.3.2. In this situation, we will give a very explicit description of Γ_r (see Lemma 5.4.1).

We use the notation of Section 3.1.1, and we assume that the projection $\varphi : X \to \mathbb{A}^n = Y$ is generically étale (i.e., the restriction of Q to X is not a zero divisor). Let $U \subset X$ be the locus $Q \neq 0$.

Recall that $X \subset \mathbb{A}^{n+l}$ and the coordinates in $\mathbb{A}^{n+l}$ are denoted by

$$x_1, \ldots, x_n, y_1, \ldots, y_l.$$

We have $X \times X \subset \mathbb{A}^{n+l} \times \mathbb{A}^{n+l}$. The coordinates in $\mathbb{A}^{n+l} \times \mathbb{A}^{n+l}$ will be denoted by

$$x_i, y_j, \tilde{x}_i, \tilde{y}_j, \quad 1 \le i \le n, \ 1 \le j \le l.$$

By definition, $\Gamma_r := \mathrm{Spec}\, A$, where $A \subset H^0(U \times U)$ is the subalgebra generated by all regular functions on $X \times X$ and also the following ones:

(5.4) $$\xi_i := \frac{\tilde{x}_i - x_i}{Q(x,y)^r}, \quad \eta_j := \frac{\tilde{y}_j - y_j}{Q(x,y)^{r-1}},$$

(5.5) $$\tilde{\xi}_i := \frac{\tilde{x}_i - x_i}{Q(\tilde{x},\tilde{y})^r}, \quad \tilde{\eta}_j := \frac{\tilde{y}_j - y_j}{Q(\tilde{x},\tilde{y})^{r-1}}.$$

Let us now give an explicit description of the scheme Γ_r. Consider the morphism

(5.6) $$\Gamma_r \to X \times \mathbb{A}^{n+l},$$

where the map $\Gamma_r \to X$ is given by x_i's and y_j's, and the map $\Gamma_r \to \mathbb{A}^{n+l}$ is given by ξ_i's and η_j's.

LEMMA 5.4.1. (i) *The morphism* (5.6) *identifies* Γ_r *with the locally closed subscheme* $\Gamma_r' \subset X \times \mathbb{A}^{n+l}$ *defined by the equation*

(5.7) $$\eta + \hat{C}(x,y)u(x,y,\xi,\eta) = 0$$

(*which is a system of l scalar equations*) *and the inequality*

(5.8) $$v(x,y,\xi,\eta) \neq 0,$$

where $C(x,y)$ is the matrix $\frac{\partial f}{\partial y}$, $\hat{C}(x,y)$ is the matrix adjugate to $C(x,y)$, and

$$u(x,y,\xi,\eta) := -\frac{\begin{array}{c} f(x + Q(x,y)^r\xi, y + Q(x,y)^{r-1}\eta) \\ -f(x,y) - Q(x,y)^{r-1}C(x,y)\eta \end{array}}{Q(x,y)^r},$$

(5.9) $$v(x,y,\xi,\eta) := \frac{Q(x + Q(x,y)^r\xi, y + Q(x,y)^{r-1}\eta)}{Q(x,y)}.$$

(ii) *The morphism* $\Gamma_r \to X$ *given by x_i's and y_j's is flat.*

Note that $u = (u_1, \dots, u_l)$ is a vector function. Also note that $v(x,y,\xi,\eta)$ and $u_j(x,y,\xi,\eta)$ are polynomials (not merely rational functions).

PROOF. By (5.4), we have

(5.10) $$\tilde{x}_i = x_i + Q(x,y)^r\xi_i, \quad \tilde{y}_j = y_j + Q(x,y)^{r-1}\eta_j.$$

So, formula (5.9) shows that

(5.11) $$v(x,y,\xi,\eta) = Q(\tilde{x},\tilde{y})/Q(x,y).$$

Thus, $Q(\tilde{x},\tilde{y})/Q(x,y)$ is a regular function on Γ_r. By symmetry, $Q(x,y)/Q(\tilde{x},\tilde{y})$ is also a regular function on Γ_r. Therefore, the inequality (5.8) holds on Γ_r. It is easy to check that the equality (5.7) also holds on Γ_r (use the definition of u from the formulation of the lemma and the equalities $f(\tilde{x},\tilde{y}) = f(x,y) = 0$).

The coordinate ring of Γ_r is generated by $x_i, y_j, \xi_i, \eta_j, v^{-1}$; this is clear from (5.10) and the formulas

$$\tilde{\xi}_i = v^{-r}\xi_i, \quad \tilde{\eta}_j = v^{1-r}\eta_j,$$

which follow from (5.4)–(5.5) and (5.11). So, the morphism (5.6) identifies Γ_r with a closed subscheme of the locally closed subscheme $\Gamma_r' \subset X \times \mathbb{A}^{n+l}$ defined by (5.7) and (5.8). On the other hand, $\Gamma_r \supset U \times U$. So to prove statement (i), it remains to be shown that $U \times U$ is schematically dense in Γ_r'. This follows from flatness of the morphism $\Gamma_r' \to X$, which we are going to prove.

Note that (5.7) is a system of l equations for a point in $X \times \mathbb{A}^{n+l}$, so it suffices to check that the fiber of Γ_r' over any point $(x_0, y_0) \in X$ has dimension $\leq n$. Since $\Gamma_r' \times_X U = U \times U$, we can assume that $(x_0, y_0) \notin U$, which means that $Q(x_0, y_0) = 0$. Under this condition, we have to show that the set of solutions to the equation

$$(5.12) \qquad \eta + \hat{C}(x_0, y_0) u(x_0, y_0, \xi, \eta) = 0$$

(which is a system of l equations for $n + l$ unknowns) has dimension $\leq n$. To see this, note that $\det C(x_0, y_0) = Q(x_0, y_0) = 0$, so $\hat{C}(x_0, y_0)$ has rank ≤ 1. Also note that $u(x_0, y_0, \xi, \eta)$ is a sum of a quadratic form in η and a function of ξ (this is clear from the definition of u given in the formulation of the lemma). These two facts imply that for any ξ, there are at most two values of η satisfying (5.12). $\qquad \square$

5.4.2. *The composition law in Γ_r*

We have described Γ_r as a subscheme of $X \times \mathbb{A}^{n+l}$. In these terms, one can write an explicit formula for the composition map

$$(5.13) \qquad \Gamma_r \times_X \Gamma_r \to \Gamma_r.$$

A point of $\Gamma_r \times_X \Gamma_r$ is a collection $(x, y, \xi, \eta, \tilde{\xi}, \tilde{\eta})$, where $(x, y) \in X$, ξ and η satisfy conditions (5.7)–(5.8), and $\tilde{\xi}, \tilde{\eta}$ satisfy similar conditions

$$\tilde{\eta} + \hat{C}(\tilde{x}, \tilde{y}) u(\tilde{x}, \tilde{y}, \tilde{\xi}, \tilde{\eta}) = 0, \quad v(\tilde{x}, \tilde{y}, \tilde{\xi}, \tilde{\eta}) \neq 0;$$

here $\tilde{x}, \tilde{y}$ are given by (5.10). It is straightforward to check that the map (5.13) is as follows:

$$(5.14) \qquad (x, y, \xi, \eta, \tilde{\xi}, \tilde{\eta}) \mapsto (x, y, \xi + v(x, y, \xi, \eta)^r \tilde{\xi}, \eta + v(x, y, \xi, \eta)^{r-1} \tilde{\eta}),$$

where $v(x, y, \xi, \eta)$ is defined by (5.9).

5.5. Smoothness and the group schemes $(\Gamma_r)_z$, $z \in \Delta_X$

PROPOSITION 5.5.1. (i) *The groupoid Γ_r is smooth.*

(ii) *For $z \in \Delta_X$, the group scheme $(\Gamma_r)_z := \Gamma_r \times_X z$ is as described in Section 4.7.*

PROOF. By Lemma 5.4.1(ii), Γ_r is flat over X. So, it suffices to prove statement (ii).

Let $z \in \Delta_X$. Let $\mathrm{Fib}_z := X \times_Y z$ be the corresponding fiber[7] of $\varphi :$ $X \to Y$. Since the question is local, we can assume that X and Y are as in Section 5.4. Moreover, we can assume that l (i.e., the number of the variables y_j) is equal to the dimension of the tangent space of Fib_z at z. This means that $C(x_0, y_0) = 0$, where C is the matrix $\frac{\partial f}{\partial y}$ and $(x_0, y_0) = z$.

By Lemma 5.4.1 and formula (5.14), $(\Gamma_r)_z$ is the subscheme of $\mathbb{A}^{n+l}$ defined by the conditions

$$(5.15) \qquad \eta + \hat{C}(x_0, y_0)u(x_0, y_0, \xi, \eta) = 0, \quad v(x_0, y_0, \xi, \eta) \neq 0$$

and equipped with the group operation

$$(5.16) \qquad (\tilde{\xi}, \tilde{\eta}) \cdot (\xi, \eta) = (\xi + v(x_0, y_0, \xi, \eta)^r \tilde{\xi}, \eta + v(x_0, y_0, \xi, \eta)^{r-1} \tilde{\eta}),$$

where u and v are as in the formulation of Lemma 5.4.1. Since $C(x_0, y_0) = 0$, we see that $\hat{C}(x_0, y_0) = 0$ if $l > 1$ and $\hat{C}(x_0, y_0) = 1$ if $l = 1$.

If $r \geq 3$, then it is easy to see that $u(x_0, y_0, \xi, \eta) = 0$, $v(x_0, y_0, \xi, \eta) = 1$, so $(\Gamma_r)_z \simeq \mathbb{G}_a^n$.

Now, suppose that $r = 2$. If $l > 1$, then $\hat{C}(x_0, y_0) = 0$, and it is easy to check[8] that $v(x_0, y_0, \xi, \eta) = 1$, so $(\Gamma_r)_z \simeq \mathbb{G}_a^n$. This agrees with Section 4.7(i): indeed, l is the dimension of the tangent space of Fib_z at z, so if $l > 1$, then the multiplicity of z in Fib_z is greater than 2.

Now let $r = 2$, $l = 1$. Recall that $\frac{\partial f}{\partial y}$ vanishes at (x_0, y_0), so the Taylor expansion of $f(x_0, y_0 + \eta)$ looks as follows:

$$f(x_0, y_0 + \eta) = f(x_0, y_0) + a\eta^2 + \cdots$$

It is easy to check that

$$u(x_0, y_0, \xi, \eta) = a\eta^2 + \frac{\partial f}{\partial x}(x_0, y_0) \cdot \xi, \quad v(x_0, y_0, \xi, \eta) = 1 + 2a\eta,$$

so $(\Gamma_r)_z$ is the subscheme of $\mathbb{A}^{n+1}$ defined by the conditions

$$(5.17) \qquad \eta + a\eta^2 + \frac{\partial f}{\partial x}(x_0, y_0) \cdot \xi = 0, \quad 1 + 2a\eta \neq 0$$

and equipped with the group operation

$$(5.18) \qquad (\tilde{\xi}, \tilde{\eta}) \cdot (\xi, \eta) = (\xi + (1 + 2a\eta)^2 \tilde{\xi}, \eta + \tilde{\eta} + 2a\eta\tilde{\eta}).$$

If $a = 0$, this group scheme is isomorphic to $\mathbb{G}_a^n$, which agrees with Section 4.7(i). If $a \neq 0$, the group scheme $(\Gamma_r)_z$ depends on whether $\frac{\partial f}{\partial x}(x_0, y_0) = 0$ (i.e., on whether z is a singular point of X), and it is straightforward to check that $(\Gamma_r)_z$ is as described in Section 4.7(ii–iv). □

---

[7]If z is a k-point, we can write this fiber as $\varphi^{-1}(\varphi(z))$.

[8]Use formula (5.9) and note that Q has zero differential at (x_0, y_0) because $Q = \det C$, $C(x_0, y_0) = 0$, and $l > 1$.

5.6. Verifying the claim of Section 4.3.3(iii)

For a k-algebra A, let $F(A)$ be the set of triples $(I, \bar{x}, \bar{y})$, where $I \subset A$ is an ideal, $\bar{x} \in (A/I^r)^n$, $\bar{y} \in (A/I^{r-1})^l$. Consider the map

$$(5.19) \qquad X(A) \to F(A),$$

which takes $(x, y) \in X(A) \subset A^n \times A^l$ to $(I, \bar{x}, \bar{y})$, where I is the ideal generated by $Q(x, y)$, $\bar{x} \in (A/I^r)^n$ is the image of $x \in A^n$, and $\bar{y} \in (A/I^{r-1})^l$ is the image of $y \in A^l$. It is easy to check that the map (5.19) factors through the quotient set $X(A)/\Gamma_r(A)$. Since F is an fppf sheaf, we get a map

$$(X/\Gamma_r)(A) \to F(A).$$

Applying the above construction to $A = R[t]$, one gets a morphism

$$(5.20) \qquad \mathrm{Maps}_N^{\circ}(\mathbb{A}^1, X/\Gamma_r) \to Z_r,$$

where $\mathrm{Maps}_N^{\circ}(\mathbb{A}^1, X/\Gamma_r)$ is as in Section 4.2.6 and Z_r is as in Section 3.2.1. Moreover, the morphism (5.20) factors through $Z_r' := \mathrm{Im}\,(Z_{r+1} \to Z_r)$ (recall that Z_r' is an open subscheme of Z_r and if $r \geq 3$, then $Z_r' = Z_r$). It is easy to check that the morphism (5.20) is a monomorphism.

It remains to be proved that if $r \geq 3$, then for every triple $(q, \bar{x}, \bar{y}) \in Z_r(R)$, there exists a faithfully flat finitely presented $R[t]$-algebra A, such that the triple $(qA, \bar{x}, \bar{y}) \in F(A)$ belongs to the image of the map (5.19). Choose $x_0 \in R[t]^n$, $y_0 \in R[t]^l$ mapping to $\bar{x} \in (R[t]/(q^r))^n$, $\bar{y} \in (R[t]/(q^{r-1}))^l$. It suffices to find a flat finitely presented morphism $\mathrm{Spec}\,A' \to \mathrm{Spec}\,R[t]$ with non-empty fibers over points of $\mathrm{Spec}\,R[t]/(q)$ and an element $\eta \in (A')^l$, such that

$$(5.21) \qquad f(x_0, y_0 + q^{r-1}\eta) = 0,$$

$$(5.22) \qquad q^{-1}Q(x_0, y_0 + q^{r-1}\eta) \in (A')^{\times}.$$

By Section 3.2.1(6) and the assumption $r \geq 3$, the element $q^{-1}Q(x_0, y_0)$ is invertible modulo q. So, (5.21) is the only essential condition for η: one can always achieve (5.22) by modifying A' slightly.

Let $C_0 := \frac{\partial f}{\partial y}(x_0, y_0)$ and let $\hat{C}_0$ be the adjugate matrix, so $\hat{C}_0 C_0 = Q(x_0, y_0)$. Write

$$f(x_0, y_0 + t\eta) = f(x_0, y_0) + tC_0\eta + t^2 g(t, \eta),$$

and then rewrite (5.21) as

$$(5.23) \qquad q^{-1}Q(x_0, y_0)\eta + q^{-r}\hat{C}_0 f(x_0, y_0) + q^{r-2}\hat{C}_0 g(q^r, \eta) = 0,$$

Note that $q^{-1}Q(x_0, y_0) \in A$ and $q^{-r}\hat{C}_0 f(x_0, y_0) \in A^l$ by conditions (4)-(5) from Section 3.2.1.

Let W be the A-scheme whose A'-points are solutions to (5.23). It suffices to show that the fiber of W over any point of $\mathrm{Spec}\,A/(q)$ is non-empty and the morphism $W \to \mathrm{Spec}\,A$ is flat at each point of W where q vanishes. Since (5.23) is a system of l equations for l unknowns, it is enough to show that the fiber of W over any point of $\operatorname{Spec} A/(q)$ is finite and non-empty. In fact, it has exactly one point because $q^{-1}Q(x_0, y_0)$ is invertible modulo q and $r - 2 > 0$.

5.7. The Lie algebroid of Γ_r

Let $r \geq 2$. As already mentioned in Section 4.5.3, we have the Lie subalgebroid $\mathfrak{a}_r := (\varphi^*\Theta_Y)(-r\Delta_X) \subset \Theta_X$. Using the definition of the anchor map $\tau : \operatorname{Lie}(\Gamma_r) \to \Theta_X$ given in Section 4.5.2, one checks that τ induces an isomorphism $\operatorname{Lie}(\Gamma_r) \xrightarrow{\sim} \mathfrak{a}_r$ (e.g., one can use Lemma 5.4.1).

Acknowledgment

This work is partially supported by NSF grant DMS-1303100.

Bibliography

[1] M. Crainic and R. L. Fernandes, Lectures on integrability of Lie brackets. In *Lectures on Poisson Geometry*, eds. T. Ratiu, A. Weinstein and N. T. Zung, Geometry & Topology Monographs, Vol. 17, Geometry & Topology Publications, Coventry, 2011. Also available as e-print arXiv:math/0611259.

[2] V. Drinfeld, On the Grinberg–Kazhdan formal arc theorem, preprint, 2002, arXiv:math-AG/0203263.

[3] M. Finkelberg and I. Mirković, Semiinfinite flags. I. Case of global curve P^1. In *Differential Topology, Infinite-Dimensional Lie Algebras, and Applications*, pp. 81–112, Amer. Math. Soc. Transl. Ser. 2, Vol. 194, Amer. Math. Soc., Providence, RI, 1999. See also e-print alg-geom/9707010.

[4] M. Grinberg and D. Kazhdan, Versal deformations of formal arcs, *Geometric and Functional Analysis*, 10(3):543–555, 2000. See also e-print math.AG/9812104.

[5] F. F. Knudsen and D. Mumford, The projectivity of the moduli space of stable curves. I. Preliminaries on "det" and "Div", *Math. Scand.* 30(1):19–55, 1976.

[6] G. Laumon and L. Moret-Bailly, *Champs Algébriques*, Ergebnisse der Mathematik und ihrer Grenzgebiete (3 Folge, A Series of Modern Surveys in Mathematics), Vol. 39, Springer-Verlag, Berlin, 2000.

[7] M. Olsson, *Algebraic Spaces and Stacks*, American Mathematical Society Colloquium Publications, Vol. 62, American Mathematical Society, Providence, RI, 2016.

[8] *Schémas en groupes, I: Propriétés générales des schémas en groupes*, Séminaire de Géométrie Algébrique du Bois Marie 1962/64 (SGA 3), Dirigé par M. Demazure et A. Grothendieck, Lecture Notes in Mathematics, Vol. 151, Springer-Verlag, Berlin, 1970. Reedited by P. Gille and P. Polo, Documents Mathématiques (Paris), **7**, Société Mathématique de France, Paris, 2011.

[9] Stacks project (run by A. J. de Jong), Available online at http://stacks.math.columbia.edu/.

4

Non-complete Completions

4

Mercedes Haiech

4

Institut de recherche mathématique de Rennes
UMR 6625 du CNRS
Université de Rennes 1
Campus de Beaulieu
35042 Rennes cedex, France

The completion of a topological ring R is, by definition, endowed with a topology for which the completion is complete. However, when R is endowed with an adic topology, there are various natural topologies that can be defined on the completion. When R is a noetherian ring, these various topologies coincide on the completion, but this is no longer true when R is not noetherian. We present such an example in this chapter.

1. Introduction

The notion of completion is a well-known one in analysis on metrizable spaces. In an algebro-geometric setting, completions arise naturally when we look at formal neighborhoods of a point. The main definitions of analysis can be adapted in a way compatible with the algebraic structures. In our framework, we will use the following definition of completion.

DEFINITION 1.1 (Completion). Let R be a topological ring, whose topology is adic, say given by the filtration $(I^n)_{n \in \mathbf{N}^*}$. Let M be a topological R-module whose topology is given by a filtration of submodules, say $M_1 \subset M_2 \subset \cdots$, such that for every integer $n \in \mathbf{N}^*$, we have $I^n M \subset M_n$.
The completion $\widehat{M}$ of the topological R-module M is defined as the inverse limit in the category of topological R-modules:

$$\widehat{M} = \varprojlim_{n} M/M_n.$$

In particular, we will say that M is complete if the canonical map $M \to \widehat{M}$ is an isomorphism in the category of topological R-modules.

Even though the definition is given for R-modules, we can also define the completion of a ring A, endowed with a filtration, by seeing A as an A-module and applying the previous definition.

EXAMPLE 1.2. Let us give two examples to showcase this definition. In these examples, k will be a field.

(1) The first one is the completion of the polynomial ring $k[x]$ endowed with the filtration given by $(\langle x \rangle^n)_{n \in \mathbf{N}^*}$. This ring is also the ring $k[[x]]$ of formal power series in one variable. An element f of this ring is a formal sum of the form $f = \sum_{i \geq 0} a_i x^i$, where $a_i \in k$.

(2) The second example is a generalization of the first one. Let us consider the polynomial ring $k[(x_l)_{l \in \mathbf{N}}]$ with countably infinitely many variables. We denote by $I = \langle (x_l)_{l \in \mathbf{N}} \rangle$ the ideal of $k[(x_l)_{l \in \mathbf{N}}]$ generated by all the variables and endow $k[(x_l)_{l \in \mathbf{N}}]$ with the filtration $(I^n)_{n \in \mathbf{N}}$. The completion $k\widehat{[(x_l)_{l \in \mathbf{N}}]}$ is also denoted by $k[[(x_l)_{l \in \mathbf{N}}]]$.

The completion $\widehat{M}$ of a topological R-module M has a natural neighborhood basis of 0 which is explicitly described by the filtration

$$(K_n = \mathrm{Ker}(\widehat{M} \to M/I^n M))_{n \in \mathbf{N}^*},$$

where the maps $\pi_n \colon \widehat{M} \to M/I^n M$ are given by the universal property of inverse systems. This leads to the following question.

QUESTION 1.3. Do the two topologies, given respectively by the filtration $(I^n \widehat{M})_{n \in N^*}$ and $(K_n)_{n \in N^*}$, coincide on $\widehat{M}$?

The answer is well known if R is a noetherian ring and M a finitely generated R-module (see, for example, [5, 23.L, Corollary 4]).

THEOREM 1.4. Let R be a noetherian ring, I an ideal of R and M a finitely generated R-module endowed with the I-adic topology. If $\widehat{M} = \varprojlim_n M/I^n M$ denotes the completion of M and $(K_n)_{n \in \mathbf{N}^*}$ the filtration on $\widehat{M}$ where $K_n = \mathrm{Ker}(\widehat{M} \to M/I^n M)$, then, for every integer $n \geq 1$, we have $K_n = \widehat{I^n M} = I^n \widehat{M} = (\widehat{I})^n \widehat{M}$.

The answer is more complicated when it comes to non-noetherian rings. In particular, does Theorem 1.4 still hold?

Yekutieli gives in the introduction of [8] a counterexample in the non-noetherian case. More precisely, he proves that, if A denotes the ring $k[[(x_l)_{l \in \mathbf{N}}]]$ of formal power series and $\mathfrak{m} = \langle (x_l)_{l \in \mathbf{N}} \rangle$ the ideal of the polynomial ring $k[(x_l)_{l \in \mathbf{N}}]$ generated by the elements $(x_l)_{l \in \mathbf{N}}$, then A is not complete for the topology induced by the filtration $\mathfrak{m}^n A$.

If we go back to the general case, as $\widehat{M}$ is also an $\widehat{R}$-module, where $\widehat{R}$ denotes the completion of the ring R with respect to the I-adic topology, we could wonder if it is possible for $\widehat{M}$ to be $\widehat{I}$-adically complete, where $\widehat{I}$ is the completion (as a R-module) of the ideal I. It is well known that if R is a noetherian ring and if M is finitely generated over R, the $\widehat{I}$-adic topology coincides with the natural topology on $\widehat{M}$ (see [**3, 5**]). However, in the general case, this turns out to be false, as suggested by Exercise 12 in [**1**, Chapter III, §2] and explained in Stacks Project [**7**]. In this chapter, we have organized the arguments of the previous reference and proved that the result given in [**7**] can be generalized as follows.

THEOREM 1.5. *Let k be a field, $k[(x_l)_{l\in\mathbf{N}}]$ a polynomial ring and $\mathfrak{m} = \langle(x_l)_{l\in\mathbf{N}}\rangle$ the ideal generated in $k[(x_l)_{l\in\mathbf{N}}]$ by the elements $(x_l)_{l\in\mathbf{N}}$. Then, the completion $k[[(x_l)_{l\in\mathbf{N}}]]$ of the ring $k[(x_l)_{l\in\mathbf{N}}]$, when the ring $k[(x_l)_{l\in\mathbf{N}}]$ is endowed with the $\mathfrak{m}$-adic topology, is neither $\mathfrak{m}$-adically complete nor $\widehat{\mathfrak{m}}$-adically complete. More generally, for every $n \geq 1$, the completion $k[[(x_l)_{l\in\mathbf{N}}]]$ is not $\widehat{\mathfrak{m}^n}$-adically complete.*

2. A Necessary and Sufficient Condition to be Adically Complete

PROPOSITION 2.1. *Let R be a ring and $(K_i)_{i\in\mathbf{N}}$, $(L_i)_{i\in\mathbf{N}}$, $(M_i)_{i\in\mathbf{N}}$ be three inverse systems of R-modules whose transition maps are respectively denoted by $k_i\colon K_{i+1} \to K_i$, $l_i\colon L_{i+1} \to L_i$ and $m_i\colon M_{i+1} \to M_i$. Assume that*

$$0 \to K_i \to L_i \to M_i \to 0$$

is an exact sequence of R-modules, such that the following diagram of R-modules morphisms is commutative:

$$
\begin{array}{ccccccccc}
0 & \longrightarrow & K_i & \xrightarrow{\tau_i} & L_i & \xrightarrow{\mu_i} & M_i & \longrightarrow & 0 \\
& & {\scriptstyle k_i}\big\uparrow & & {\scriptstyle l_i}\big\uparrow & & {\scriptstyle m_i}\big\uparrow & & \\
0 & \longrightarrow & K_{i+1} & \xrightarrow{\tau_{i+1}} & L_{i+1} & \xrightarrow{\mu_{i+1}} & M_{i+1} & \longrightarrow & 0
\end{array}
$$

If every morphism k_i is surjective, then the following sequence is exact:

$$0 \longrightarrow \varprojlim_i K_i \xrightarrow{\tau} \varprojlim_i L_i \xrightarrow{\mu} \varprojlim_i M_i \longrightarrow 0$$

REMARK 2.2. Proposition 2.1 is a special case of what happens when the directed inverse system $(K_i, k_i)_{i\geq 1}$ is a Mittag-Leffler inverse system (see, for example, [**4**, III §10 Proposition 10.3]).

PROOF. Recall that the inverse limit functor is left exact [**4**, III §10, p. 164].

Thus, we only have to prove that the map μ is surjective. Let $y = (y_1, y_2, \dots) \in \varprojlim_i M_i$. Since, for every integer $i \geq 1$, the morphism μ_i is surjective, we can find $x_i \in L_i$ such that $\mu_i(x_i) = y_i$. But the element $(x_1, x_2, \ldots)$ is possibly not in $\varprojlim_i L_i$. Nevertheless, we have

$$\mu_i(l_i(x_{i+1})) = m_i(\mu_{i+1}(x_{i+1})) = m_i(y_{i+1}) = y_i = \mu_i(x_i).$$

So, $l_i(x_{i+1}) - x_i \in \mathrm{Ker}(\mu_i)$. Let us construct by induction a sequence $(\tilde{x}_i)_{i \in \mathbf{N}} \in \varprojlim_i L_i$ as follows: let $\tilde{x}_1 = x_1$. As $l_2(x_2) - x_1 \in \mathrm{Ker}(\mu_1) = \mathrm{Im}(\tau_1)$, there is an element $c_1 \in K_1$, such that $l_1(x_2) - x_1 = \tau_1(c_1)$. As the map k_1 is surjective, we can find $\tilde{c}_2 \in K_2$, such that $k_1(\tilde{c}_2) = c_1$. Let $\tilde{x}_2 = x_2 - \tau_2(\tilde{c}_2)$. So, we have

$$\begin{aligned}
l_1(\tilde{x}_2) &= l_1(x_2 - \tau_2(\tilde{c}_2)) \\
&= l_1(x_2) - l_1(\tau_2(\tilde{c}_2)) \\
&= x_1 + \tau_1(c_1) - l_1(\tau_2(\tilde{c}_2)) \\
&= x_1 + \tau_1(c_1) - \tau_1(k_1(\tilde{c}_2)) \\
&= x_1 \\
&= \tilde{x}_1.
\end{aligned}$$

Suppose we have constructed the elements $(\tilde{x}_1, \tilde{x}_2, \ldots, \tilde{x}_r) \in \prod_{i=1}^r L_i$, such that, for every integer $i \in \{1, \ldots, r-1\}$, we have $l_i(\tilde{x}_{i+1}) = \tilde{x}_i$ and $l_r(x_{r+1}) - \tilde{x}_r \in \mathrm{Ker}(\mu_r)$ and, for every integer $i \in \{1, \ldots, r\}$, we also have $\mu_i(\tilde{x}_i) = y_i$. There is an element $c_r \in K_r$, such that $l_r(x_{r+1}) - \tilde{x}_r = \tau_r(c_r)$. As the map k_r is surjective, we can find an element $\tilde{c}_{r+1} \in K_{r+1}$, such that $k_r(\tilde{c}_{r+1}) = c_r$. Let $\tilde{x}_{r+1} = x_{r+1} - \tau_{r+1}(\tilde{c}_{r+1})$. It is clear that $l_{r+1}(x_{r+2}) - \tilde{x}_{r+1} \in \mathrm{Ker}(\mu_{r+1})$ and that $\mu_{r+1}(\tilde{x}_{r+1}) = y_r$. Furthermore,

$$\begin{aligned}
l_r(\tilde{x}_{r+1}) &= l_r(x_{r+1} - \tau_2(\tilde{c}_{r+1})) \\
&= l_r(x_{r+1}) - l_r(\tau_{r+1}(\tilde{c}_{r+1})) \\
&= \tilde{x}_r + \tau_r(c_r) - l_r(\tau_{r+1}(\tilde{c}_{r+1})) \\
&= \tilde{x}_r + \tau_r(c_r) - \tau_r(k_r(\tilde{c}_{r+1})) \\
&= \tilde{x}_r.
\end{aligned}$$

Thus, we have constructed by induction an element $x = (\tilde{x}_1, \tilde{x}_2, \ldots) \in \varprojlim_i L_i$, which satisfies $\mu(x) = y$. So, the morphism μ is surjective. $\square$

PROPOSITION 2.3. *With the notations and under the assumptions of Proposition 2.1, the following assertions are equivalent:*

(1) *The map $\mu \colon \varprojlim_i L_i \to \varprojlim_i M_i$ is an isomorphism of R-modules.*
(2) *We have $\varprojlim_i K_i \simeq 0$.*
(3) *For every integer $i \geq 0$, we have $K_i \simeq 0$.*

PROOF. The implications $(3) \Rightarrow (2) \Rightarrow (1)$ are clear.
Assuming (1), we will prove (2). Since the sequence

$$0 \longrightarrow \varprojlim_i K_i \overset{\tau}{\longrightarrow} \varprojlim_i L_i \overset{\mu}{\longrightarrow} \varprojlim_i M_i \longrightarrow 0$$

is exact, then τ is injective, hence $\varprojlim_i K_i \simeq \mathrm{Im}(\tau)$. Furthermore, $\mathrm{Im}(\tau) = \mathrm{Ker}(\mu) = 0$, which proves $\varprojlim_i K_i \simeq 0$.

Now, assume (2), we will prove (3). Suppose, by *reductio ad absurdum*, that we can find an integer $n \in \mathbf{N}^*$, such that $K_n \neq 0$. In particular, let $x_n \in K_n \setminus \{0\}$. Since the transition maps $k_i \colon K_{i+1} \to K_i$ are supposed to be surjective, we can construct by induction a sequence $(x_i)_{i \geq n}$, such that, for every integer $i \geq n$, $k_i(x_{i+1}) = x_i$. This means in particular that $x_i \neq 0$ for every $i \geq n$. For $1 \leq i < n$, we define by induction x_i by $x_i := k_i(x_{i+1})$. Finally, the element $(x_i)_{i \in \mathbf{N}^*}$ is in $\varprojlim_i K_i$ by construction and is non-zero. This is a contradiction; thus, for every integer $i \geq 0$, we have $K_i \simeq 0$. $\square$

PROPOSITION 2.4. *Let R be a ring and I an ideal of R. Let M be an R–module. Let $\widehat{M} := \varprojlim_n M/I^n M$. For $n \geq 1$, set*

$$K_n = \mathrm{Ker}(\widehat{M} \to M/I^n M).$$

Let $m \in \mathbf{N} \setminus \{0\}$. For every integer $n \geq 1$, we have $K_{(n+1)m} + I^{nm}\widehat{M} = K_{nm}$.

PROOF. For every integer $k \geq 1$, we denote by π_k the canonical morphism $\widehat{M} \to M/I^k M$. Hence, $K_k = \mathrm{Ker}(\pi_k)$.

Let $m \geq 1$ be an integer. Since the following diagram of R–modules morphisms is commutative

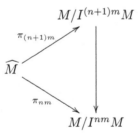

we get $\mathrm{Ker}(\pi_{(n+1)m}) \subset \mathrm{Ker}(\pi_{nm})$. Besides, it is clear that $I^{nm}\widehat{M} \subset K_{nm}$. Thus, we deduce the inclusion $K_{(n+1)m} + I^{nm}\widehat{M} \subset K_{nm}$.

Now, let $x = (x_j)_{j \in \mathbf{N}^*} \in \widehat{M}$, where $x_j \in M/I^j M$. For the purpose of notations, x_j may designate an element in $M/I^j M$ or one of its fibers in M. Assume that $x \in K_{nm}$, hence, for every integer $j \leq nm$, we have $x_j = 0$. We set $u = (u_j)_{j \in \mathbf{N}^*} \in \widehat{M}$ defined from x in the following way:

$$\begin{cases} u_j = 0 & \forall j,\ 1 \leq j \leq mn, \\ u_j = x_i & \forall j,\ mn+1 \leq j < m(n+1), \\ u_j = x_{m(n+1)} & \forall j,\ j \geq m(n+1). \end{cases}$$

By construction, $u \in I^{nm}\widehat{M}$. Furthermore, $x - u \in K_{(n+1)m}$ since, for every $j \leq (n+1)m$, we have $x_j - u_j = 0$. So, we have proved that $x = y + u \in K_{(n+1)m} + I^{nm}\widehat{M}$, which leads us to the conclusion. $\square$

PROPOSITION 2.5. *Let R be a ring, I an ideal in R. Let M be an R-module. Denote $\widehat{M} := \varprojlim_n M/I^n M$. Let $m \in \mathbf{N} \setminus \{0\}$. Then, we have an isomorphism of R-modules $\widehat{M} \simeq \varprojlim_n M/I^{nm} M$.*

PROOF. For $a \geq b$, let $\pi_{a,b}$ be the canonical map $\pi_{a,b} \colon M/I^a M \to M/I^b M$.

Then, the map

$$\widehat{M} \to \varprojlim_n M/I^{nm} M$$
$$(x_j)_{j \in \mathbf{N}^*} \mapsto (x_{jm})_{j \in \mathbf{N}^*}$$

is an isomorphism of R-modules, whose inverse is given by

$$\varprojlim_n M/I^{nm} M \to \widehat{M}$$
$$(x_{jm})_{j \in \mathbf{N}^*} \mapsto (y_j)_{j \in \mathbf{N}^*}$$

where, if for some integer k, we have $km < j \leq (k+1)m$, then

$$y_j = \pi_{(k+1)m,j}(x_{(k+1)m}). \qquad \square$$

THEOREM 2.6. *Let R be a ring and I an ideal in R. Let M be an R-module. Set $\widehat{M} := \varprojlim_n M/I^n M$ and $K_n = \mathrm{Ker}(\widehat{M} \to M/I^n M)$. Let $\mathfrak{b}$ be an ideal of $\widehat{R}$, such that there exists $m \in \mathbf{N}^*$, such that $I^m \widehat{M} \subset \mathfrak{b} \widehat{M} \subset K_m$. Then, the completion $\widehat{M}$ is $\mathfrak{b}$-adically complete if and only if, for every integer $n \geq 1$, the equality $K_{nm} = \mathfrak{b}^n \widehat{M}$ holds.*

PROOF. As for every integer $n \geq 1$, the inclusions $\mathfrak{b}^n \widehat{M} \subset (K_m)^n \subset K_{nm}$ hold, we will deduce an exact sequence:

$$0 \longrightarrow K_{nm}/\mathfrak{b}^n \widehat{M} \xrightarrow{\tau_n} \widehat{M}/\mathfrak{b}^n \widehat{M} \xrightarrow{\mu_n} M/I^{nm} M \longrightarrow 0.$$

Let us describe the morphisms τ_n and μ_n to prove the exactness of the previous sequence. Let $i_n : K_{nm} \to \widehat{M}$ be the including map. So, the induced morphism $\tau_n : K_{nm}/\mathfrak{b}^n \widehat{M} \to \widehat{M}/\mathfrak{b}^n \widehat{M}$ is also injective. As $\mathfrak{b}^n \widehat{M} \subset K_{nm}$, we have $\widehat{M}/\mathfrak{b}^n \widehat{M} \to \widehat{M}/K_{nm} \cong M/I^{nm} M$. This morphism is surjective since it is induced from the surjective morphism $\widehat{M} \twoheadrightarrow \widehat{M}/K_{nm}$. We have to prove that $\mathrm{Ker}(\mu_n) = \mathrm{Im}(\tau_n)$. It is clear since $\mathrm{Ker}(\mu_n) = K_{nm}/\mathfrak{b}^n \widehat{M} = \mathrm{Im}(\tau_n)$.

Now, we will prove that the sequence

$$0 \to \varprojlim_n K_{nm}/\mathfrak{b}^n \widehat{M} \to \varprojlim_n \widehat{M}/\mathfrak{b}^n \widehat{M} \to \varprojlim_n M/I^{nm} M \to 0$$

is exact.

We have three inverse systems: $(K_{nm}/\mathfrak{b}^n \widehat{M}, k_n)_{n \in \mathbf{N}^*}$, $(\widehat{M}/\mathfrak{b}^n \widehat{M}, l_n)_{n \in \mathbf{N}^*}$ and $(M/I^{nm} M, m_n)_{n \in \mathbf{N}^*}$. For every integer $n \geq 1$, the transition maps $l_n \colon \widehat{M}/\mathfrak{b}^n \widehat{M} \to \widehat{M}/\mathfrak{b}^{n+1} \widehat{M}$ and $m_n \colon M/I^{nm} \to M/I^{(+1)nm}$ are defined as projection maps. The definition of k_n also comes from the universal property of quotient. Since $K_{(n+1)m} \subset K_{nm}$, we get a map $\varphi_n \colon K_{(n+1)m} \to$

$K_{nm} \to K_{nm}/\mathfrak{b}^n \widehat{M}$. Moreover, $\mathfrak{b}^{n+1} \subset \mathfrak{b}^n$, then φ_n induced, by the universal property of quotient, the map

$$k_n \colon K_{(n+1)m}/\mathfrak{b}^{n+1}\widehat{M} \to K_{nm}/\mathfrak{b}^n\widehat{M}.$$

Furthermore, we have to prove that, for every integer $n \geq 1$, the transition maps k_n are surjective, i.e., every map φ_n is surjective, which is to say that $K_{(n+1)m} + \mathfrak{b}^n\widehat{M} = K_{nm}$. Thanks to Proposition 2.4, we have $K_{(n+1)m} + I^{nm}\widehat{M} = K_{nm}$ and by hypothesis $I^m\widehat{M} \subset \mathfrak{b}$, we thus get the wanted equality.

All the necessary requirements have been met to apply Proposition 2.1, thus the sequence

$$0 \to \varprojlim_n K_{nm}/\mathfrak{b}^n\widehat{M} \to \varprojlim_n \widehat{M}/\mathfrak{b}^n\widehat{M} \to \varprojlim_n M/I^{nm}M \to 0$$

is exact.

But we have seen in Proposition 2.5 that $\varprojlim_n M/I^{nm}M \simeq \widehat{M}$. So, the conclusion comes from the fact that the following assertions are equivalences which is an application of Proposition 2.3:

(1) The ring $\widehat{M}$ is $\mathfrak{b}$-adically complete.
(2) The map $\varprojlim_n \widehat{M}/\mathfrak{b}^n\widehat{M} \to \widehat{M}$ is bijective.
(3) $\varprojlim_n K_{nm}/\mathfrak{b}^n\widehat{M} = 0$.
(4) For every integer $n \geq 1$, we have $K_{nm} = \mathfrak{b}^n\widehat{M}$. $\qquad\square$

3. A Not I-adically Complete Completion

We will now construct an example of a *non-noetherian* ring R and an ideal I whose completion, with respect to the filtration $(I^n)_{n\geq 1}$, is neither I-adically complete nor K_1-adically complete. With these notations, we recall that $K_n := \mathrm{Ker}(\widehat{R} \to R/I^n)$. Our aim is to prove the following theorem.

THEOREM 3.1. *There exist a non-noetherian ring R and a proper ideal I of R, such that, if $\widehat{R}$ denotes the completion of the ring R whose topology is given by the filtration $(I^n)_{n\geq 1}$ and $K_n := \mathrm{Ker}(\widehat{R} \to R/I^n)$, then*

(1) $K_1 \neq I\widehat{R}$;
(2) $K_{2m} \neq (K_m)^2$ *for every $m \geq 1$.*

In particular, the completion $\widehat{R}$ of R is neither I-adically complete nor K_m-adically complete.

This theorem is a consequence of Propositions 3.3 and 3.4.

Let k be a field, $R = k[(x_l)_{l\in\mathbf{N}}]$ the polynomial ring with countably infinitely many variables, and $\mathfrak{m} = \langle(x_l)_{l\in\mathbf{N}}\rangle$, the ideal generated by the x_l. The ring R is non-noetherian and the ideal $\mathfrak{m}$ is not finitely generated. The completion $\widehat{R}$ of R, with respect to the topology given by the filtration $(\mathfrak{m}^n)_{n\geq 1}$, is the ring $k[[(x_l)_{l\in\mathbf{N}}]]$ of formal power series as introduced in

Example 1.2. More specifically, an element $f \in k[[(x_l)_{l \in \mathbf{N}}]]$ can be seen as a formal series

$$f = \sum_{\nu \in \mathbf{N}^{\mathbf{N}}} a_\nu x^\nu,$$

where

(1) $\nu \colon \mathbf{N} \to \mathbf{N}$ has finite support,
(2) the notation x^ν stands for $\prod_{l \in \mathbf{N}} x_l^{\nu(l)}$,
(3) for every integer $d \geq 0$, the set $\{a_\nu \mid a_\nu \neq 0, \sum_{l \in \mathbf{N}} \nu(l) = d\}$ is finite.

NOTATION 3.2. From now and for the remainder of this section, the notation R will stand for the polynomial ring $k[(x_l)_{l \in \mathbf{N}}]$ and $\widehat{R} = k[[(x_l)_{l \in \mathbf{N}}]]$ for its completion, when R is endowed with the topology given by the filtration $(\mathfrak{m}^n)_{n \geq 1}$, where $\mathfrak{m} = \langle (x_l)_{l \in \mathbf{N}} \rangle$.

PROPOSITION 3.3. *The completion $\widehat{R}$ of the ring R is not complete for the $\mathfrak{m}$-adic topology.*

PROOF. Thanks to Theorem 2.6 which gives us a condition to be $\mathfrak{m}$-adically complete, it suffices to show that $K_1 \neq \mathfrak{m}\widehat{R}$ to prove our proposition. To reach this goal, let us consider the element $f = \sum_{l \geq 1} x_l^l$ in the ring $k[[(x_l)_{l \in \mathbf{N}}]]$. It is clear that $f \in K_1$. If we suppose that $f \in \mathfrak{m}\widehat{R}$, there exists an integer $r \in \mathbf{N} \setminus \{0\}$, $u_1, \ldots, u_r \in \mathfrak{m}$, and $h_1, \ldots, h_r \in \widehat{R}$, such that

$$f = \sum_{i=1}^r u_i \cdot h_i.$$

Since every u_i belongs to $\mathfrak{m}$, only a finite number of variables are involved in the expression of u_i, so there exists an integer $n_0 \in \mathbf{N} \setminus \{0\}$, such that for every integer $i \in \{1, \ldots, r\}$, the variable x_{n_0} does not appear in the polynomial u_i.

Consider the morphism of rings $\psi \colon \widehat{R} \to k[[x_{n_0}]]$, induced by the morphism of polynomial rings $\varphi \colon R \to k[x_{n_0}]$, defined by $\varphi(x_l) = 0$ if $i \neq n_0$ and $\varphi(x_{n_0}) = x_{n_0}$. In particular, the map ψ also satisfies $\psi(x_l) = 0$ if $i \neq n_0$ and $\psi(x_{n_0}) = x_{n_0}$. Hence, $\psi(\sum_{i=1}^r u_i \cdot h_i) = 0$, but $\psi(\sum_{l \geq 1} x_l^l) = x_{n_0}^{n_0}$. This is a contradiction, thus $f \notin \mathfrak{m}\widehat{R}$. □

THEOREM 3.4. *Let $m \in \mathbf{N} \setminus \{0\}$. The completion $\widehat{R}$ of the ring R is not complete for the K_m-adic topology.*

Before we start the proof, recall some propositions about the dimension of universally catenary rings, see [2, part 2 Dimension theory] for definitions.

Let A be a ring, then a chain of ideals of the form $\mathfrak{p}_r \supsetneq \mathfrak{p}_{r-1} \supsetneq \cdots \supsetneq \mathfrak{p}_0$ is said of length r. We recall that if E is a subset of A, then

$$V(E) := \{\mathfrak{p} \in \mathrm{Spec}(A) \mid E \subset \mathfrak{p}\}$$

and that the dimension of A, denoted by $\dim(A)$, is the supremum of the length of prime chains. Furthermore, if I is an ideal of A, the dimension of I, denoted by $\dim(I)$, is defined as $\dim(A/I)$. If we set $\sqrt{I}$ to be the radical of the ideal I, a consequence of the definition is the equality

$$\dim(A/I) = \dim(A/\sqrt{I}).$$

Finally, we denote by $\dim(V(E))$ the dimension of the ideal generated by E. This last definition makes sense because if I and J are two ideals of A, such that $V(I) = V(J)$, then $\sqrt{I} = \sqrt{J}$, and in particular, $\dim(I) = \dim(J)$.

PROPOSITION 3.5. *Let k be a field and $k[[x_1, \ldots, x_n]]$ the ring of formal power series with n variables. Let $(a_i)_{1 \leq i \leq n}$ be non-zero elements in k and m a non-zero integer. Let E be the subset $\{a_1 x_1^m, \ldots, a_n x_n^m\}$ of $k[[x_1, \ldots, x_n]]$. Then $\dim(V(E)) = 0$.*

PROOF. Set $I = \langle E \rangle$. Since the elements $(a_i)_{1 \leq i \leq n}$ are non-zero elements in k, they have an inverse. In particular, $\sqrt{I} = \langle x_1, x_2, \ldots, x_n \rangle$. Then

$$\dim(V(E)) = \dim(k[[x_1, \ldots, x_n]]/\sqrt{I}) = \dim(k).$$

But $\dim(k) = 0$, hence $\dim(V(E)) = 0$. $\qquad\square$

LEMMA 3.6. *Let A be a ring and I, J be two ideals of A, such that $J \subset I$. Then $\dim(A/I) \leq \dim(A/J)$.*

PROOF. Let $\mathfrak{p}_r \supsetneq \mathfrak{p}_{r-1} \supsetneq \cdots \supsetneq \mathfrak{p}_0 \supset I$ be a chain of prime ideals containing I. Then, it is certainly a chain of prime ideals containing J. In particular, if $r = \dim(A/I)$, we get $\dim(A/I) \leq \dim(A/J)$. $\qquad\square$

PROPOSITION 3.7. *Let $(A, \mathfrak{m})$ be a noetherian domain and a catenary commutative ring. Suppose that $I = \langle f_1, \ldots, f_r \rangle$ is a proper ideal of A generated by r elements. Thus, we have $\dim(A/I) \geq \dim(A) - r$.*

PROOF. See the discussion [**2**, part II, §13, p. 286] after Corollary 13.5 and [**5**, 12.I, Theorem 18, p. 77]. $\qquad\square$

Finally, we need a last proposition before we begin the proof of Theorem 3.4.

PROPOSITION 3.8. *The ring of formal power series in n variables over a field k, namely, $k[[x_1, \ldots, x_n]]$, is a universally catenary commutative ring. Furthermore, $\dim(k[[x_1, \ldots, x_n]]) = n$.*

PROOF. Since a field is a universally catenary commutative ring and since $k[[x_1, \ldots, x_n]]$ is a noetherian ring, the fact that $k[[x_1, \ldots, x_n]]$ is universally catenary is a consequence from [**6**, Theorem 1.12, p. 24].

Let us prove the assertion about the dimension. The sequence of primes ideals

$$\langle x_1, \ldots, x_n \rangle \supset \langle x_1, \ldots, x_{n-1} \rangle \supset \cdots \supset \langle x_1 \rangle \supset \{0\}$$

is of length n, hence $\dim(k[[x_1, \ldots, x_n]]) \geq n$. On the other hand, we deduce from Proposition 3.7 that

$$\dim(k[[x_1, \ldots, x_n]]) \leq n + \dim(k[[x_1, \ldots, x_n]]/\langle x_1, \ldots, x_n \rangle)$$
$$= n + \dim(k)$$
$$= n.$$

Hence, $\dim(k[[x_1, \ldots, x_n]]) = n$. $\square$

PROOF OF THEOREM 3.4. We want to prove that $\widehat{R} = k[[(x_l)_{l \in \mathbf{N}^*}]]$ is not K_m-adically complete. Let K_m play the role of $\mathfrak{b}$ of Theorem 2.6. Since the mentioned theorem states an equivalence, if we find an integer $n \geq 1$, such that $K_{nm} \neq K_m^n$, then $\widehat{R}$ will not be K_m-adically complete. In particular, we will prove that $(K_m)^2 \neq K_{2m}$.

Note that an element of $(K_m)^2$ can be written as a finite sum,

$$(3.9) \qquad\qquad\qquad \sum_{j=1,\ldots,t} f_j g_j,$$

with $f_j, g_j \in K_m$. Let us construct polynomials $z_i \in k[(x_l)_{l \in \mathbf{N}^*}]$ and sequences of integers $2m \leq d_1 < d_2 < \cdots$ and $0 < l_1 < l_2 < \cdots$, such that

$$z_i \in k[x_{l_i}, x_{l_i+1}, \ldots, x_{l_{i+1}-1}]$$

and, in the ring $k[[x_{l_i}, x_{l_i+1}, \ldots, x_{l_{i+1}-1}]]$, the element z_i cannot be written as a sum like (3.9) with $t \leq i$.

Set $l_1 = 1$, and for every integer $i \geq 2$, take $l_i = l_{i-1} + 2(i-1) + 1$. Let $(d_i)_{i \geq 1}$ be an increasing sequence of integers (such that $d_i < d_{i+1}$), such that, for every integer $i \geq 1$, the integer d_i is coprime with the characteristic of the field k. Define $z_i = \sum_{k=l_i}^{l_{i+1}-1} x_k^{d_i}$ which clearly lies in $k[x_{l_i}, x_{l_i+1}, \ldots, x_{l_{i+1}-1}]$. Suppose that, for one integer $t \leq i$, the polynomial z_i can be written as a sum

$$(3.10) \qquad\qquad\qquad z_i = \sum_{i=1,\ldots,t} f_i g_i,$$

with $f_i, g_i \in k[[x_{l_i}, x_{l_i+1}, \ldots, x_{l_{i+1}-1}]]$ and such that the polynomials f_i, g_i have no constant terms. Then, let us set

$$E = \left(\frac{\partial z_i}{\partial x_{n_i}}, \ldots, \frac{\partial z_i}{\partial x_{l_{i+1}-1}} \right) = (d_i x_{l_i}^{d_i-1}, \ldots, d_i x_{l_{i+1}-1}^{d_i-1}).$$

Let us consider

$$V(E) \subset \mathrm{Spec}(k[[x_{l_i}, x_{l_i+1}, \ldots, x_{l_{i+1}-1}]]).$$

Then $\dim(V(E)) = 0$ because $V(E)$ is a point (see Proposition 3.5). On the contrary, we have that, for every integer $j \in \{n_i, \ldots, n_{i+1} - 1\}$,

$$\frac{\partial(\sum_{i=1,\ldots,t} f_i g_i)}{\partial x_j} \in \langle f_1, \ldots, f_t, g_1, \ldots, g_t \rangle.$$

Let us set $E' = \left\{ \frac{\partial(\sum_{i=1,\ldots,t} f_i g_i)}{\partial x_j} \right\}_{l_i \leq j \leq l_{i+1}-1}$. Actually, $E = E'$ by equation (3.10), but here we will use the notation E' to specify that we use form $\sum_{i=1,\ldots,t} f_i g_i$ of z_i for the calculus. Then, by Lemma 3.6,

$$\dim(V(E')) \geq \dim(V(f_1, \ldots, f_t, g_1, \ldots, g_t)).$$

Besides, the dimension of $V(f_1, \ldots, f_t, g_1, \ldots, g_t)$ is at least

$$l_{i+1} - 1 - l_i + 1 - 2t = 2i + 1 - 2t \geq 1,$$

by Proposition 3.7 applied to the ring $k[[x_{l_i}, \ldots, x_{l_{i+1}-1}]]$ (which is catenary of dimension $n_{i+1} - n_i$, as explained in Proposition 3.8) and the ideal $\langle f_1, \ldots, f_t, g_1, \ldots, g_t \rangle$ which is proper since every polynomials f_i or g_i have no constant terms. Thus, $\dim(V(E')) \geq 1$. Here is a contradiction since $0 = \dim(V(E)) = \dim(V(E')) \geq 1$.

Set $z = \sum_{i \geq 1} z_i \in k[[(x_l)_{l \in \mathbf{N}^*}]]$. It is clear that $z \in K_{2m}$. If $z \in (K_m)^2$, we could find $t > 0$ and $f_j, g_j \in k[[(x_l)_{l \in \mathbf{N}^*}]]$, such that

$$z = \sum_{j=1,\ldots,t} f_j g_j.$$

Let us set

$$f_{j,t} = f_j(0, 0, \ldots, 0, x_{l_t}, \ldots, x_{l_{t+1}-1}, 0, 0, \ldots),$$
$$g_{j,t} = g_j(0, 0, \ldots, 0, x_{l_t}, \ldots, x_{l_{t+1}-1}, 0, 0, \ldots).$$

Then, since $z_t = z(0, 0, \ldots, 0, x_{l_t}, \ldots, x_{l_{t+1}-1}, 0, 0, \ldots)$, we have

$$(3.11) \qquad z_t = \sum_{j=1,\ldots,t} f_{j,t} g_{j,t}.$$

Thus, $f_{j,t}$ and $g_{j,t}$ are elements of $k[[x_{l_t}, x_{n_i+1}, \ldots, x_{l_{t+1}-1}]]$. However, by construction, z_t cannot be written as a sum like equation (3.9) since $t \leq t$. Hence, equality (3.11) is impossible. Thus, $z \notin (K_m)^2$. $\square$

Acknowledgments

I would like to thank D. Bourqui and J. Sebag for introducing me to this question and for their very careful reading of my work. I am also grateful to the referee for her/his extensive comments, which improved the presentation of this chapter.

Bibliography

[1] N. Bourbaki, *Algebre Commutative*, Chapitres 1–4. Springer, 2006.

[2] D. Eisenbud, *Commutative Algebra: With a View Toward Algebraic Geometry*, Graduate Texts in Mathematics, Vol. 150. Springer New York, 1995.

[3] J. P. Lafon, *Algèbre commutative: langages g'eom'etrique et alg'ebrique*, Hermann edition. Collection Enseignement des sciences, Vol. 24, Hermann, 1977.

[4] S. Lang, *Algebra*, 3rd edn. Springer-Verlag, 2002.

[5] H. Matsumura, *Commutative Algebra*, 2nd edn. Mathematics Lecture Note Series, Vol. 56, Benjamin-Cummings, 1980.

[6] H. Seydi, Anneaux henséliens et conditions de chaînes. *Bull. Soc. Math. France*, 98: 9–31, 1970.

[7] The Stacks Project Authors. Tag 05ja. https://stacks.math.columbia.edu/tag/05JA, 2017.

[8] A. Yekutieli, On flatness and completion for infinitely generated modules over noetherian rings. *Commun. Algebra*, 39(11):4221–4245, 2011. doi: 10.1080/00927872.2010.522159.

5

The Local Structure of Arc Schemes

David Bourqui* and Julien Sebag[†]

Institut de recherche mathématique de Rennes
UMR 6625 du CNRS, Université de Rennes 1
Campus de Beaulieu, 35042 Rennes cedex, France
* david.bourqui@univ-rennes1.fr
† julien.sebag@univ-rennes1.fr

In this chapter, we present an important result due to Drinfeld, Grinberg and Kazhdan (see [17] in the present volume), which deals with the local structure of arc scheme at rational non-degenerate arcs, and various applications in the direction of singularity theory.

1. Introduction

The local structure of arc schemes, and especially the completions of their local rings, was first investigated by Grinberg and Kazhdan in [19] for rational arcs, and by Reguera in [33] for a class a schematic points of arc scheme, called the *stable points* of arc schemes. Whereas [33] relates the study to (and in fact is motivated by) singularity theory and the so-called *Nash problem*, the exegesis of the main theorem of [19], with singularity theory as a guideline, had not appeared in the literature and is the subject of the present chapter which is broadly based on [4,6–9].

The main result of [19] is a crucial statement in the direction of the understanding of the local structure of arc schemes. It was conjectured, under a weaker form, by Drinfeld in private communication (see [19, Introduction]). In [19], it is proved under the assumption that the base field is of characteristic zero, but it was subsequently extended by Drinfeld himself (see [**17**, Theorem 0.1]) to a version valid over an arbitrary field, which we now state.

THEOREM 1.1. *Let k be a field. Let V be a k-variety with $v \in V(k)$. We assume that $\dim_v(V) \geq 1$. Let $\gamma \in \mathscr{L}_\infty(V)(k)$ be a rational point of the associated arc scheme, not contained in $\mathscr{L}_\infty(V_{\text{sing}})$, and whose origin $\gamma(0)$ is v. If $\mathscr{L}_\infty(V)_\gamma$ denotes the formal neighborhood of the k-scheme $\mathscr{L}_\infty(V)$ at the point γ, there exists an affine k-scheme S of finite type, with $s \in S(k)$, and an isomorphism of formal k-schemes:*

$$(1.2) \qquad \mathscr{L}_\infty(V)_\gamma \cong S_s \hat{\otimes}_k k[[(T_i)_{i \in \mathbf{N}}]].$$

Very recently, this statement has been used in [**13**, **12**] for specific purposes in the study of automorphic forms.

After explaining Drinfeld's original arguments of the proof of Theorem 1.1 (and its natural generalization, see Theorem 3.21) in Section 3.6, we provide an alternative way of proving such a result and various illustrations of this statement (see Section 3). Thanks to Theorem 1.1, we define the notion of finite-dimensional formal models of formal neighborhoods of arc schemes at rational arcs in Definition 3.1, and begin the study of the Drinfeld–Grinberg–Kazhdan theorem with respect to singularity theory. Thanks to a cancellation lemma due to Gabber stated in Section 4, we introduce the notion of minimal formal models of formal neighborhoods of arc schemes at rational arcs in Definition 5.1 and explain how this new geometric object gives rise to a (local) measure of the singularity, in particular in Theorem 5.5. We also show how to deduce from it various numerical invariants of the singularities in the particular cases of curve and toric singularities (see Sections 7.3 and 7.5). In the end, we open discussions on the genericness of Theorem 1.1 (see Section 7.6) and introduce open questions in Section 8.1.

2. Conventions and Notations

2.1. In this chapter, k is a field of arbitrary characteristic (unless explicitly stated otherwise); $k[[T]]$ is the ring of power series over the field k. For every ring R, an R-*variety* is an R-scheme of finite type. The category of R-schemes is denoted by $\mathfrak{Sch}_R$. In this chapter, $R = k$ or $R = k[[T]]$, and $R_n = k[[T]]/\langle T^{n+1} \rangle$ for every integer $n \in \mathbf{N}$. The singular locus V_{sing} of V is defined as the (unique) reduced closed subscheme associated with the non-smooth locus of V.

2.2. If V is a $k[[T]]$-variety and $n \in \mathbf{N}$, the restriction *à la* Weil of the $k[T]/\langle T^{n+1} \rangle$-scheme $V \otimes_R R_n$ with respect to the morphism of k-algebras $k \hookrightarrow k[[T]]/\langle T^{n+1} \rangle$ exists; it is a k-scheme of finite type which is called the n-*jet scheme* of V and that we denote by $\mathscr{L}_n(V)$. The projective limit $\varprojlim_n (\mathscr{L}_n(V))$ exists in the category of k-schemes; it is the *arc* *scheme* associated with V and we denote it by $\mathscr{L}_\infty(V)$. For every integer $n \in \mathbf{N}$, the canonical morphism of k-schemes $\pi_n^\infty : \mathscr{L}_\infty(V) \to \mathscr{L}_n(V)$ is called the *truncation morphism* of level n. Let A be a k-algebra. As proved in [1], there exists a natural bijection

$$(2.1) \qquad \mathrm{Hom}_{\mathfrak{Sch}_k}(\mathrm{Spec}(A), \mathscr{L}_\infty(V)) \cong \mathrm{Hom}_{\mathfrak{Sch}_{k[[T]]}}(\mathrm{Spec}(A[[T]]), V).$$

An arc of V, i.e., a point of the k-scheme $\mathscr{L}_\infty(V)$, which is *not* contained in the singular locus V_{sing} of V, is called a *non-degenerate* arc. In other words, the subset $\mathscr{L}_\infty^\circ(V) := \mathscr{L}_\infty(V) \setminus \mathscr{L}_\infty(V_{\mathrm{sing}})$ is the set of non-degenerate arcs.

REMARK 2.2. If V is a k-variety, then one can apply this construction to the $k[[T]]$-variety $V_T := V \otimes_k k[[T]]$. Then, the obtained k-scheme $\mathscr{L}_\infty(V_T)$ (resp. $\mathscr{L}_n(V)$) is nothing but the usual arc scheme (resp. jet scheme of level n) associated with V.

2.3. We denote the following category by $\mathbf{Lacp}_k$. The objects are the topological local k-algebras, which are topologically isomorphic to $\mathfrak{m}$-adic completions of local k-algebras (the latter being endowed with the projective limit topology) and whose residue fields are k-algebras isomorphic to k. In particular, the objects of $\mathbf{Lacp}_k$ are topologically complete k-algebras. The morphisms in $\mathbf{Lacp}_k$ are the continuous morphisms of k-algebras; note that every element x of the maximal ideal of an object of $\mathbf{Lacp}_k$ is topologically nilpotent, i.e., the sequence $(x^n)_{n \in \mathbf{N}}$ converges to 0, and thus cannot be mapped to a unit by a continuous morphism; hence, such a morphism is necessarily local. We denote by $\mathbf{Tes}_k$ the full subcategory of $\mathbf{Lacp}_k$ whose objects are *test-rings* (over k), i.e., local k-algebras in $\mathbf{Lacp}_k$ with nilpotent maximal ideal and residue field isomorphic to k. We stress that, in particular, a test-ring is not necessarily noetherian. We also denote by $\mathbf{Lncp}_k$ the full subcategory of $\mathbf{Lacp}_k$ whose objects are the complete local noetherian k-algebras with residue field k-isomorphic to k. As a crucial example of object in $\mathbf{Lacp}_k$ (resp. $\mathbf{Lncp}_k$), we have, for every set I (resp. for every *finite* set I), the local k-algebra $k[[(T_i)_{i \in I}]]$ which is the completion of the polynomial k-algebra $k[(T_i)_{i \in \mathbf{N}}]$ with respect to the maximal ideal $\langle (T_i)_{i \in I} \rangle$.

REMARK 2.3. When the set I is infinite, the topological ring $k[[(T_i)_{i \in I}]]$ must be considered carefully. See [22] in the present volume for a justification of this remark.

We denote by $\mathbf{D}_k^I := \mathrm{Spf}(k[[(T_i)_{i \in I}]])$ the associated k-formal scheme. For every object $A \in \mathbf{Lacp}_k$ (resp. $A \in \mathbf{Lncp}_k$), we set $A[[(T_i)_{i \in I}]] := A \hat{\otimes}_k k[[(T_i)_{i \in I}]]$, which is again an object of $\mathbf{Lacp}_k$ (resp. $\mathbf{Lncp}_k$). Two objects A, B of $\mathbf{Lacp}_k$ (resp. $\mathbf{Lncp}_k$) are said to be *stably isomorphic* if there exist sets I and J (resp. finite sets I and J) and an isomorphism $A[[(T_i)_{i \in I}]] \cong B[[(T_i)_{i \in J}]]$.

2.4. Let $(A, \mathfrak{m}_A)$ be a complete local ring with residue field k. For $f \in A[[T]]$, denote by $\bar{f}$ the image of f by the natural quotient morphism $A[[T]] \to k[[T]]$. Let d be a non-negative integer. An element $f \in A[[T]]$ is said to be *d-regular* if $\mathrm{ord}_t(\bar{f}) = d$. It is said to be *d-Weierstrass* (*or d-distinguished*) if it is a monic polynomial of $A[T]$ with degree T which is d-regular. We denote by $\mathscr{W}(A, d)$ the set of the d-Weierstrass elements of $A[T]$. By the uniqueness in Weierstrass division theorem (see [**29**, Theorem 9.1]), every $f \in A[[T]]$ such that $\bar{f} \neq 0$ is not a zero divisor in $A[[T]]$.

3. The Drinfeld–Grinberg–Kazhdan Theorem

In this section, we first give some examples illustrating Theorem 1.1 (without proof, but with adequate references). Then, we give a short account of Drinfeld's arguments showing the result. An important feature of the proof is the interpretation of the functor of points of formal neighborhoods in arc scheme in terms of deformation of arcs. We illustrate on some examples another approach for proving the theorem in case the variety is defined by binomial equations, and we compare it with Drinfeld's approach. Finally, we describe some extensions and variants of Theorem 1.1. Prior to all of this, we recall a useful terminology, introduced in [**13**].

DEFINITION 3.1. Let k be a field. Let V be a k-variety and $\gamma \in \mathscr{L}_\infty^\circ(V)(k)$ such that $\dim_{\gamma(0)}(V) \geq 1$ (see Section 2.2 for the notation $\mathscr{L}_\infty^\circ(V)$). Every algebraizable noetherian affine formal k-scheme $\mathscr{S}$ such that there exists an isomorphism of formal k-schemes

$$\mathscr{L}_\infty(V)_\gamma \to \mathscr{S} \,\hat{\times}_k \mathbf{D}_k^{\mathbf{N}}$$

is called a *finite-dimensional formal model* (or simply a *finite formal model*) of the pair $(\mathscr{L}_\infty(V), \gamma)$.

Thus, Theorem 1.1 may be reformulated as follows: under the above hypotheses, the pair $(\mathscr{L}_\infty(V), \gamma)$ always admits a finite formal model. In fact, we shall see that Drinfeld's proof of Theorem 1.1 is effective in the sense that it provides a procedure to compute explicitly a presentation of a finite formal model in terms of a presentation of V and the datum of (sufficiently many coefficients of) the arc γ.

3.1. Examples

We give some examples illustrating Theorem 1.1, providing in particular explicit examples of finite formal models.

EXAMPLE 3.2. Let k be a field. Let V be a k-variety and $v \in V(k)$ be a smooth point. For example, by [**35**, Lemme 3.4.2], we easily deduce that $\mathscr{L}_\infty(V)_\gamma \cong \mathbf{D}_k^{\mathbf{N}}$ for every arc $\gamma \in \mathscr{L}_\infty(V)(k)$ with origin $\gamma(0) = v$. For the converse, see Section 5.2.

EXAMPLE 3.3 (see [**17**]). Let k be a field. Let $f \in k[X_1, \ldots, X_n]$ be a polynomial such that $f(0) = 0$. Let H be the hypersurface of $\mathrm{Spec}(k[X_1, \ldots, X_n, Y, Z])$ with equation $YZ + F(X_1, \ldots, X_n) = 0$. Let $\gamma \in \mathscr{L}_\infty(H)$ be the k-arc given by

$$\gamma_{X_1}(T) = \cdots = \gamma_{X_n}(T) = 0, \quad \gamma_Y(T) = 0, \quad \gamma_Z(T) = T.$$

Then, one has an isomorphism of formal k-schemes:

$$\mathscr{L}_\infty(H)_\gamma \cong \mathrm{Spf}(k[[X_1, \ldots, X_n]]/\langle f \rangle) \hat{\times}_k \mathbf{D}_k^{\mathbf{N}}.$$

EXAMPLE 3.4. The previous example by Drinfeld may easily be generalized as follows. Let k be a field. Let $f_1, \ldots, f_r \in k[X_1, \ldots, X_n]$ be r polynomials such that $f_1(0) = \cdots = f_r(0) = 0$. Let V be the subvariety of $\mathrm{Spec}(k[X_1, \ldots, X_n, Y_1, \ldots, Y_r, Z])$ with equations

$$Y_i Z + F_i(X_1, \ldots, X_n) = 0, \quad 1 \le i \le r.$$

Let $\gamma \in \mathscr{L}_\infty(V)$ be the k-arc given by

$$\gamma_{X_1}(T) = \cdots = \gamma_{X_n}(T) = 0, \quad \gamma_{Y_1}(T) = \cdots = \gamma_{Y_r}(T) = 0, \quad \gamma_Z(T) = T.$$

Then, one has an isomorphism of formal k-schemes:

$$\mathscr{L}_\infty(V)_\gamma \cong \mathrm{Spf}(k[[X_1, \ldots, X_n]]/\langle f_1, \ldots, f_r \rangle) \hat{\times}_k \mathbf{D}_k^{\mathbf{N}}.$$

REMARK 3.5. Let S be a k-scheme of finite type and $s \in S(k)$. By Example 3.4, we observe that there exist a k-variety V and $\gamma \in \mathscr{L}_\infty(V)(k)$ a non-degenerate arc on V such that S_s is a finite formal model of $(\mathscr{L}_\infty(V), \gamma)$. By Section 3.5, one may even assume that V is reduced and irreducible. Thus, if we put no particular restriction on V and/or on the non-degenerate arc γ, any algebraizable noetherian affine formal k-scheme can appear as a finite formal model.

EXAMPLE 3.6 (see [**8**]). Let $m \ge 1$ be an even integer. Let k be a field with characteristic zero or greater than $2(m+1)$. Let $(\mathscr{C}, x)$ be a pointed curve formally isomorphic to $X_1^{m+1} - X_2^2$. Let $\gamma \in \mathscr{L}_\infty(\mathscr{C})(k)$ be a primitive k-parameterization at x. Recall that a non-primitive k-parameterization η of $\mathscr{C}$ at x is a formal parameterization $\eta \colon \widehat{\mathcal{O}_{\mathscr{C},x}} \to k[[T]]$ of $\mathscr{C}$ at x such that there exist a power series $\sigma \in T^2 k[[T]]$ and a formal parameterization $\eta' \colon \widehat{\mathcal{O}_{\mathscr{C},x}} \to k[[T]]$ satisfying $\eta = \sigma \circ \eta'$. Then, one has an isomorphism of formal k-schemes

$$\mathscr{L}_\infty(V)_\gamma \cong \mathrm{Spf}(k[[X]]/\langle X^{\frac{m}{2}+1} \rangle) \hat{\times}_k \mathbf{D}_k^{\mathbf{N}}.$$

3.2. Formal neighborhoods of arc scheme at rational arcs and arc deformations

One of the key points in the proof of Theorem 1.1 is to identify the functor of points associated with the formal neighborhoods of arc schemes at rational points in terms of *arc deformations*. Let us explain this point.

3.3. If $\widehat{\mathbf{Tes}_k}$ is the category of pre-cosheaves on the category $\mathbf{Tes}_k$ (i.e., covariant functors from the category $\mathbf{Tes}_k$ to the category of sets), we define the functor

$$F \colon \mathbf{Lacp}_k \ \rightarrow \ \widehat{\mathbf{Tes}_k}$$
$$\widehat{\mathcal{O}} \ \rightarrow \ \mathrm{Hom}_{\mathbf{Lacp}_k}(\widehat{\mathcal{O}}, \cdot).$$

One has the following seemingly standard observation (see [**17**, **7**]).

OBSERVATION 3.7. *The functor F is fully faithful.*

One will use the following consequence of the observation: let S and S' be k-schemes, let $s \in S(k)$ and $s' \in S'(k)$, let S_s and $S'_{s'}$ be the associated formal neighborhoods and let $f_A \colon S_s(A) \to S'_{s'}(A)$ be a natural map defined for every test-ring $(A, \mathfrak{m}_A)$; then there exists a unique morphism of formal k-schemes $f \colon S_s \to S'_{s'}$ inducing f_A for every test-ring A; moreover, f is an isomorphism if and only if f_A is bijective for every A.

3.4. Let V be a k-variety. Let $\gamma \in \mathscr{L}_\infty(V)(k)$. Then, in the sense of Observation 3.7, the formal k-scheme $\mathscr{L}_\infty(V)_\gamma$ is *uniquely* determined by the functor $F(\mathcal{O}_{\mathscr{L}_\infty(V),\gamma})$. Let A be a test-ring. Let $\gamma_A \in \mathscr{L}_\infty(V)_\gamma(A)$. The datum of γ_A corresponds to one of the following (equivalent) commutative diagrams:

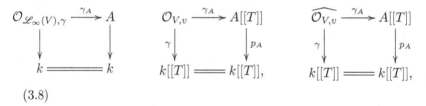

(3.8)

where we denote by $p_A \colon A[[T]] \to k[[T]]$ the unique local morphism which extends the projection $A \to A/\mathfrak{m}_A \cong k$. The set $\mathscr{L}_\infty(V)_\gamma(A)$ parameterizes the elements $\gamma_A \in V(A[[T]])$ whose reduction modulo $\mathfrak{m}_A$ coincides with γ.

DEFINITION 3.9. *Every morphism $\gamma_A \in \mathscr{L}_\infty(V)_\gamma(A)$ is called an A-deformation of γ.*

3.5. Reduction to reduced and irreducible germs

The following remark is originally made but not used in Drinfeld's argument establishing Theorem 1.1. It is interesting to note that one can always reduce to the case where the variety is reduced and irreducible. More precisely, let V be a k-variety, $\gamma \in \mathscr{L}_\infty(V)^\circ(k)$ and $v = \gamma(0)$. The arc γ corresponds to a morphism of local k-algebra $\gamma \colon \mathcal{O}_{V,v} \to k[[T]]$. The localization $(\mathcal{O}_{V,v})_{\mathrm{Ker}(\gamma)}$ is isomorphic to the local ring of the image of the generic point of $\mathrm{Spec}(k[[T]])$ in V; by assumption, the latter is a regular local ring, in particular $(\mathcal{O}_{V,v})_{\mathrm{Ker}(\gamma)}$ is a domain, which shows that there exists a unique minimal prime ideal $\mathfrak{p}$ of $\mathcal{O}_{V,v}$ contained in $\mathrm{Ker}(\gamma)$. Now consider a test-ring A and an A-deformation $\gamma_A \colon \mathcal{O}_{V,v} \to A[[T]]$ of γ. Let $x \in \mathfrak{p}$. Since $\mathfrak{p}$ vanishes in the localization $(\mathcal{O}_{V,v})_{\mathrm{Ker}(\gamma)}$, there exists $y \notin \mathrm{Ker}(\gamma)$ such that $xy = 0$. In particular, $\gamma_A(x)\gamma_A(y) = 0$. Since $\gamma(y) \neq 0$, $\gamma_A(y)$ is not a zero divisor in $A[[T]]$, thus $\gamma_A(x) = 0$. Summing up, there exists a unique minimal prime ideal $\mathfrak{p}$ of $\mathcal{O}_{V,v}$ contained in $\mathrm{Ker}(\gamma)$, and it is also contained in the kernel of every A-deformation of γ, for every test-ring A. Letting (W, w) be a germ such that $\mathcal{O}_{W,w} = \mathcal{O}_{V,v}/\mathfrak{p}$ and $\eta \colon \mathcal{O}_{W,w} \to k[[T]]$ be the arc induced by γ shows that $\mathscr{L}_\infty(V)_\gamma \cong \mathscr{L}_\infty(W)_\eta$.

3.6. The proof of the Drinfeld–Grinberg–Kazdhan theorem, following Drinfeld

The argument described here follows those of Drinfeld in the original proof (see [7] for a detailed argument). As pointed out before, one essential ingredient, already used in [19], lies in the identification of the formal k-scheme $\mathscr{L}_\infty(X)_\gamma$ with its functor of points and the observation that this functor is completely determined by its restriction to the category of test-rings (see Section 3.3). In this way, we are reduced to analyze the functorial behavior of the family of sets $\mathscr{L}_\infty(X)_\gamma(A)$ when A runs over the test-rings over k. Precisely, this study will define an affine k-variety S, a k-point $s \in S$, and a functorial family of bijective maps $\mathscr{L}_\infty(X)_\gamma(A) \to S_s(A) \times \mathfrak{m}_A^{\mathbf{N}}$. Now, we sketch how to construct this morphism of functors.

First, by a standard argument using "Elkik's trick" and the fact that for any test-ring A, $A[[t]] \backslash \mathfrak{m}_A[[T]]$ contains no zero divisor, we may assume that $X = \mathrm{Spec}(k[X_1, \ldots, X_n, Y_1, \ldots, Y_r]/\langle f_1, \ldots, f_r \rangle)$, where $f = (f_1, \ldots, f_r)$ is a family of polynomials vanishing at the origin, that $\gamma(0)$ is the origin and that the determinant of the matrix $D = (\partial f_i/\partial Y_j)_{1 \leq i,j \leq r}$ does not vanish at $\gamma = (x(T), y(T)) \in k[[T]]^n \times k[[T]]^r$. Let C be the adjugate matrix of D and $J = \det(D)$; one thus has $CD = DC = J\mathrm{I}_r$.

Let $d := \mathrm{ord}_t(J(x(T), y(T)))$. By assumption, this is a non-negative integer.

Let $(A, \mathfrak{m}_A)$ be a test-ring.

Now, there exist $\tilde{x}(T) \in (k[T]_{\leq 2d-1})^n$, $\tilde{y}(T) \in (k[T]_{\leq 2d-1})^r$, $z(T) \in k[[T]]^n$ and $w(T) \in k[[T]]^r$ such that $x(T) = T^{2d}z(T) + \tilde{x}(T)$ and $y(T) = T^d w(T) + \tilde{y}(T)$.

Let $\tilde{\mathcal{B}}(A)$ be the set whose elements are the triples

$$(x_A(T), y_A(T), q_A(T)) \in \mathfrak{m}_A[T]_{\leq 2d-1}^n \times \mathfrak{m}_A[T]_{\leq d-1}^r \times \mathscr{W}(A, d)$$

which satisfy the relations:

$$\begin{cases} q_A(T) \text{ divides } J(\tilde{x}(T) + \tilde{x}_A(T), \tilde{y}(T) + \tilde{y}_A(T)); \\ q_A(T)^2 \text{ divides } C \cdot f(\tilde{x}(T) + \tilde{x}_A(T), \tilde{y}(T) + \tilde{y}_A(T)). \end{cases}$$

One easily observes that the set $\widetilde{\mathcal{B}}(A)$ is in natural bijection with the set of A-points $S_s(A)$ of a noetherian formal k-scheme isomorphic to the formal neighborhood at the origin of a closed k-subscheme S of the k-affine space of dimension $2dn + rd + d$ (see the explicit Examples 3.14 and 3.18).

Let $\mathcal{B}(A) := \widetilde{\mathcal{B}}(A) \times \mathfrak{m}_A[[T]]^n$. Note that $\mathcal{B}(A)$ is in natural bijection with $\widetilde{\mathcal{B}}(A) \times (\mathfrak{m}_A^n)^{\mathbf{N}}$.

The end of the proof consists of constructing a natural bijection

$$\theta_A \colon \mathscr{L}_\infty(V)_\gamma(A) \xrightarrow{\sim} \mathcal{B}(A).$$

By the previous observations and Section 3.3, this will show that there exists an isomorphism of formal k-schemes

$$\theta_k \colon \mathscr{L}_\infty(V)_\gamma \to S_s \hat{\times}_k \mathbf{D}_k^{\mathbf{N}}.$$

Let us just explain here how θ_A is defined. Let γ_A be an A-deformation of γ; it corresponds to a pair $(x_A, y_A) \in \mathfrak{m}_A[[T]]^n \times \mathfrak{m}_A[[T]]^r$ such that $f(x+x_A, y+y_A) = 0$ in A^r. Note that the power series $J(x+x_A, y+y_A) \in A[[T]]$ is d-regular. Let $J(x + x_A, y + y_A) = q_A \cdot u_A$ be the Weierstrass factorization of $J(x + x_A, y + y_A)$, where $q_A \in A[T]$ is d-Weierstrass and $u_A \in A[[T]]$ is an invertible power series.

Let us write

$$x(T) + x_A(T) = (z(T) + z_A(T))q_A(T)^2 + \tilde{x}(T) + \tilde{x}_A(T),$$

where

$$\tilde{x}_A(T) \in \mathfrak{m}_A[T]_{\leq 2d-1}^n \text{ and } \tilde{z}_A(T) \in \mathfrak{m}_A[[T]]^n,$$

and

$$y(T) + y_A(T) = (w(T) + w_A(T))\tilde{q}_A(T) + \tilde{x}(T) + \tilde{y}_A(T),$$

where

$$\tilde{y}_A(T) \in \mathfrak{m}_A[T]_{\leq d-1}^r \text{ and } w_A(T) \in \mathfrak{m}_A[[T]]^r.$$

One sets $\theta_A(x_A, y_A) = (\tilde{x}_A, \tilde{y}_A, q_A, z_A)$. Then, one can show that θ_A takes its values in $\mathcal{B}(A)$, is functorial in A and is bijective.

REMARK 3.10. As already pointed out, Drinfeld's method allows to compute explicitly a finite formal model of the pair $(\mathscr{L}_\infty(V), \gamma)$. A presentation of such a model is obtained by making the conditions defining $\widetilde{\mathcal{B}}(A)$ explicit (see the explicit Examples 3.14 and 3.18).

3.7. An alternative approach for binomial equations

In this section, we present an alternative proof of Theorem 1.1 which applies to varieties defined by binomial equations, e.g., (normal) toric varieties or monomial curves. When applicable, it also produces effectively a finite formal model, though in general the obtained model is different from the one given by Drinfeld's approach (see Remark 3.16). As in Drinfeld's argument, we crucially use the fact that the formal neighborhood is determined by its points with value in test-rings and the Weierstrass factorization. But contrary to the original proof, we do not use the Taylor formula. We shall content ourselves by giving an illustration for two specific examples. The general case can be obtained by adapting the arguments described here.

EXAMPLE 3.11. Let us consider $\mathscr{C} = \mathrm{Spec}(k[X,Y]/\langle X^3 - Y^2 \rangle)$ and the primitive arc $\gamma(T) = (T^2, T^3)$. For simplicity, we assume that the characteristic of the field k is different from 2. For every test-ring $(A, \mathfrak{m}_A)$, we denote by $\widetilde{\mathcal{C}}(A)$ the subset of $\mathscr{W}(A, 2) \times \mathscr{W}(A, 3)$ (see Section 2.4 for the notation) formed by the triples $(r_A(T), s_A(T))$ such that $r_A(T)^3 = s_A(T)^2$. We also set $\mathcal{C}(A) := \widetilde{\mathcal{C}}(A) \times (1 + \mathfrak{m}_A[[T]])$. Note that $\mathcal{C}(A)$ is in natural bijection with $\widetilde{\mathcal{C}}(A) \times \mathfrak{m}_A^{\mathbb{N}}$. Let us identify $\mathscr{L}_\infty(\mathscr{C})_\gamma(A)$ with the set of elements $(x_A(T), y_A(T)) \in \mathfrak{m}_A[[T]]^2$ such that $(T^2 + x_A(T))^3 = (T^3 + y_A(T))^2$. For every $(x_A(T), y_A(T)) \in \mathscr{L}_\infty(\mathscr{C})_\gamma(A)$ let $T^2 + x_A(T) = r_A(T)u_A(T)$, with $r_A(T) \in \mathscr{W}(A, 2)$, and $T^3 + y_A(T) = s_A(T)v_A(T)$, with $s_A(T) \in \mathscr{W}(A, 3)$, be the Weierstrass factorization of $T^2 + x_A(T)$ and $T^3 + y_A(T)$, respectively. Using the uniqueness of the Weierstrass factorization, one easily checks that the map

$$\theta_A \colon (x_A(T), y_A(T)) \mapsto (r_A(T), s_A(T), u_A(T))$$

induces a bijection, functorial in A, from $\mathscr{L}_\infty(\mathscr{C})_\gamma(A)$ to $\mathcal{C}(A)$, whose inverse bijection is given by

$$(r_A(T), s_A(T), u_A(T)) \mapsto (r_A(T)u_A(T) - T^2, s_A(T)u_A(T)^{\frac{3}{2}} - T^3).$$

We also observe that, for every test-ring A, the set $\widetilde{\mathcal{C}}(A)$ coincides with the A-points of the formal k-scheme

$$\mathrm{Spf}(k[[X]]/\langle X^2 \rangle) \hat{\times}_k \mathbf{D}_k.$$

which is therefore by Section 3.3 a finite formal model of $(\mathscr{L}_\infty(\mathscr{C}), \gamma(T))$. Indeed, the condition $r_A(T)^3 = s_A(T)^2$ defining $\widetilde{\mathcal{C}}(A)$ means concretely that $\widetilde{\mathcal{C}}(A)$ is the set of $(r_{A,1}, r_{A,0}, s_{A,2}, s_{A,1}, s_{A,0}) \in \mathfrak{m}_A^5$ such that

$$(3.12) \qquad (T^2 + r_{A,1}T + r_{A,0})^3 = (T^3 + s_{A,2}T^2 + s_{A,1}T + s_{A,0})^2.$$

Let Δ be the discriminant of $T^2 + r_{A,1}T + r_{A,0}$. Killing the T-coefficient of this polynomial by the usual translation and denoting $s'_{A,2}, s'_{A,1}, s'_{A,0} \in \mathfrak{m}_A$ as the resulting variables on the right-hand side, we see that equation (3.12) is equivalent to

$$(3.13) \qquad (T^2 + \Delta)^3 = (T^3 + s'_{A,2}T^2 + s'_{A,1}T + s'_{A,0})^2.$$

By [9, Proposition 5.4], this implies that $\Delta^2 = 0$ and that the map $(r_A(T), s_A(T)) \mapsto (\Delta, r_{A,1})$ is a natural bijection

$$\{(r_A(T), s_A(T)), \ r_A(T)^3 = s_A(T)^2\} \xrightarrow{\sim} \mathrm{Spf}(k[[\Delta]]/\langle \Delta^2 \rangle)(A) \times \mathfrak{m}_A.$$

By an analogous argument, the authors prove in [9] the formula in Example 3.6. We finally observe that, in this case, this alternative method provides a finite formal model which is of dimension 1 and which simplifies naturally to a finite formal model of dimension 0.

EXAMPLE 3.14. Let us now make explicit the finite formal model of $(\mathcal{L}_\infty(\mathscr{C}), \gamma)$ obtained by following the computational algorithm produced by Drinfeld's argument. Keeping the notation used in the proof, one can take $d = 3$, and for every test-ring A, $\tilde{\mathcal{B}}(A)$ is the set whose elements are the triples

$$(x_A(T), y_A(T), q_A(T)) \in (\mathfrak{m}_A[T]_{\leq 5} \times \mathfrak{m}_A[T]_{\leq 2} \times \mathscr{W}(A, 3)),$$

which satisfy the relations

$$\begin{cases} q_A(T) \text{ divides } y_A(T); \\ q_A(T)^2 \text{ divides } (T^2 + x_A(T)^3) - y_A(T)^2. \end{cases}$$

Since $q_A(T)$ is a Weierstrass polynomial of degree 3 and $\deg(y_A(T)) \leq 2$, the condition that $q_A(T)$ divide $y_A(T)$ imposes that $y_A(T) = 0$. In order to give an explicit set of algebraic relations defining the finite formal model determined by $\tilde{\mathcal{B}}$, one has to express algebraically in the variables $(x_0, x_1, x_2, x_3, x_4, x_5, q_0, q_1, q_2)$ the condition

$$(q_0 + q_1 T + q_2 T^2 + T^3)^2 \text{ divides } \left(T^2 + \sum_{i=0}^{5} x_i T^i\right)^3$$

It suffices to perform the Euclidian division of $(T^2 + \sum_{i=0}^5 x_i T^i)^3$ by the monic polynomial $(q_0 + q_1 T + q_2 T^2 + T^3)^2$. The remainder is a polynomial in $k[(q_i)_{0 \leq i \leq 2}, (x_i)_{0 \leq i \leq 5}, T]$ with T-degree ≤ 5. The T^α-coefficients for $\alpha \in \{0, \ldots, 5\}$ of this polynomial give six algebraic relations $F_\alpha \in k[(q_i)_{0 \leq i \leq 2}, (x_i)_{0 \leq i \leq 5}]$ and finally the finite formal model determined by $\tilde{\mathcal{B}}(A)$ is

(3.15) $S_s = \mathrm{Spf}(k[[(q_i)_{0 \leq i \leq 2}, (x_i)_{0 \leq i \leq 5}]]/\langle F_\alpha\rangle_{0 \leq \alpha \leq 5}.$

The computation gives, for example,

$$F_0 = 10x_5^3 q_0^2 q_2^9 - 72x_5^3 q_0^2 q_1 q_2^7 - 27x_4 x_5^2 q_0^2 q_2^8 + 168x_5^3 q_0^2 q_1^2 q_2^5 + 56x_5^3 q_0^3 q_2^6$$

$$+ 168x_4 x_5^2 q_0^2 q_1 q_2^6 + 24x_4^2 x_5 q_0^2 q_2^7 + 24x_3 x_5^2 q_0^2 q_2^7 - 140x_5^3 q_0^2 q_1^3 q_2^3$$

$$- 210x_5^3 q_0^3 q_1 q_2^4 - 315x_4 x_5^2 q_0^2 q_1^2 q_2^4 - 126x_4 x_5^2 q_0^3 q_2^5 - 126x_4^2 x_5 q_0^2 q_1 q_2^5$$

$$- 126x_3x_5^2q_0^2q_1q_2^5 - 7x_4^3q_0^2q_2^6 - 42x_3x_4x_5q_0^2q_2^6 - 21x_2x_5^2q_0^2q_2^6$$

$$+ 30x_5^3q_0^2q_1^4q_2 + 180x_5^3q_0^3q_1^2q_2 + 180x_4x_5^2q_0^2q_1^3q_2 + 60x_5^3q_0^4q_2^3$$

$$+ 360x_4x_5^2q_0^3q_1q_2^3 + 180x_4^2x_5q_0^2q_1^2q_2^3 + 180x_3x_5^2q_0^2q_1^2q_2^3 + 90x_4^2x_5q_0^3q_2^4$$

$$+ 90x_3x_5^2q_0^3q_2^4 + 30x_4^3q_0^2q_1q_2^4 + 180x_3x_4x_5q_0^2q_1q_2^4 + 90x_2x_5^2q_0^2q_1q_2^4$$

$$+ 18x_3x_4^2q_0^2q_2^5 + 18x_3^2x_5q_0^2q_2^5 + 36x_2x_4x_5q_0^2q_2^5 + 18x_1x_5^2q_0^2q_2^5$$

$$- 21x_5^2q_0^2q_2^6 - 20x_5^3q_0^3q_1^3 - 15x_4x_5^2q_0^2q_1^4 - 60x_5^3q_0^4q_1q_2 - 180x_4x_5^2q_0^3q_1^2q_2$$

$$- 60x_4^2x_5q_0^2q_1^3q_2 - 60x_3x_5^2q_0^2q_1^3q_2 - 90x_4x_5^2q_0^4q_2^2 - 180x_4^2x_5q_0^3q_1q_2^2$$

$$-180x_3x_5^2q_0^3q_1q_2^2 - 30x_4^3q_0^2q_1^2q_2^2 - 180x_3x_4x_5q_0^2q_1^2q_2^2 - 90x_2x_5^2q_0^2q_1^2q_2^2$$

$$- 20x_4^3q_0^3q_2^3 - 120x_3x_4x_5q_0^3q_2^3 - 60x_2x_5^2q_0^3q_2^3 - 60x_3x_4^2q_0^2q_1q_2^3$$

$$- 60x_3^2x_5q_0^2q_1q_2^3 - 120x_2x_4x_5q_0^2q_1q_2^3 - 60x_1x_5^2q_0^2q_1q_2^3 - 15x_3^2x_4q_0^2q_2^4$$

$$- 15x_2x_4^2q_0^2q_2^4 - 30x_2x_3x_5q_0^2q_2^4 - 30x_1x_4x_5q_0^2q_2^4 - 15x_0x_5^2q_0^2q_2^4$$

$$+ 90x_5^2q_0^2q_1q_2^4 + 36x_4x_5q_0^2q_2^5 + 4x_5^3q_0^5 + 36x_4x_5^2q_0^4q_1 + 36x_4^2x_5q_0^3q_1^2$$

$$+ 36x_3x_5^2q_0^3q_1^2 + 4x_4^3q_0^2q_1^3 + 24x_3x_4x_5q_0^2q_1^3 + 12x_2x_5^2q_0^2q_1^3 + 36x_4^2x_5q_0^4q_2$$

$$+ 36x_3x_5^2q_0^4q_2 + 24x_4^3q_0^3q_1q_2 + 144x_3x_4x_5q_0^3q_1q_2 + 72x_2x_5^2q_0^3q_1q_2$$

$$+ 36x_3x_4^2q_0^2q_1^2q_2 + 36x_3^2x_5q_0^2q_1^2q_2 + 72x_2x_4x_5q_0^2q_1^2q_2 + 36x_1x_5^2q_0^2q_1^2q_2$$

$$+ 36x_3x_4^2q_0^3q_2^2 + 36x_3^2x_5q_0^3q_2^2 + 72x_2x_4x_5q_0^3q_2^2 + 36x_1x_5^2q_0^3q_2^2$$

$$+ 36x_3^2x_4q_0^2q_1q_2^2 + 36x_2x_4^2q_0^2q_1q_2^2 + 72x_2x_3x_5q_0^2q_1q_2^2 + 72x_1x_4x_5q_0^2q_1q_2^2$$

$$+ 36x_0x_5^2q_0^2q_1q_2^2 - 90x_5^2q_0^2q_1^2q_2^2 + 4x_3^3q_0^2q_2^3 + 24x_2x_3x_4q_0^2q_2^3$$

$$+ 12x_1x_4^2q_0^2q_2^3 + 12x_2^2x_5q_0^2q_2^3 + 24x_1x_3x_5q_0^2q_2^3 + 24x_0x_4x_5q_0^2q_2^3$$

$$- 60x_5^2q_0^3q_2^3 - 120x_4x_5q_0^2q_1q_2^3 - 15x_4^2q_0^2q_2^4 - 30x_3x_5q_0^2q_2^4 - 3x_4^3q_0^4$$

$$- 18x_3x_4x_5q_0^4 - 9x_2x_5^2q_0^4 - 18x_3x_4^2q_0^3q_1 - 18x_3^2x_5q_0^3q_1 - 36x_2x_4x_5q_0^3q_1$$

$$-18x_1x_5^2q_0^3q_1 - 9x_3^2x_4q_0^2q_1^2 - 9x_2x_4^2q_0^2q_1^2 - 18x_2x_3x_5q_0^2q_1^2$$

$$- 18x_1x_4x_5q_0^2q_1^2 - 9x_0x_5^2q_0^2q_1^2 + 12x_5^2q_0^2q_1^3 - 18x_3^2x_4q_0^3q_2 - 18x_2x_4^2q_0^3q_2$$

$$- 36x_2x_3x_5q_0^3q_2 - 36x_1x_4x_5q_0^3q_2 - 18x_0x_5^2q_0^3q_2 - 6x_3^3q_0^2q_1q_2$$

$$- 36x_2x_3x_4q_0^2q_1q_2 - 18x_1x_4^2q_0^2q_1q_2 - 18x_2^2x_5q_0^2q_1q_2 - 36x_1x_3x_5q_0^2q_1q_2$$

$$- 36x_0x_4x_5q_0^2q_1q_2 + 72x_5^2q_0^3q_1q_2 + 72x_4x_5q_0^2q_1^2q_2 - 9x_2x_3^2q_0^2q_2^2$$

$$- 9x_2^2x_4q_0^2q_2^2 - 18x_1x_3x_4q_0^2q_2^2 - 9x_0x_4^2q_0^2q_2^2 - 18x_1x_2x_5q_0^2q_2^2$$

$$- 18x_0x_3x_5q_0^2q_2^2 + 72x_4x_5q_0^3q_2^2 + 36x_4^2q_0^2q_1q_2^2 + 72x_3x_5q_0^2q_1q_2^2$$

$$+ 24x_3x_4q_0^2q_2^3 + 24x_2x_5q_0^2q_2^3 + 2x_3^3q_0^3 + 12x_2x_3x_4q_0^3 + 6x_1x_4^2q_0^3$$

$$+ 6x_2^2x_5q_0^3 + 12x_1x_3x_5q_0^3 + 12x_0x_4x_5q_0^3 - 9x_5^2q_0^4 + 6x_2x_3^2q_0^2q_1$$

$$+ 6x_2^2 x_4 q_0^2 q_1 + 12 x_1 x_3 x_4 q_0^2 q_1 + 6 x_0 x_4^2 q_0^2 q_1 + 12 x_1 x_2 x_5 q_0^2 q_1$$

$$+ 12 x_0 x_3 x_5 q_0^2 q_1 - 36 x_4 x_5 q_0^3 q_1 - 9 x_4^2 q_0^2 q_1^2 - 18 x_3 x_5 q_0^2 q_1^2$$

$$+ 6 x_2^2 x_3 q_0^2 q_2 + 6 x_1 x_3^2 q_0^2 q_2 + 12 x_1 x_2 x_4 q_0^2 q_2 + 12 x_0 x_3 x_4 q_0^2 q_2$$

$$+ 6 x_1^2 x_5 q_0^2 q_2 + 12 x_0 x_2 x_5 q_0^2 q_2 - 18 x_4^2 q_0^3 q_2 - 36 x_3 x_5 q_0^3 q_2$$

$$- 36 x_3 x_4 q_0^2 q_1 q_2 - 36 x_2 x_5 q_0^2 q_1 q_2 - 9 x_3^2 q_0^2 q_2^2 - 18 x_2 x_4 q_0^2 q_2^2$$

$$- 18 x_1 x_5 q_0^2 q_2^2 + 12 x_5 q_0^2 q_2^3 - x_2^3 q_0^2 - 6 x_1 x_2 x_3 q_0^2 - 3 x_0 x_3^2 q_0^2$$

$$- 3 x_1^2 x_4 q_0^2 - 6 x_0 x_2 x_4 q_0^2 - 6 x_0 x_1 x_5 q_0^2 + 12 x_3 x_4 q_0^3 + 12 x_2 x_5 q_0^3$$

$$+ 6 x_3^2 q_0^2 q_1 + 12 x_2 x_4 q_0^2 q_1 + 12 x_1 x_5 q_0^2 q_1 + 12 x_2 x_3 q_0^2 q_2 + 12 x_1 x_4 q_0^2 q_2$$

$$+ 12 x_0 x_5 q_0^2 q_2 - 18 x_5 q_0^2 q_1 q_2 - 9 x_4 q_0^2 q_2^2 - 3 x_2^2 q_0^2 - 6 x_1 x_3 q_0^2$$

$$- 6 x_0 x_4 q_0^2 + 6 x_5 q_0^3 + 6 x_4 q_0^2 q_1 + 6 x_3 q_0^2 q_2 + x_0^3 - 3 x_2 q_0^2 - q_0^2$$

The other F_α have the same level of complexity.

REMARK 3.16. It transpires clearly that the finite formal model S_s of the pair $\mathcal{L}_\infty(\mathscr{C}, \gamma)$ provided by Drinfeld's arguments is different from the one given using the binomial structure of the cusp. By the computation using the binomial approach and Gabber's theorem (see Theorem 4.1), one knows that for some non-negative integer n, one has $S_s \cong$ $\mathrm{Spf}(k[[X]]/\langle X^2 \rangle) \hat{\times}_k \mathbf{D}_k^n$ but to guess and show this directly from description in Section 3.15 seems rather challenging. In fact, here, one can prove that $S_s \cong \mathrm{Spf}(k[[X]]/\langle X^2 \rangle) \hat{\times}_k \mathbf{D}_k^4$ but the proof is essentially a matter of rewriting the data used in Drinfeld's approach in terms of those used in the alternative approach.

EXAMPLE 3.17. We now consider the toric surface

$$\mathscr{S} = \mathrm{Spec}(k[X_0, X_1, X_2, X_3]/\langle X_0 X_2 - X_1^2, X_1 X_3 - X_2^2, X_0 X_3 - X_1 X_2^2 \rangle$$

and the arc $\gamma(T) = (T, T, T, T^2)$. For simplicity, we assume that the characteristic of the field k is different from 2.

For every test-ring $(A, \mathfrak{m}_A)$, we denote by $\widetilde{\mathcal{C}}(A)$ the subset of $\mathscr{W}(A,1)^3 \times \mathscr{W}(A,2)$ consisting of the quadruples $(q_{A,i}(T))_{i \in \{0,\ldots,3\}}$ such that

$$q_{A,0}(T) q_{A,2}(T) = q_{A,1}(T)^2 \quad \text{and} \quad q_{A,1}(T) q_{A,3}(T) = q_{A,2}(T)^3.$$

Note that the third relation one might be tempted to consider, namely $q_{A,0}(T) q_{A,3}(T) = q_{A,1}(T) q_{A,2}(T)^2$, is a consequence of the two previous relations. As pointed out by the referee, this illustrates the general fact that the computation of formal neighborhood of non-degenerate arcs may be reduced to the case of arcs on complete intersections.

For every A-deformation $(x_{A,i}(T))_{i \in \{0,\ldots,3\}} \in \mathcal{L}_\infty(\mathscr{S})_\gamma(A)$, and for every $i \in \{0,\ldots,3\}$ let $x_{A,i}(T) = q_{A,i}(T) u_{A,i}(T)$ be the Weierstrass factorizations of $x_{A,i}(T)$. Using the uniqueness of the Weierstrass factorization, one easily checks that the map

$$\theta_A : (x_{A,i}(T))_{i\in\{0,\dots,3\}} \mapsto (q_{A,i}(T))_{i\in\{0,\dots,3\}}, u_{A,1}(T), u_{A,2}(T))$$

induces a bijection, functorial in A, from $\mathscr{L}_\infty(\mathscr{S})_\gamma(A)$ to $\widetilde{\mathcal{C}}(A) \times (1 + \mathfrak{m}_A[[T]])^2$, whose inverse bijection is given by

$$(q_{A,i}(T)_{i\in\{0,\dots,3\}}, u_{A,1}(T), u_{A,2}(T))$$

$$\mapsto (q_{A,0}(T)u_{A,1}(T)^2 u_{A,2}(T)^{-1}, q_{A,1}(T)u_{A,1}(T), q_{A,2}(T)u_{A,2}(T),$$

$$q_{A,3}(T)u_{A,1}(T)^{-1}u_{A,2}(T)^3).$$

We also observe that, for every test-ring A, the set $\widetilde{\mathcal{C}}(A)$ is in natural bijection with the A-points of the formal k-scheme

$$\mathrm{Spf}(k[[X]]/\langle X^2 \rangle)\hat{\times}_k\mathbf{D}_k,$$

which is therefore by Section 3.3 a finite formal model of $(\mathscr{L}_\infty(\mathscr{S}), \gamma)$.

Indeed, setting $q_{A,0}(T) = q_{0,0} + T$, $q_{A,1}(T) = q_{1,0} + T$, $q_{A,2}(T) = q_{2,0} + T$ and $q_{A,3}(T) = q_{3,0} + q_{3,1}T + T^2$, with the $q_{i,j} \in \mathfrak{m}_A$, one obtains (after the change of variable $T \mapsto T - q_{2,0}$) that a presentation in $\mathrm{Spf}(k[[q_{0,0}, q_{1,0}, q_{3,1}, q_{3,0}]])$ of the finite model defined by $\widetilde{C}$ simplified by $\mathbf{D}_k$ is given by the T-coefficients of the polynomials

$$T(T + q_{0,0}) - (T + q_{1,0})^2 \quad \text{and} \quad (T + q_{1,0})(T^2 + q_{3,1}T + q_{3,0}) - T^3.$$

After identification and elimination, we find that the latter finite model is isomorphic to $\mathrm{Spf}(k[[q_{1,0}]]/\langle q_{1,0}^2 \rangle)$.

EXAMPLE 3.18. We still consider the toric surface

$$\mathscr{S} = \mathrm{Spec}(k[X_0, X_1, X_2, X_3]/\langle X_0X_2 - X_1^2, X_1X_3 - X_2^3, X_0X_3 - X_1X_2^2 \rangle)$$

and the arc $\gamma(T) = (T, T, T, T^2)$ and we make explicit the finite formal model of $(\mathscr{L}_\infty(\mathscr{S}), \gamma)$ obtained by following the computational algorithm produced by Drinfeld's argument.

Keeping the notation used in the proof, one can take $n = r = 2$, $X_1 = X_0$, $X_2 = X_1$, $Y_1 = X_2$, $Y_2 = X_3$, $f_1 = X_0X_2 - X_1^2$ and $f_2 = X_1X_3 - X_2^3$. In particular one has $D = X_0X_1$, $C = \begin{pmatrix} X_1 & 0 \\ 2X_2 & X_0 \end{pmatrix}$ and $d = 2$. For every test-ring A, $\tilde{\mathcal{B}}(A)$ is the set whose elements are the triples

$$(x_{A,0}(T), x_{A,1}(T), x_{A,2}(T), x_{A,3}(T), q_A(T)) \in (\mathfrak{m}_A[T]_{\leq 3})^2$$

$$\times(\mathfrak{m}_A[T]_{\leq 1})^2 \times \mathscr{W}(A, 2),$$

which satisfy the following relations:

$$\begin{cases} q_A(T) \text{ divides } (T + x_{A,0})(T + x_{A,1}); \\ q_A(T)^2 \text{ divides } (T + x_{A,1})((T + x_{A,0})(T + x_{A,2}) - (T + x_{A,1})^2); \\ q_A(T)^2 \text{ divides } 2(T + x_{A,2})((T + x_{A,0})(T + x_{A,2}) - (T + x_{A,1})^2) \\ \quad + (T + x_{0,A})((T + x_{A,1})(T^2 + x_{A,3}) - (T + x_{A,2})^3). \end{cases}$$

In order to give an explicit set of algebraic relations defining the finite formal model determined by $\widetilde{\mathcal{B}}$, one processes similarly as in the case of Example 3.14. One obtains a presentation

$$S_s = \mathrm{Spf}(k[[(q_i)_{0 \leq i \leq 1}, (x_{j,i})_{j \in \{0,1\}, 0 \leq i \leq 3}, (x_{j,i})_{j \in \{2,3\}, 0 \leq i \leq 1}]]/\langle F_\alpha \rangle_{0 \leq \alpha \leq 7}),$$

(3.19)

where the expressions of the F_α are rather complicated.

REMARK 3.20. From a computational point of view, and as could be guessed by comparing the models provided by the two approaches in the above examples, the binomial approach, when applicable, has the advantage to provide a much more efficient algorithm for computing a presentation of a finite formal model. Needless to say, the hypothesis that the variety is defined by binomial equations is very restrictive. An interesting question in this direction would be to find an efficient effective method to compute a finite formal model and applicable in full generality. Another interesting feature of the binomial approach, already pointed out in remark 3.16, is that it seems to provide more directly a finite formal model of minimal dimension (see [5], and Section 4 for the unicity up to isomorphism of this minimal formal model). We are not aware of any general efficient method (either from the computational or the theoretical point of view) to deduce information on the minimal finite formal model starting from the description provided by Drinfeld's argument.

3.8. Some extensions and variants of the Drinfeld–Grinberg–Kazhdan theorem

As proved in [7], the Drinfeld–Grinberg–Kazhdan can be slightly extended and made precise under the following form.

THEOREM 3.21. *Let k be a field. Let V be a $k[[T]]$-scheme of finite type, with no connected component isomorphic to $\mathrm{Spec}(k[[T]])$. Let $\gamma \in \mathscr{L}_\infty^\circ(V)(k)$. There exist an affine k-scheme S of finite type, with $s \in S(k)$, and an isomorphism of formal k-schemes:*

$$\theta_k \colon \mathscr{L}_\infty(V)_\gamma \to S_s \hat{\times}_k \mathbf{D}_k^{\mathbf{N}}.$$

Besides, for every separable field extension K of k, the isomorphism θ_k is compatible with base change to K.

REMARK 3.22. In [7], it is also shown that this theorem holds in case V is topologically of finite-type formal $k[[T]]$-scheme, though one does not necessarily obtain an *algebraizable* finite formal model. It would be an interesting problem to investigate whether such a result could be extended to the case of topologically of finite-type formal R-schemes when R is complete discrete valuation ring of mixed characteristics. (In that case, arc schemes have to be replaced by Greenberg schemes; see, e.g., [35, §3].) To the best of our knowledge, the latter question is open.

REMARK 3.23. From the point of view of differential algebra, arc schemes are defined by differential ideals generated by order 0 equations. One can also consider "differential arc schemes" defined by differential ideals generated by "genuine" differential equations. As proved by the authors in [6], one can obtain DGK-like results in the context of differential arcs.

3.9. Let V be a k-variety and γ be a rational non-degenerate arc. Instead of considering the formal neighborhood of γ in $\mathscr{L}_\infty(V)$, it should also be interesting to consider formal neighborhood in some "naturally relevant" closed subschemes of $\mathscr{L}_\infty(V)$. For example, one can look for any non-negative integer n at the formal neighborhood of γ in $\mathscr{L}_\infty(V)^{\gamma,n} := \pi_n^{-1}(\pi_n(\gamma))$. Here again, Drinfeld's argument may be extended to show that the analog of the DGK theorem remains true. Moreover, one obtains that for n large enough $\mathscr{L}_\infty(V)^{\gamma,n}_\gamma$ becomes isomorphic to $\mathbf{D}_k^{\mathbf{N}}$, a result that may be seen as a Denef–Loeser's fibration theorem-type result in the context of formal neighborhoods.

EXAMPLE 3.24. Let us illustrate the proof of this result in the case of the cusp $\mathscr{C} = \mathrm{Spec}(k[X,Y]/\langle X^3 - Y^2\rangle$ and $\gamma(T) = (T^2, T^3)$. We assume that k is of characteristic zero. For every test-ring A, one has to understand the set of pair $(x_A(T), y_A(T)) \in \mathfrak{m}_A[[T]]^2$ such that

$$(T^2 + T^n x_A(T))^3 = (T^3 + T^n y_A(T))^2.$$

For $n \geq 3$, this boils down to the relation $(1 + T^{n-2}x_A(T))^2 = (1 + T^{n-3}y_A(T))^3$. Thus, T divides $x_A(T)$ and, once having chosen arbitrarily $y_A(T)$, $x_A(T)$ is entirely determined. We infer that in this case $\mathscr{L}_\infty(\mathscr{C})^{\gamma,n}_\gamma \cong \mathbf{D}_k^{\mathbf{N}}$.

For $n = 2$, one obtains $T^2(1 + x_A(T))^3 = (T + y_A(T))^2$. If $(T + y_A(T)) = (T + \alpha)u_A(T)$ is the Weierstrass factorization of $T + y_A(T)$ (with $\alpha \in \mathfrak{m}_A$), this is equivalent to $(T + \alpha)^2 = 0$ and $(1 + x_A(T))^3 = u_A(T)^2$. Thus, we obtain $\mathscr{L}_\infty(\mathscr{C})^{\gamma,n}_\gamma \cong \mathbf{D}_k^{\mathbf{N}}$.

For $n = 1$, one obtains $T(T + x_A(T))^3 = (T^2 + y_A(T))^2$. If $(T + x_A(T)) = (T + \alpha)u_A(T)$ is the Weierstrass factorization of $T + x_A(T)$ (with $\alpha \in \mathfrak{m}_A$) and $(T^2 + y_A(T)) = (T + \beta_1 T + \beta_0)v_A(T)$ is the Weierstrass factorization of $T^2 + y_A(T)$ (with $\beta_1, \beta_0 \in \mathfrak{m}_A$), the above relation is equivalent to

$$T(T + \alpha)^3 = (T^2 + \beta_1 T + \beta_0)^2 \quad u_A(T)^3 = v_A(T)^2.$$

Thus, $v_A(T)$ may be expressed in terms of $u_A(T)$, and expanding the first relation, one may find after suitable eliminations that $\mathscr{L}_\infty(\mathscr{C})^{\gamma,n}_\gamma \cong \mathrm{Spf}(k[[\beta_1]]/\beta_1^3)\hat{\times}_k\mathbf{D}_k^{\mathbf{N}}$.

3.10. There also exist more global variants of Theorem 1.1: see [13, 12, 24, 32] as well as the survey paper [11]. These global variants are expected to play an important part in representation theory and the Langlands program.

4. A Simplification Lemma in Formal Geometry

We retain the notation and assumptions of Theorem 1.1. Recall from definition 3.1 the notion of *finite formal model* of the pair $(\mathscr{L}_\infty(V), \gamma)$. An elementary observation is that for a given pair $(\mathscr{L}_\infty(V), \gamma)$, a finite formal model is not unique, even up to isomorphism: indeed, if S_s satisfies the statement of Theorem 1.1, so does $S_s \hat{\times} \mathbf{D}_k$. More generally, every representative of the stable isomorphism class in $\mathbf{Lncp}_k$ of S_s provides a finite formal model of $(\mathscr{L}_\infty(V), \gamma)$. Nevertheless, we have the following simplification theorem of Gabber. Let k be a field. One says that a k-algebra $A \in \mathbf{Lncp}_k$ is *cancellable* (in $\mathbf{Lncp}_k$) if there exists a k-algebra $B \in \mathbf{Lncp}_k$ such that A is isomorphic to $B[[T]]$. Let us note that for every k-algebra $A \in \mathbf{Lncp}_k$, there exist an integer $N \in \mathbf{N}$ and a non-cancellable k-algebra $A_{\min} \in \mathbf{Lncp}_k$ such that A is isomorphic to $A_{\min}[[T_1, \ldots, T_N]]$.

THEOREM 4.1 (O. Gabber). *Let $A, B \in \mathbf{Lncp}_k$, and let I, J be sets (possibly infinite). Assume that $A[[(T_i)_{i \in I}]]$ and $B[[(U_j)_{j \in J}]]$ are isomorphic (in $\mathbf{Lacp}_k$). Then, up to exchanging A and B, there exists a finite subset $I' \subset I$ such that $A[[(T_i)_{i \in I'}]]$ and B are isomorphic (in $\mathbf{Lncp}_k$). In particular, if both A and B are non-cancellable, then they are isomorphic.*

PROOF. We sketch the arguments of the complete proof which is presented in the appendix of [7].

Let $A, B \in \mathbf{Lncp}_k$. Let $\mathfrak{M}_A$ (resp. $\mathfrak{M}_B$) be the maximal ideal of A (resp. B). Clearly, we may assume that A, B are non-cancellable. We set $A' := A[[(T_i)_{i \in I}]]$, $B' := B[[(U_j)_{i \in J}]]$, and let $\varphi : A' \to B'$ be an isomorphism.

We have a natural injective morphism $\iota_A : A \to A'$ admitting a retraction ρ_A given by $T_i \mapsto 0$. We define analogously ι_B and ρ_B.

Let $\mathfrak{M}_{A'}$ (resp. $\mathfrak{M}_{B'}$) be the maximal ideal of A' (resp. B') and $\overline{\mathfrak{M}_{A'}^2}$ (resp. $\overline{\mathfrak{M}_{B'}^2}$) be the closure of $\mathfrak{M}_{A'}^2$ (resp. $\mathfrak{M}_{B'}^2$).

Identifying, via ι_A, $\mathfrak{M}_A/\mathfrak{M}_A^2$ with a subvector space of $\mathfrak{M}_{A'}/\overline{\mathfrak{M}_{A'}^2}$, we have a decomposition

$$(4.2) \qquad \mathfrak{M}_{A'}/\overline{\mathfrak{M}_{A'}^2} \cong \mathfrak{M}_A/\mathfrak{M}_A^2 \bigoplus_{i \in I} k \, t_i$$

(where we denote by t_i the class of T_i modulo $\overline{\mathfrak{M}_{A'}^2}$) and a similar decomposition for B'.

Using Lemma 4.3, we obtain that the composition

$$h : A \xrightarrow{\iota_A} A' \xrightarrow[\cong]{\varphi} B' \xrightarrow{\rho_B} B$$

induces an isomorphism $\mathfrak{M}_A/\mathfrak{M}_A^2 \cong \mathfrak{M}_B/\mathfrak{M}_B^2$. By a straightforward induction, we infer that, for every integer $n \in \mathbf{N}$, the morphism h induces a surjection of finite dimensional k-vector spaces:

$$h_n : \mathfrak{M}_A^n/\mathfrak{M}_A^{n+1} \to \mathfrak{M}_B^n/\mathfrak{M}_B^{n+1}.$$

Thus, one has $\dim(\mathfrak{M}_B^n/\mathfrak{M}_B^{n+1}) \leq \dim(\mathfrak{M}_A^n/\mathfrak{M}_A^{n+1})$. Exchanging the roles of A in B, we get the opposite inequalities, hence the equality of the dimensions for all n. Thus, the morphism h_n is an isomorphism for every integer $n \in \mathbf{N}$; hence, the morphism h is an isomorphism. $\qquad\square$

LEMMA 4.3. *Identifying, via* φ, *$\mathfrak{M}_{A'}/\overline{\mathfrak{M}_{A'}^2}$ and $\mathfrak{M}_{B'}/\overline{\mathfrak{M}_{B'}^2}$, the images of $\mathfrak{M}_A/\mathfrak{M}_A^2$ by ι_A and $\mathfrak{M}_B/\mathfrak{M}_B^2$ by ι_B coincide.*

REMARK 4.4. The proof of this lemma is detailed in [**7**]. Theorem 4.1 was previously known under the extra assumption that the sets I and J are finite (see [**23**]). Note that the theorem shows in particular that two objects in $\mathbf{Lncp}_k$, which are stably isomorphic in $\mathbf{Lacp}_k$, are in fact already stably isomorphic in $\mathbf{Lncp}_k$. On the other hand, there is also a version of the theorem where A and B are only assumed to be objects of $\mathbf{Lacp}_k$ (see [**4**]).

5. The Minimal Formal Model of a Rational Non-degenerate Arc

Combining Theorems 4.1 and 1.1, one introduces a *new* geometric object in the study of singularities. Theorem 5.5 shows in particular that it may provide a relevant measure of the complexity of singular points of algebraic varieties.

5.1. The minimal finite-dimensional formal model

Let k be a field, V be a k-variety and $\gamma \in \mathscr{L}_\infty^\circ(V)(k)$ be a non-degenerate rational arc. By Theorem 1.1, $(\mathscr{L}_\infty(V), \gamma)$ admits finite-dimensional formal models (see Definition 3.1). By Theorem 4.1, all the finite-dimensional formal models of $(\mathscr{L}_\infty(V), \gamma)$ are stably isomorphic, and there exists a unique (up to isomorphism) *non-cancellable* complete noetherian algebraizable local k-algebra $A_{\min}(\gamma)$ such that $\mathscr{S}_V(\gamma) := \mathrm{Spf}(A_{\min}(\gamma))$ (simply denoted by $\mathscr{S}(\gamma)$) is a finite-dimensional formal model of $(\mathscr{L}_\infty(V), \gamma)$.

DEFINITION 5.1. Let k be a field. Let V be a k-variety. Let $\gamma \in \mathscr{L}_\infty^\circ(V)(k)$ be a non-degenerate rational arc. We call $\mathscr{S}(\gamma)$ (or more precisely its isomorphism class) *the* minimal formal model of $(\mathscr{L}_\infty(V), \gamma)$.

REMARK 5.2. The minimal formal model of a non-degenerate arc is also the finite formal model of minimal dimension among all the finite formal models of this arc.

REMARK 5.3. In general, Drinfeld's proof of Theorem 1.1 seems to be far from providing information about the *minimal* formal model (see Remark 3.20). An interesting question would be to determine an effective algorithm whose output would be the minimal formal model of $\mathscr{L}_\infty(V)_\gamma$ from the datum of the equations of an embedding of the affine k-variety V in an affine space and that of a sufficiently large truncation of the arc γ.

REMARK 5.4. Considering, e.g., the non-cancellable complete local k-algebra

$$k[[x_0, \ldots, x_n]]/\langle x_0^2, x_0 x_1, \ldots, x_0 x_n \rangle$$

(the formal neighborhood of the origin in the affine n-space with an embedded point at the origin) and using Example 3.4, we see that if we put no particular restriction on V and/or on the non-degenerate arc γ, the dimension of a minimal finite formal model may be arbitrarily large.

5.2. The minimal formal model is a measure of the singularities

Let V be a k-variety and $v \in V(k)$ be a rational point. A challenging problem is to understand the geometric information on (V, v) contained in the minimal formal model $\mathscr{S}(\gamma)$ of $(\mathscr{L}_\infty(V), \gamma)$, where γ is an non-degenerate arc with $\gamma(0) = v$. Of course, this should involve understanding how $\mathscr{S}(\gamma)$ varies when γ runs over the families of non-degenerate arcs with origin v (see Section 7). Disregarding this matter for the moment, a first step in this direction is given by the following result obtained in [10].

THEOREM 5.5. *Let V be a k-variety and $v \in V(k)$ such that $\dim_v(V) \geq 1$. Let $\gamma \in \mathscr{L}_\infty(V)(k)$ be a non-degenerate rational arc such that $\gamma(0) = v$. Then, the following conditions are equivalent:*

(1) *The unique formal branch containing γ is smooth.*
(2) *The formal neighborhood $\mathscr{L}_\infty(V)_\gamma$ is isomorphic to $\mathbf{D}_k^{\mathbf{N}}$.*
(3) *The minimal finite formal model $\mathscr{S}(\gamma)$ is isomorphic to $\mathrm{Spf}(k)$.*

A word of explanation is in order regarding the first condition. Under the assumption of the statement of Theorem 5.5, one can show that γ induces a continuous morphism of complete local k-algebras $\hat{\gamma} \colon \widehat{\mathcal{O}_{V,v}} \to k[[T]]$ whose kernel contains a unique minimal prime ideal $\mathfrak{p}$ of the ring $\widehat{\mathcal{O}_{V,v}}$. Roughly speaking, the non-degenerate arc γ factorizes through a unique irreducible component of $\mathrm{Spec}\,(\widehat{\mathcal{O}_{V,v}})$. Moreover, one can show that the kernel of every deformation of γ contains $\mathfrak{p}$; in other words, it is contained in this irreducible component. Then, the first condition means that $\widehat{\mathcal{O}_{V,v}}/\mathfrak{p}$ is formally smooth. In particular, if $\widehat{\mathcal{O}_{V,v}}$ is a domain, it means that v is a smooth point of V.

REMARK 5.6. Under the assumption of the statement of Theorem 5.5 and the extra assumption that $\mathcal{O}_{V,v}$ is reduced, if there is a unique formal branch passing through v and (5.2) holds, then v is a smooth point of V. Let us stress that we cannot relax the hypothesis that $\mathcal{O}_{V,v}$ is reduced. This is clear by Section 3.5. Let us give an explicit example by considering the case where $V = \mathrm{Spec}(k[X, Y]/\langle X^2, XY \rangle)$ and v is the origin. In this case, we observe that property (5.2) in the statement of Theorem 5.5 is satisfied for the arc $\gamma = (0, T)$, since, by the very definitions, one has

$$\mathscr{L}_\infty(V)_\gamma \cong \mathscr{L}_\infty(\mathrm{Spec}(k[Y]))_\alpha \cong \mathbf{D}_k^{\mathbf{N}},$$

where $\gamma = T$, but $\widehat{\mathcal{O}_{V,v}}$ is not formally smooth.

There is also a version of Theorem 5.5 in the context of constant arcs. Unlike the proof of Theorem 5.5, this case can be more formally deduced from the usual arguments of formal smoothness.

PROPOSITION 5.7. *Let V be a k-variety and $v \in V(k)$ such that $\dim_v(V) \geq 1$. Then, the following conditions are equivalent:*

(1) *The k-variety V is smooth at v.*
(2) *The formal neighborhood $\mathscr{L}_\infty(V)_{\sigma(v)}$ is isomorphic to $\mathbf{D}_k^{\mathbf{N}}$.*

In other words, smooth constant arcs on V correspond to smooth points of V.

PROOF. We only have to show implication $2 \Rightarrow 1$. By [**21**, 17.5.1, 17.5.3], it suffices to show that the local k-algebra $\widehat{\mathcal{O}_{V,v}}$ is formally smooth for the $\mathfrak{m}_v$-adic topology (which coincides here with the projective limit topology). By [**20**, 19.3.3, 19.3.6] and the hypothesis, the k-algebra $\widehat{\mathcal{O}_{\mathscr{L}_\infty(V),\sigma(v)}}$ is formally smooth for the projective limit topology. Since the continuous morphism $\widehat{\mathcal{O}_{V,v}} \rightarrow \widehat{\mathcal{O}_{\mathscr{L}_\infty(V),\sigma(v)}}$ induced by the projection $\mathscr{L}_\infty(V) \rightarrow V$ admits a continuous retraction (induced by σ), we may conclude the proof by the very definition of formal smoothness. □

As already pointed out, Definition 5.1 and Theorem 5.5 suggest the following question:

QUESTION 5.8. What piece of information on the singularities of V does the minimal formal model of $(\mathscr{L}_\infty(V), \gamma)$ encode (for a "suitable" choice of γ)?

Section 7 provides the first elements of answer in this direction.

6. The Case of Degenerate Arcs

In [**8**], it is shown that Theorem 1.1 can not be extended to the case of constant arcs. Precisely, we prove the following statement.

THEOREM 6.1. *Let k be a field of characteristic zero which does not contain a root of the equation $T^2 + 1 = 0$. Let $f \in k[X, Y]$ be the polynomial $X^2 + Y^2$. Let us denote by $\mathscr{C}$ the affine plane curve defined by the datum of f, and by $\mathfrak{o} \in \mathscr{C}(k)$ the origin of $\mathbf{A}_k^2$. We still denote by $\mathfrak{o}$ the induced constant arc in $\mathscr{L}_\infty(\mathscr{C})(k)$. Then, the arc $\mathfrak{o}$ does not satisfy the statement of Theorem 1.1.*

Let us note that, by fiber products, one can provide such examples in every dimension. The basic idea of the proof relies on results of differential algebra. Recall that, when the base-field k is of characteristic zero, one can describe the arc scheme associated with an affine k-variety thanks to this algebraic theory due to J. Ritt and E. Kolchin. Indeed, if V is defined in $\mathbf{A}_k^N$ by the datum of the ideal I of $k[X_1, \ldots, X_N]$, then one knows that

$$V \cong \mathrm{Spec}(k[X_{i,j}; i \in [\![1, N]\!], j \in \mathbf{N}]/[I]),$$

where, as usual, we denote by $[I]$ the differential closure of I under the action of the k-derivation Δ defined by the formula $\Delta(X_{i,j}) = X_{i,j+1}$ for every integer $i \in [\![1, N]\!]$, and every integer $j \in \mathbf{N}$ (see [**3**] in the present volume for more details). Then, the proof of Theorem 6.1 can be deduced from the non-existence of a strong basis for the differential ideal $[X]^2$ (see [**34**] for an introduction of differential algebra and [**34**, pp. 11–13] for details on strong basis). More recently, Chiu and Hauser in [**14**] have completed this first answer by showing the following statement which in particular positively answers Question 1 and Question 2 of [**8**].

THEOREM 6.2. *Let k be a field of characteristic zero. Let V be an algebraic variety. Let $x \in V_{\mathrm{sing}}(k)$. We still denote by x the induced constant arc in $\mathscr{L}_\infty(V)(k)$. Then, the arc x does not satisfy the statement of Theorem 1.1.*

The proof is based on the following lemma which is a characterization for non-noetherian formal neighborhoods to be *cylinders*. Let us stress that this statement is an extension in the non-noetherian case of [**36**, Lemma 4].

LEMMA 6.3. *Let k be a field of characteristic zero. Let V be a k-scheme with $x \in V(k)$. Then, the ring $\widehat{\mathcal{O}_{V,x}}$ is a cylinder if and only if it admits a regular k-derivation.*

Recall that $\widehat{\mathcal{O}_{V,x}}$ is a cylinder if there exist a k-scheme W, with $y \in W$, and a continuous isomorphism $\widehat{\mathcal{O}_{V,x}} \xrightarrow{\sim} \widehat{\mathcal{O}_{W,y}}[[T]]$. A regular derivation D of $\widehat{\mathcal{O}_{V,x}}$ is a continuous k-derivation of $\widehat{\mathcal{O}_{V,x}}$ such that there exists an element $x \in \mathfrak{m}_x$ with $D(x) \in (\widehat{\mathcal{O}_{V,x}})^\times$.

SKETCH OF PROOF OF THEOREM 6.2. Since the question is local, we may assume that the k-variety V is affine. We also may assume that the completion $\widehat{\mathcal{O}_{V,x}}$ is non-cancellable. If there exists an isomorphism

$$\mathscr{L}_\infty(V)_x \cong S_s \hat{\times}_k \mathbf{D}_k^{\mathbf{N}},$$

then, by Lemma 6.3, there exists a regular k-derivation $\mathscr{D}$ on $\mathscr{L}_\infty(V)_x$. The key point of the proof of Chiu and Hauser is to show that such a k-derivation induces a regular k-derivation on V. Then, by Lemma 6.3 (or more simply by [**36**, Lemma 4]), the ring $\widehat{\mathcal{O}_{V,x}}$ is cancellable which is a contradiction of our assumption. □

7. Dependency on the Arc

7.1. Consider a smooth rational point v on a k-variety V. Then, the formal neighborhood of an arc with center v does not depend on the choice of the arc and is trivial. But in case v is singular, in general, changing the involved arc while keeping the center fixed will modify the isomorphism class of the formal neighborhood. This may happen, for example, when reparameterizing the arc by a non-invertible endomorphism of $k[[T]]$ (see Example 7.2). In the case of curves, we prove in Section 7.2 that primitive arcs (at a given singularity and branch) provide a class of rational arcs for which the formal neighborhood is invariant. In higher dimension, prompted by the observations in the toric case (see Section 7.4), one may wonder whether the formal neighborhood is constant among generic rational arcs lying on a given Nash component of the variety, which would constitute a generalization of the case of curves.

EXAMPLE 7.1. Let k be a field. Let $\mathscr{C} = \mathrm{Spec}(k[X,Y]/\langle X^3 - Y^2 \rangle)$ be the affine plane cusp and $\mathfrak{o}$ be the origin. If γ is any primitive k-parameterization of the singular point $\mathfrak{o}$, then we have shown in Section 3.7 that the formal neighborhood $\mathscr{L}_\infty(V)_\gamma$ is isomorphic to the formal k-scheme $\mathrm{Spf}(k[X]/\langle X^2 \rangle) \hat{\times}_k \mathbf{D}_k^{\mathbf{N}}$.

EXAMPLE 7.2. Let k be a field whose characteristic is not 2 or 3. Let $\mathscr{C} = \mathrm{Spec}(k[X,Y]/\langle X^3 - Y^2 \rangle)$ be the affine plane cusp and $\mathfrak{o}$ be the origin. Let $\gamma \in \mathscr{L}_\infty(\mathscr{C})(k)$ be a non-degenerate arc. As a consequence of Theorem 1.1, it is meaningful to introduce the nilpotence index of $\mathscr{L}_\infty(V)_\gamma$, that we denote by $\mathrm{nil}_\gamma(V)$ (see Section 8 and Theorem 8.1 for a complete justification of this fact). Now, let $\gamma(T) = (T^2, T^3)$ and $\eta(T) = (T^4, T^6)$. One can show that

$$\mathrm{nil}_\eta(\mathscr{C}) \geq 3 > 2 = \mathrm{nil}_\gamma(\mathscr{C}).$$

Hence, the formal k-schemes $\mathscr{L}_\infty(\mathscr{C})_\gamma$ and $\mathscr{L}_\infty(\mathscr{C})_\eta$ are not isomorphic (see [**7**]).

7.2. Invariance of the formal neighborhood in the case of unibranch germs of curves

Let $\mathscr{C}$ be an integral k-curve, geometrically unibranch at $c \in \mathscr{C}(k)$. A *primitive k-parameterization* $\gamma \in \mathscr{L}_\infty(\mathscr{C})(k)$ at c^1 is an arc γ on $\mathscr{C}$, with $\gamma(0) = c$, satisfying the following property: for every morphism $\gamma' : \mathcal{O}_{\mathscr{C},c} \to k[[T]]$ of local k-algebras, there exists a morphism of formal k-schemes $p_k : \mathrm{Spf}(k[[T]]) \to \mathrm{Spf}(k[[T]])$ such that we have $\gamma' = \gamma \circ p_k$.

---

[1]If k is assumed to be perfect, the assumption that c is geometrically unibranch guarantees the existence of primitive k-parameterizations at c.

PROPOSITION 7.3. *Let* V, V' *be two* k-*varieties. Let* $\gamma \in \mathscr{L}_\infty(V)(k)$, $\gamma' \in \mathscr{L}_\infty(V')(k)$, $v = \gamma(0)$ *and* $v' = \gamma'(0)$. *Assume that there exist isomorphisms* $f : \widehat{\mathcal{O}_{V,v}} \to \widehat{\mathcal{O}_{V',v'}}$ *and* $p : k[[T]] \to k[[T]]$ *such that* $p \circ \gamma = \gamma' \circ f$ *Then, there exists an isomorphism of formal* k-*schemes* $\mathscr{L}_\infty(V)_\gamma \cong \mathscr{L}_\infty(V')_{\gamma'}$.

PROOF. One may assume $\gamma = \gamma' \circ f$. Then, for every test-ring A, the composition by f induces the following diagram of maps, functorial in A:

$$
\begin{array}{ccc}
\mathrm{Hom}_k^{\mathrm{lcp}}(\widehat{\mathcal{O}_{V,v}}, A[[T]]) & \xrightarrow[\sim]{\circ f} & \mathrm{Hom}_k^{\mathrm{lcp}}(\widehat{\mathcal{O}_{V',v'}}, A[[T]]) \\
\uparrow & & \uparrow \\
\mathscr{L}_\infty(V)_\gamma(A) & \longrightarrow & \mathscr{L}_\infty(V')_{\gamma'}(A)
\end{array}
$$

where the map $\mathscr{L}_\infty(V)_\gamma(A) \to \mathscr{L}_\infty(V')_{\gamma'}(A)$, obtained by restriction, is also a bijection. Hence, the formal k-schemes $\mathscr{L}_\infty(V)_\gamma$ and $\mathscr{L}_\infty(V')_{\gamma'}$ are isomorphic. $\square$

COROLLARY 7.4. *Let* k *be a field. Let* $\mathscr{C}$ *be a* k-*curve. Let* $x \in \mathscr{C}(k)$. *Assume that there exists a primitive* k-*parameterization* $\gamma_x \in \mathscr{L}_\infty(\mathscr{C})(k)$ *of* $\mathscr{C}$ *at* x. *Then, the isomorphism class of the formal* k-*scheme* $\mathscr{L}_\infty(\mathscr{C})_{\gamma_x}$ *does not depend on the choice of the involved primitive* k-*parameterization. In particular, it is a formal invariant of the curve singularity* $(\mathscr{C}, x)$.

7.3. A new geometric invariant for curve singularities

Let us state an important consequence of Theorem 4.1, which shows the relevance of minimal formal models in the study of curve singularities as a *new geometric formal invariant* (see [**9**]).

COROLLARY 7.5. *Let* k *be a field. Let* $\mathscr{C}$ *be a* k-*curve, with* $x \in \mathscr{C}(k)$. *Assume that there exists a primitive* k-*parameterization* $\gamma \in \mathscr{L}_\infty(\mathscr{C})(k)$ *of* $\mathscr{C}$ *at* x. *Let* (S, s) *be a pointed* k-*scheme such that* S_s *is a finite-dimensional formal model of* $(\mathscr{L}_\infty(\mathscr{C}), \gamma)$. *Then, the stable isomorphism class of the formal neighborhood* S_s, *and the complete* k-*algebra* $(\widehat{\mathcal{O}_{S,s}})_{\mathrm{min}}$, *are formal invariants of the curve singularity* $(\mathscr{C}, x)$.

EXAMPLE 7.6. Let $m \geq 1$ be an even integer. Let k be a field with characteristic zero or greater than $2(m+1)$. Let $(\mathscr{C}, x)$ be the simple A_m-plane curve singularity. Let $\gamma \in \mathscr{L}_\infty(\mathscr{C})(k)$ be a primitive k-parameterization at x. Then, the formal k-scheme $\mathrm{Spf}(k[[X]]/\langle X^{m/2+1}\rangle)$ is the minimal formal model of $(\mathscr{C}, x)$ (see [**9**]).

In the case of monomial curve singularities, we have the following result on the minimal formal model (see [**9, 5**]).

THEOREM 7.7. *Let k be a field of characteristic zero. Let $(\mathscr{C}, x)$ be a germ of monomial k-curve singularity, with multiplicity m. Then, its minimal formal model is of dimension 0 and effectively computable once one knows a primitive parameterization. Moreover, the embedding dimension of the minimal formal model equals $m - 1$.*

As an application of the explicit knowledge of the minimal formal model, one shows in [**9**] for the family of curve singularities $(\mathrm{Spec}(k[X, Y]/\langle X^N - Y^M \rangle, \mathfrak{o})$ that Theorem 1.1 is, in a strong sense, not compatible with the truncation morphism $\pi_n^\infty \colon \mathscr{L}_\infty(V) \to \mathscr{L}_n(V)$. Roughly speaking, although Theorem 1.1 is a finiteness result, it is not a stability-like result with respect to truncations.

7.4. Invariance of the formal neighborhood in the toric case

We point out an important feature of the behavior of formal neighborhoods of non-degenerate arcs on toric varieties. This description will be connected to the general notion of *Nash components* (also called *Nash families*) of a k-variety. Let V be a k-variety. Recall that the associated arc scheme $\mathscr{L}_\infty(V)$ is endowed with a canonical projection $\pi_0 \colon \mathscr{L}_\infty(V) \to V$. A *Nash component* of V is an irreducible component of the subset $\pi_0^{-1}(V_{\mathrm{sing}}) \setminus \mathscr{L}_\infty(V_{\mathrm{sing}})$.

DEFINITION 7.8. *Let k be a field of characteristic zero and let V be an integral k-scheme. One says that a divisorial valuation ν on V is essential (resp. strongly essential) if, for every resolution of singularities $h \colon W \to V$, the center of ν on W is the generic point of an irreducible component (resp. an irreducible component of codimension 1 in W) of $h^{-1}(V_{\mathrm{sing}})$. If $h \colon W \to V$ is a resolution of the singularities of V, we say that a divisor D of W is an essential divisor of V (resp. a strongly essential divisor of V) if the associated valuation is essential (resp. strongly essential).*

Roughly speaking, an essential divisor of V is a divisor that "appears" in every resolution of the singularities of V. If the field k is assumed to be of characteristic zero, the *Nash map* injects the set of Nash components into the set of essential divisors of V; the *Nash problem* is then to understand when this injection is a surjection, see [**31**]. By [**25**], one knows that the Nash problem admits a positive answer in the case of toric singularities. In particular, every essential divisor D of V is associated to a unique Nash component $\mathcal{N}_D$ in $\mathscr{L}_\infty(V)$. In [**5**], the authors obtained the following statement, the proof of which is crucially based on the study of arc schemes associated with toric varieties made in [**26, 25**].

THEOREM 7.9. *Let k be a field of characteristic zero and V be a toric k-variety. Let D be an essential divisor of V. Then, there exists a Zariski non-empty open subset U_D of $\mathcal{N}_D$ such that $\mathcal{O}_{\widehat{\mathscr{L}_\infty(V)}, \gamma}$ and thus the minimal formal model of $(\mathscr{L}_\infty(V), \gamma)$ are invariant (up to isomorphism) for $\gamma \in U_D(k)$.*

7.5. The minimal formal model of toric singularities

Let V be a toric k-variety and $\gamma \in \mathscr{L}_\infty(V)(k)$ be a k-rational arc, non-contained in the complement of the open orbit. With every such arc, one can associate a divisorial toric valuation ν_γ (see [26]). A crucial ingredient of the proof of Theorem 7.9 is the fact that $\widehat{\mathcal{O}_{\mathscr{L}_\infty(V),\gamma}}$ depends only on the valuation ν_γ. This follows from [26] but it may also be recovered from the alternative approach of the proof of Theorem 1.1 for binomial equations. Thus, one may speak of the minimal formal model of a toric valuation. In [5], one obtains the following theorem.

THEOREM 7.10. *Let k be a field of characteristic zero. Let V be a toric k-variety and ν be a strongly essential toric valuation. Then, the minimal formal model of ν is of dimension 0 and effectively computable. Besides, its embedding dimension coincides with the Mather discrepancy $\hat{k}_\nu$ of the toric valuation ν.*

For the definition and the basic properties of the notion of Mather discrepancy, see, e.g., [27, 28, 18, 16].

REMARK 7.11. More generally, one obtains in [5] an explicit description of the minimal formal model, as well as its dimension and embedding dimension, for other classes of toric valuations. This allows for example to show that even when restricting to arcs on toric three-dimensional varieties, the dimension of the minimal formal model may be arbitrarily large. For an (not necessarily strongly) essential toric valuation ν, one obtains a presentation of an irreducible finite formal model with dimension $\hat{k}_\nu$ which we conjecture to be the minimal finite formal model.

7.6. The genericness of formal neighborhoods of non-degenerate arcs and the relation with stable points

As pointed out in Section 7.1, the formal neighborhoods of arc schemes at rational arcs vary with respect to the involved arc even when keeping the origin fixed. But, as illustrated in Sections 7.3 and 7.4, in some cases, a "generic behavior" (which may be described in terms of Nash components) seems to appear with a generic choice of the considered arcs. Let us make this basic idea precise. (For simplicity, we assume that the field k is algebraically closed.)

• If $\mathscr{C}$ is a singular curve, classical results assert that there exists a bijection between the set of isomorphism classes of k-parameterizations and the set of Nash components.[2] In particular, every Nash component contains, up to isomorphism, a unique k-parameterization, and the set of k-parameterizations of a given Nash component $\mathcal{N}$ are the rational points of an open subset U of $\mathcal{N}$. Thus, by Corollary 7.4, we observe that the

---

[2]This bijection solves the Nash problem in dimension 1.

formal neighborhood $\widehat{\mathscr{L}_\infty(\mathscr{C})_\gamma}$ is invariant (up to isomorphisms) when γ runs over $U(k)$.

• Theorem 7.9 is, by its very statement, the analog of the former remark in the case of toric singularities.

These observations lead us to ask the following question.

QUESTION 7.12. Let k be a field of characteristic zero. Let V be a k-variety. Let $\mathcal{N}$ be a Nash component of V. Does there exist an open subscheme U of $\mathcal{N}$ such that the formal neighborhood $\widehat{\mathscr{L}_\infty(\mathscr{C})_\gamma}$ is invariant (up to isomorphisms) when γ runs over $U(k)$?

7.7. We also believe that when the answer to the above question is positive, the generic behavior of the formal neighborhood in a Nash component should be related to the formal neighborhood at the generic point of this component, which is an example of the notion of *stable point* introduced by Reguera in [**33**], precisely in relation with the Nash problem. By definition, a stable point of an arc scheme is the generic point of an irreducible constructible subset of this scheme (not contained in $\mathscr{L}_\infty(V_{\mathrm{sing}})$; see also [**15**, §11]). In particular, we note that there is a strong similarity between Theorem 7.10 and results in [**30**] computing the embedding dimension of the formal neighborhoods of stable points. One may wonder whether there may exist a kind of specialization property relates to the formal neighborhoods of stable points on the one hand, and of non-degenerate rational arcs on the other hand. A possible more precise question could be stated as follows.

QUESTION 7.13. Keep the notation and assumptions of Question 7.12 and assume that the answer is positive. Let $\gamma \in U(k)$ and $\mathfrak{p}$ be the generic point of $\mathcal{N}$, with residue field $\kappa(\mathfrak{p})$. Choose a section of the quotient morphism $\mathcal{O}_{\widehat{\mathscr{L}_\infty(V),\mathfrak{p}}} \to \kappa(\mathfrak{p})$. Are the complete local $\kappa(\mathfrak{p})$-algebras $\mathcal{O}_{\widehat{\mathscr{L}_\infty(V),\mathfrak{p}}}$ and $\mathcal{O}_{\widehat{\mathscr{L}_\infty(V),\gamma}} \widehat{\otimes}_k \kappa(\mathfrak{p})$ stably isomorphic?

CONJECTURE 7.14. Let k be a field of characteristic zero. Let $(V, \mathfrak{o})$ be an A_n-singularity of dimension 1 or 2 (or more generally one might assume that V is a binomial variety). Then, Question 7.13 admits a positive answer. A recent work of Mario Morán Cañón and the authors (see [**2**]) gives positive elements of answer in the direction of this conjecture.

8. Nilpotency in Formal Neighborhoods

If A is a ring whose nilradical $\mathcal{N}(A)$ is nilpotent, then the *nilpotence index* of A is the smallest positive integer m such that $\mathcal{N}(A)^m = \{0\}$. As a consequence of Theorem 1.1, one obtains the following theorem.

THEOREM 8.1. *Let V be a k-variety, and let $\gamma \in \mathscr{L}_\infty^\circ(V)(k)$. Then, the nilradical of the ring $\mathcal{O}_{\widehat{\mathscr{L}_\infty(V),\gamma}}$ is finitely generated, and in particular* nilpotent. Moreover, the nilpotence index of the ring $\mathcal{O}_{\widehat{\mathscr{L}_\infty(V)},\gamma}$ equals the nilpotence index of $\widehat{\mathcal{O}_{S,s}}$, for every pointed k-scheme (S,s) such that S_s is a finite formal model of $(\mathscr{L}_\infty(V),\gamma)$.

PROOF. Let (S,s) be a pointed k-scheme such that S_s is a finite formal model of $(\mathscr{L}_\infty(V),\gamma)$. One has an isomorphism of k-formal schemes $S_s \cong \mathrm{Spf}(A/I)$ where A is a power series ring in a finite number of variables over k and I is a proper ideal of A. Since S_s is a finite formal model of $(\mathscr{L}_\infty(V),\gamma)$, there exists an isomorphism of formal k-schemes

$$\mathscr{L}_\infty(V)_\gamma \xrightarrow{\sim} \mathrm{Spf}(A[[(T_i)_{i\in\mathbf{N}}]]/I[[(T_i)_{i\in\mathbf{N}}]]).$$

Since the ideal I is finitely generated, we observe that $I \cdot A[[(T_i)_{i\in\mathbf{N}}]] = I[[(T_i)_{i\in\mathbf{N}}]]$. Let us set $B = A[[(T_i)_{i\in\mathbf{N}}]]$. One can check that $\sqrt{I \cdot B} = (\sqrt{I}) \cdot B$, which shows that $\sqrt{I \cdot B}$ is finitely generated. Let m be the nilpotence index of A/I. Then, one has $(\sqrt{I \cdot B})^m = (\sqrt{I})^m \cdot B \subset I \cdot B$. On the other hand, if n is a positive integer satisfying $(\sqrt{I \cdot B})^n \subset I \cdot B$, one has $(\sqrt{I})^n \cdot B \cap A \subset I \cdot B \cap A$, hence $(\sqrt{I})^n \subset I$. That concludes the proof. $\square$

Let V be a k-variety, and let $\gamma \in \mathscr{L}_\infty^\circ(V)(k)$. We define $\mathrm{nil}_\gamma(V)$ to be the nilpotence index of $\mathscr{L}_\infty(V)_\gamma$. If $v \in V(k)$, we denote by $\mathscr{L}_\infty^\circ(V,v)$ the subset of $\mathscr{L}_\infty^\circ(V)(k)$ formed by the rational arcs γ on V with base-point $\gamma(0) = v$. This definition suggests to introduce the following invariant.

DEFINITION 8.2. Let k be a field. Let V be a k-variety, with $v \in V(k)$. The *absolute nilpotence index* of the pair (V,v) is the integer

$$\mathrm{nil}_{\mathrm{abs}}(V,v) := \inf_{\gamma \in \mathscr{L}_\infty^\circ(V,v)} (\mathrm{nil}_\gamma(V)).$$

The absolute nilpotence index of (V,v) only depends on the pointed k-variety (V,v) by construction. Besides, there exists an arc $\gamma_{\mathrm{abs}} \in \gamma \in \mathscr{L}_\infty^\circ(V)(k)$ with origin v such that $\mathrm{nil}_{\mathrm{abs}}(V,v) = \mathrm{nil}_{\gamma_{\mathrm{abs}}}(V)$. We do not know, in general, whether such arcs are or not unique, up to isomorphism.

THEOREM 8.3. *Let k be a field. Let V be a k-variety, with $v \in V(k)$. Then, the absolute nilpotence index $\mathrm{nil}_{\mathrm{abs}}(V,v)$ is a formal invariant of the singularity (V,v).*

PROOF. Let (V',v') be a pointed k-variety endowed with an isomorphism of complete local k-algebras: $f : \widehat{\mathcal{O}_{V',v'}} \xrightarrow{\sim} \widehat{\mathcal{O}_{V,v}}$. Let $\gamma_v \in \mathscr{L}_\infty(V)(k)$ be an arc with origin v such that $\mathrm{nil}_{\gamma_v}(V) = \mathrm{nil}_{\mathrm{abs}}(V,v)$ and $\gamma_{v'} \in \mathscr{L}_\infty(\mathscr{C}')(k)$ its image by f, i.e., $\gamma_{v'} = \gamma_v \circ f$. For every test-ring A, the composition by f induces the following diagram of maps, functorial in A:

$$
\begin{array}{ccc}
\mathrm{Hom}_k^{\mathrm{loct}}(\widehat{\mathcal{O}_{V,v}}, A[[T]]) & \xrightarrow[\sim]{\circ f} & \mathrm{Hom}_k^{\mathrm{loct}}(\widehat{\mathcal{O}_{V',v'}}, A[[T]]) \\
\uparrow & & \uparrow \\
\mathscr{L}_\infty(\mathscr{C})_{\gamma_v}(A) & \longrightarrow & \mathscr{L}_\infty(\mathscr{C}')_{\gamma_{v'}}(A)
\end{array}
$$

where the map $\mathscr{L}_\infty(\mathscr{C})_{\gamma_v}(A) \rightarrow \mathscr{L}_\infty(\mathscr{C}')_{\gamma_{v'}}(A)$ is obtained by restriction and also is a bijection. Hence, the formal k-schemes, $\mathscr{L}_\infty(\mathscr{C})_{\gamma_v}$ and $\mathscr{L}_\infty(\mathscr{C})_{\gamma_{v'}}$ are isomorphic. By this way, we deduce that $\mathrm{nil}_{\mathrm{abs}}(V, v) \geq \mathrm{nil}_{\mathrm{abs}}(V', v')$. We conclude, by symmetry, by applying the same arguments to f^{-1}. □

8.1. Further comments and open questions concerning nilpotency

Let us introduce various open problems concerning nilpotency and its interpretation in the special case of curves.

DEFINITION 8.4. Let k be a field. Let $\mathscr{C}$ be an integral k-curve unibranch at $x \in \mathscr{C}(k)$. Let $\gamma \in \mathscr{L}_\infty(\mathscr{C})(k)$ be a primitive k-parameterization of $\mathscr{C}$ at x. We call the *nilpotence degree* of $\mathscr{C}$ at x the integer $\mathrm{nil}_\gamma(\mathscr{C})$. We denote it by $\mathrm{nil}_x(\mathscr{C})$.

8.2. It is clear from Corollary 7.4 that the integer $\mathrm{nil}_x(\mathscr{C})$ does not depend on the choice of the primitive k-parameterization at x and is a formal invariant of the singularity $(\mathscr{C}, x)$. In an obvious way, we have $\mathrm{nil}_x(\mathscr{C}) \geq \mathrm{nil}_{\mathrm{abs}}(\mathscr{C}, x)$.

QUESTION 8.5. Let k be a field. Let $\mathscr{C}$ be an integral k-curve geometrically unibranch at $x \in \mathscr{C}(k)$. Does the formula $\mathrm{nil}_x(\mathscr{C}) = \mathrm{nil}_{\mathrm{abs}}(\mathscr{C}, x)$ hold?

REMARK 8.6. The nilpotence degree is an important tool used in [8].

8.3. Nilpotency conjecture

If $\mathscr{C}$ is a k-curve geometrically unibranch at $x \in \mathscr{C}(k)$, we denote by $\delta(\mathscr{C}, x)$ the *degree of the singularity of $\mathscr{C}$ at x*. Recall if $\bar{\mathscr{C}}$ is the normalization of $\mathscr{C}$ and $\bar{x}$ the unique lifting of x, one has $\bar{\mathscr{C}}_{\bar{x}} \cong \mathrm{Spf}(k[[T]])$. The integer $\delta(\mathscr{C}, x)$ then is defined as the cardinal of the set $\mathbf{N} \setminus \Gamma$ where Γ is formed by the T-adic valuations associated with the functions on $\mathscr{C}_x$. We introduce the following question.

QUESTION 8.7. Let k be a field of characteristic zero. Does $\mathrm{nil}_x(\mathscr{C})$ equal $\delta(\mathscr{C}, x) + 1$? In particular, if the field k is algebraically closed, if $\mathscr{C}$ is an affine plane curve with Milnor number $\mu_x(\mathscr{C})$ at x, does $\mu_x(\mathscr{C})$ equal $2(\mathrm{nil}_x(\mathscr{C}) - 1)$?

The result described in Example 3.6 provides, for example, an affirmative answer to Question 8.7 for unibranch plane curve singularities of multiplicity two. Besides, the computational method that we introduce in Section 3.7 allows us to check that the answer is also positive for the curve singularity $(\mathrm{Spec}(k[X, Y]/\langle X^N - Y^M \rangle), \mathfrak{o})$ for every pair of integers (N, M) with N, M coprime and $M = 3$ and $N \leq 100$ or $M = 4$ and $N \leq 43$ or $M = 5$ and $N \leq 21$. We also checked this for several examples of non-plane monomial curve singularities.

REMARK 8.8. It seems to us interesting to find an analogous interpretation (at least conjectural) of the nilpotency in the case of toric singularities where the situation is perhaps more tractable than in the general case.

Bibliography

[1] B. Bhatt, Algebraization and Tannaka duality. *Camb. J. Math.*, 4(4):403–461, 2016.

[2] D. Bourqui, M. Morán Cañón and J. Sebag, Comparison between formal neighborhoods of stable and rational arcs for toric singularities. Preprint, 2019.

[3] D. Bourqui, J. Nicaise and J. Sebag, Arc schemes in geometry and differential algebra. Chapter 2 in this volume.

[4] D. Bourqui and J. Sebag, Cancellation and regular derivations. To appear in *J. Algebra Appl.* https://doi.org/10.1142/S0219498819501652.

[5] D. Bourqui and J. Sebag, Finite formal models of toric singularities. To appear in *J. Math. Soc. Japan.*

[6] D. Bourqui and J. Sebag, Deformations of differential arcs. *Bull. Aust. Math. Soc.*, 94(3):405–410, 2016.

[7] D. Bourqui and J. Sebag, The Drinfeld–Grinberg–Kazhdan theorem for formal schemes and singularity theory. *Confluentes Math.*, 9(1):29–64, 2017.

[8] D. Bourqui and J. Sebag, The Drinfeld–Grinberg–Kazhdan theorem is false for singular arcs. *J. Inst. Math. Jussieu*, 16(4):879–885, 2017.

[9] D. Bourqui and J. Sebag, The minimal formal models of curve singularities. *Internat. J. Math.*, 28(10):23, 2017.

[10] D. Bourqui and J. Sebag, Smooth arcs on algebraic varieties. *J. Singul.*, 16:130–140, 2017.

[11] A. Bouthier, Théorème de structure sur les espaces d'arcs. Chapter 8 in this volume.

[12] A. Bouthier and D. Kazhdan, Faisceaux pervers sur les espaces d'arcs I: Le cas d'égales caractéristiques. Preprint, 2015, arXiv:1509.02203.

[13] A. Bouthier, B. C. Ngô and Y. Sakellaridis, On the formal arc space of a reductive monoid. *Amer. J. Math.*, 138(1):81–108, 2016.

[14] C. Chiu and H. Hauser, Triviality of the formal neighborhood of arcs. Preprint, 2016.

[15] T. de Fernex and R. Docampo, Differential on the arc space. Preprint, 2017, https://arxiv.org/abs/1703.07505.

[16] T. de Fernex, L. Ein and S. Ishii, Divisorial valuations via arcs. *Publ. Res. Inst. Math. Sci.*, 44(2):425–448, 2008.

[17] V. Drinfeld, The Grinberg–Kazhdan formal arc theorem and the Newton groupoids. Chapter 3 in this volume.

[18] L. Ein and S. Ishii, Singularities with respect to Mather–Jacobian discrepancies. In *Commutative Algebra and Noncommutative Algebraic Geometry*, Vol. II. Math. Sci. Res. Inst. Publ., Vol. 68, pp. 125–168. Cambridge University Press, New York, 2015.

[19] M. Grinberg and D. Kazhdan, Versal deformations of formal arcs. *Geom. Funct. Anal.*, 10(3):543–555, 2000.

[20] A. Grothendieck, Éléments de géométrie algébrique. IV. Étude locale des schémas et des morphismes de schémas. I. *Inst. Hautes Études Sci. Publ. Math.*, (20):259, 1964.

[21] A. Grothendieck, Éléments de géométrie algébrique. IV. Étude locale des schémas et des morphismes de schémas IV. *Inst. Hautes Études Sci. Publ. Math.*, (32):361, 1967.

[22] M. Haiech, Noncomplete completions. Chapter 4 in this volume.

[23] E. Hamann, On power-invariance. *Pacific J. Math.*, 61(1):153–159, 1975.

[24] H. Hauser and S. Woblistin, On the structure of varieties of power series in one variable. Preprint, 2016.
[25] S. Ishii and J. Kollár, The Nash problem on arc families of singularities. *Duke Math. J.*, 120(3):601–620, 2003.
[26] S. Ishii, The arc space of a toric variety. *J. Algebra*, 278(2):666–683, 2004.
[27] S. Ishii, Mather discrepancy and the arc spaces. *Ann. Inst. Fourier (Grenoble)*, 63(1):89–111, 2013.
[28] S. Ishii and A. J. Reguera, Singularities with the highest Mather minimal log discrepancy. *Math. Z.*, 275(3–4):1255–1274, 2013.
[29] S. Lang, *Algebra*, 3rd edn. Graduate Texts in Mathematics, Vol. 211. Springer-Verlag, New York, 2002.
[30] H. Mourtada and A. J. Reguera, Mather discrepancy as an embedding dimension in the space of arcs. *Publ. Res. Inst. Math. Sci.*, 54(1):105–139, 2018.
[31] J. F. Nash, Jr., Arc structure of singularities. *Duke Math. J.*, 81(1):31–38 (1996), 1995. A celebration of John F. Nash, Jr.
[32] B. C. Ngô, Weierstrass preparation theorem and singularities in the space of non-degenerate arcs. Preprint, 2017, https://arxiv.org/abs/1706.05926.
[33] A. J. Reguera, Towards the singular locus of the space of arcs. *Amer. J. Math.*, 131(2):313–350, 2009.
[34] J. F. Ritt, *Differential Algebra*. American Mathematical Society Colloquium Publications, Vol. XXXIII. American Mathematical Society, New York, 1950.
[35] J. Sebag, Intégration motivique sur les schémas formels. *Bull. Soc. Math. France*, 132(1):1–54, 2004.
[36] O. Zariski, Studies in equisingularity. I. Equivalent singularities of plane algebroid curves. *Amer. J. Math.*, 87:507–536, 1965.

6

Arc Schemes of Affine Algebraic Plane Curves and Torsion Kähler Differential Forms

David Bourqui* and Julien Sebag[†]

Institut de recherche mathématique de Rennes
UMR 6625 du CNRS, Université de Rennes 1
Campus de Beaulieu, 35042 Rennes cedex, France
**david.bourqui@univ-rennes1.fr*
[†]*julien.sebag@univ-rennes1.fr*

Let k be a field of characteristic zero. Let $\mathscr{C}$ be an affine plane k-curve, which is assumed to be integral. In this chapter, we collect various results on torsion Kähler differential form $\omega \in \Omega^1_{\mathcal{O}(\mathscr{C})/k}$ in order to study the relation between these objects and the scheme structure of the arc scheme associated with $\mathscr{C}$. Following this study, we provide a simple algorithm based on differential algebra to compute torsion Kähler differential form on plane curves.

1. Introduction

1.1. Let k be a field. For every integral k-variety $\mathscr{V}$, we denote by $\Omega^1_{\mathscr{V}/k}$ the *sheaf of differential forms (of degree 1) on* $\mathscr{V}$ and by $T_{\mathscr{V}/k} = \mathbf{Spec}(\mathrm{Sym}(\Omega^1_{\mathscr{V})/k})$ the tangent space of $\mathscr{V}$. If the k-variety $\mathscr{V}$ is smooth, the sheaf $\Omega^1_{\mathscr{V}/k}$ is locally free; in particular, it is torsion-free and the k-scheme $T_{\mathscr{V}/k}$ is integral. If $\mathscr{V}$ is not smooth, the situation is much more complicated in the neighborhood of the singular points of $\mathscr{V}$, e.g., the sheaf

$\Omega^1_{\mathscr{V}/k}$ may have torsion. The relations between the geometry of $\mathscr{V}$ and the properties of $T_{\mathscr{V}/k}, \Omega^1_{\mathscr{V}/k}$ form a classical area of investigation (see, e.g., [7, pp. 135–136; 8, 9, 3]). This survey chapter collects results mainly from [2, 6, 15–17] and connects the study of torsion Kähler differential forms with that of the scheme structure of arc scheme in the specific case of affine plane curves.

1.2. Let k be a field of characteristic zero. In this chapter, for every integral affine plane k-curve $\mathscr{C}$, we mainly characterize the torsion Kähler differential forms $\omega \in \Omega^1_{\mathcal{O}(\mathscr{C})/k}$ by various equivalent conditions. Precisely, we prove the following theorem.

THEOREM 1.1. *Let k be a field of characteristic zero. Let $\bar{k}$ be an algebraic closure of k. Let $f \in k[X,Y]$ be an irreducible polynomial. Let $\mathscr{C} = \mathrm{Spec}(k[X,Y]/\langle f \rangle)$ be the associated integral k-curve. Let $\omega = \omega_1 dX + \omega_2 dY \in \Omega^1_{k[X,Y]/k}$. Let $\bar{\omega}$ (resp. $P(\omega) = \omega_1 X_1 + \omega_2 Y_1$, resp. $\overline{P(\omega)}$) be the induced element in $\Omega^1_{\mathscr{C}/k}$ (resp. $k[X, X_1, Y, Y_1]$, resp. $\mathrm{Sym}(\Omega^1_{\mathcal{O}(\mathscr{C})/k})$). Then, the following assertions are equivalent:*

(1) *The Kähler differential form ω induces a torsion element $\bar{\omega}$ of the $\mathcal{O}(\mathscr{C})$-module $\Omega^1_{\mathcal{O}(\mathscr{C})/k}$.*

(1′) *The element $\bar{\omega}$ belongs to the kernel of the morphism $\Omega^1_{\mathcal{O}(\mathscr{V})/k} \to \Omega^1_{\mathrm{Frac}(\mathcal{O}(\mathscr{V}))/k}$.*

(2) *The Kähler differential form ω induces a nilpotent element $\overline{P(\omega)}$ of the ring $\mathrm{Sym}(\Omega^1_{\mathcal{O}(\mathscr{C})/k})$.*

(3) *For every non-constant morphism of k-schemes $\varphi \colon \mathrm{Spec}(\bar{k}[[T]]) \to \mathscr{C}$, we have $\varphi^*\omega = 0$ in $\hat{\Omega}^1_{\bar{k}[[T]]/k}$.*

(4) *The dual k-derivation $D_\omega \in \mathrm{Der}_k(k[X,Y])$, associated with ω, maps f into $\langle f \rangle$. In other words, D_ω induces a k-derivation of $\mathcal{O}(\mathscr{C})$.*

(5) *There exists a differential 2-form $\zeta \in \Omega^2_{k[X,Y]/k}$ such that $\omega \wedge df = f\zeta$.*

Besides, the following assertions are equivalent:

(6) *The Kähler differential form $\bar{\omega}$ is a non-zero torsion element of $\Omega^1_{\mathcal{O}(\mathscr{C})/k}$.*

(7) *The element $\overline{P(\omega)}$ is a non-trivial nilpotent element of $\mathrm{Sym}(\Omega^1_{\mathcal{O}(\mathscr{C})/k})$.*

(8) *The k-derivation $\overline{D_\omega}$ induced by D_ω does not belong to the submodule of $\mathrm{Der}_k(\mathcal{O}(\mathscr{C}))$ generated by the image $\bar{J}_f$ of the Jacobian derivation J_f defined to be*

$$J_f := \partial_2(f)\partial_1 - \partial_1(f)\partial_2.$$

In particular, we observe that equivalence (1) $\Leftrightarrow$ (2) of Theorem 1.1 in fact implies that the nilpotent elements of $\mathrm{Sym}(\Omega^1_{\mathcal{O}(\mathscr{C})/k})$, which are homogeneous polynomials of degree 1 in X_1, Y_1 of 1 bijectively correspond to the elements of $\mathrm{Tors}(\Omega^1_{\mathcal{O}(\mathscr{C})/k})$.

1.3. Following this characterization, we are able to deduce various consequences. In Proposition 4.1, we prove that the singular locus of the k-curve $\mathscr{C}$ is contained in the singular locus on any torsion Kähler differential form ω (see Section 4 for definitions). We explain in Sections 5 and 7 how this point of view provides a theoretic and computational approach to the study of the torsion submodule of the $\mathcal{O}(\mathscr{C})$-module $\Omega^1_{\mathcal{O}(\mathscr{C})/k}$ associated with $\mathscr{C}$. Finally, we explain in Section 6 how this statement can be translated in terms of the scheme structure of the jet/arc schemes associated with $\mathscr{C}$. (This last result is related to [11].)

2. Conventions and Notations

2.1. In this chapter, the base field k is assumed to be of characteristic zero. This chapter is devoted to the study of a property of plane curve singularities. For such an integral curve $\mathscr{C}$, there exists an irreducible polynomial $f \in k[X, Y]$ such that $\mathscr{C} = \mathrm{Spec}(k[X, Y]/\langle f \rangle)$.

2.2. We denote by ∂_1 (resp. ∂_2) the k-derivation ∂_X (resp. ∂_Y) in $\mathrm{Der}_k(k[X, Y])$. For every polynomial $f \in k[X, Y]$, we denote by δ_f the *Jacobian derivation* associated with f, i.e., the k-derivation of $k[X, Y]$ defined by the formula $\partial_2(f)\partial_1 - \partial_1(f)\partial_2$. Let $\omega \in \Omega^1_{k[X,Y]/k}$ be a Kähler differential form with $\omega = \omega_1 dX + \omega_2 dY$, $\omega_1, \omega_2 \in k[X, Y]$. Let us associate with it the k-derivation $D_\omega \in \mathrm{Der}_k(k[X, Y])$ defined by $D_\omega := \omega_2\partial_1 - \omega_1\partial_2$. The assignment $\omega \mapsto D_\omega$ defines an invertible $k[X, Y]$-linear map $\theta \colon \Omega^1_{k[X,Y]/k} \to \mathrm{Der}_k(k[X, Y])$. Furthermore, let us recall that there exist natural $k[X, Y]$-linear isomorphisms:

$$\frac{\{D \in \mathrm{Der}_k(k[X, Y]);\ D(f) \in \langle f \rangle\}}{\langle f \rangle \mathrm{Der}_k(k[X, Y])} \cong \mathrm{Der}_k(\mathcal{O}(\mathscr{C})) \cong \frac{(\langle f, \partial_2(f) \rangle : \partial_1(f))}{\langle f \rangle}.$$

(2.1)

Let us denote by $\mathrm{Der}_k(\log(\mathscr{C}))$ the submodule of $\mathrm{Der}_k(k[X, Y])$ defined to be the set

$$\{D \in \mathrm{Der}_k(k[X, Y]),\ D(f) \in \langle f \rangle\}.$$

2.3. We denote by $\mathrm{Sym}(\Omega^1_{\mathcal{O}(\mathscr{C})/k})$ the symmetric $\mathcal{O}(\mathscr{C})$-algebra associated with the $\mathcal{O}(\mathscr{C})$-module $\Omega^1_{\mathcal{O}(\mathscr{C})/k}$. We have an isomorphism:

$$\mathrm{Sym}(\Omega^1_{\mathcal{O}(\mathscr{C})/k}) \cong \frac{k[X_0, X_1, Y_0, Y_1]}{\langle f, \partial_1(f)X_1 + \partial_2(f)Y_1 \rangle}.$$

Via this isomorphism, the $\mathcal{O}(\mathscr{C})$-algebra structure on $\mathrm{Sym}(\Omega^1_{\mathcal{O}(\mathscr{C})/k})$ is provided by the natural morphism

$$\frac{k[X,Y]}{\langle f \rangle} \rightarrow \frac{k[X_0, X_1, Y_0, Y_1]}{\langle f, \partial_1(f)X_1 + \partial_2(f)Y_1 \rangle}.$$

If $\omega = \omega_1 dX + \omega_2 dY \in \Omega^1_{k[X,Y]/k}$, we denote by $P(\omega)$ the polynomial $\omega_1 X_1 + \omega_2 Y_1 \in k[X_0, X_1, Y_0, Y_1]$, and by $\overline{P(\omega)}$ its class in $\mathrm{Sym}(\Omega^1_{\mathcal{O}(\mathscr{C})/k})$.

2.4. We endow the k-algebra $k[(X_i)_{i \in \mathbf{N}}, (Y_i)_{i \in \mathbf{N}}]$ with the k-derivation Δ defined by $\Delta(X_i) = X_{i+1}$, $\Delta(Y_i) = Y_{i+1}$ for every integer $i \in \mathbf{N}$. The resulting differential k-algebra is denoted by $k\{X,Y\}$. The morphism of k-algebras $k[X,Y] \rightarrow k\{X,Y\}$, defined by $X, Y \mapsto X_0, Y_0$, gives rise to a structure of $k[X,Y]$-algebra on $k\{X,Y\}$. We adopt the identifications $X_0 := X, Y_0 := Y$. We also observe that $\Delta(f) = \partial_1(f)X_1 + \partial_2(f)Y_1$. For every subset $S \subset k\{X,Y\}$, we denote by $[S]$ the differential ideal generated by S in the differential ring $k\{X,Y\}$ and by $\{S\}$ the radical of the ideal $[S]$. If S only contains f, the notation $\{f\}$ refers to this radical differential ideal associated with S.

3. Proof of Theorem 1.1

We consider $\omega \in \Omega^1_{k[X,Y]/k}$ written as $\omega = \omega_1 dX + \omega_2 dY$, where $\omega_1, \omega_2 \in k[X,Y]$.

3.1. Equivalence (6) $\Leftrightarrow$ (7) $\Leftrightarrow$ (8)

Direct arguments of linear algebra immediately show that the negations of assertions (6), (7) and (8) of Theorem 1.1 can be translated into the existence of polynomials $a, b, c \in k[X,Y]$ such that

(3.1)
$$\begin{cases} \omega_1 = af + c\partial_1(f), \\ \omega_2 = bf + c\partial_2(f). \end{cases}$$

Then, they are all mutually equivalent.

3.2. Implication (4) $\Rightarrow$ (2)

By the Nullstellensatz, it suffices to show that the polynomial $P(\omega) = \omega_1 X_1 + \omega_2 Y_1$ belongs to every maximal ideal $\mathfrak{m}$ of $k[X_0, X_1, Y_0, Y_1]$ which contains the polynomials $f, \Delta(f)$. If $\mathfrak{m}$ also contains $\partial_1(f), \partial_2(f)$, then $\mathfrak{m} \cap k[X,Y]$ corresponds to a singular point of $\mathscr{C}$. Thus, by the assumption and Lemma 4.3, we conclude that $\omega_1, \omega_2 \in \mathfrak{m}$; hence, $P(\omega) \in \mathfrak{m}$. Otherwise, we have $\partial_1(f) \notin \mathfrak{m}$ or $\partial_2(f) \notin \mathfrak{m}$. In other words, the prime ideal $\mathfrak{m} \cap k[X,Y]$ corresponds to a smooth point of $\mathscr{C}$. Thus, the ideal $\mathfrak{m}$

corresponds to a tuple $(\alpha_0, \alpha_1, \beta_0, \beta_1) \in \bar{k}^4$ such that $f(\alpha_0 + \alpha_1 T, \beta_0 + \beta_1 T) = 0 \pmod{T^2}$. Let $\varphi(T) = (\alpha(T), \beta(T)) \in k[[T]]^2$ be a *non-constant* pair of power series such that $f(\varphi(T)) = 0$ and

$$\begin{cases} \alpha(T) \equiv \alpha_0 + \alpha_1 T \pmod{\langle T^2 \rangle}, \\ \beta(T) \equiv \beta_0 + \beta_1 T \pmod{\langle T^2 \rangle}, \end{cases}$$

Note that such a pair exists by the Hensel lemma. So, we deduce

$$(3.2) \qquad \partial_T(f(\varphi(T))) = \alpha'(T)\partial_1(f)(\varphi(T)) + \beta'(T)\partial_2(f)(\varphi(T)) = 0.$$

Since, by assumption, there exists a polynomial $a \in k[X, Y]$ such that

$$D_\omega(f) = \omega_2 \partial_1(f) - \omega_1 \partial_2(f) = af,$$

we also deduce that

$$(3.3) \qquad \omega_2(\varphi(T))\partial_1(f)(\varphi(T)) - \omega_1(\varphi(T))\partial_2(f)(\varphi(T)) = 0$$

Since $\partial_1(f)(\varphi(T)) \neq 0$ or $\partial_2(f)(\varphi(T)) \neq 0$ and $\bar{k}[[T]]$ is a domain, equations (3.2) and (3.3) imply

$$(3.4) \qquad \begin{vmatrix} \omega_2(\varphi(T)) & \alpha'(T) \\ -\omega_1(\varphi(T)) & \beta'(T) \end{vmatrix} = 0.$$

By specializing T to 0 in equation (3.4), we conclude that

$$\omega_1(\alpha_0, \beta_0)\beta_1 + \omega_2(\alpha_0, \beta_0)\alpha_1 = 0,$$

i.e., $P(\omega) \in \mathfrak{m}$. That concludes the proof of this implication.

3.3. Implication $(2) \Rightarrow (1)$

This part of the proof of Theorem 1.1 is based on the use of a particular formulation of the Rosenfeld lemma (see [5, IV, §9, Lemma 2]). Recall that if R is a ring, $h \in R$, and I an ideal of R, we denote by $(I : h^\infty)$ the ideal of R formed by those elements y such that there exists an integer $n \in \mathbf{N}$ with $yh^n \in I$. Now, the above-mentioned result reads as follows.

PROPOSITION 3.5. *Let k be a field of characteristic zero. Let $f \in k[X_1, \ldots, X_N]$ be an irreducible polynomial. Then, for every integer $i \in \{1, \ldots, N\}$, we have $\{f\} = ([f] : \partial_{X_i}(f)^\infty)$.*

Let us prove $(2) \Rightarrow (1)$. We are in case $N = 2$. Assume that $P(\omega) \in \sqrt{\langle f, \Delta(f) \rangle}$ in the ring $k[X_0, X_1, Y_0, Y_1]$. Then, $P(\omega)$ belongs to the differential ideal $\{f\}$ in the ring $k\{X, Y\}$. It follows from Proposition 3.5 that there exists an integer n such that $\partial_1(f)^n P(\omega) \in [f]$. By Section 3.1, we conclude that $\overline{\partial_1(f)}^n \bar{\omega} = 0$ in $\Omega^1_{\mathcal{O}(\mathscr{C})/k}$. Since f is irreducible and k is of characteristic zero, one has $\overline{\partial_1(f)} \neq 0$ in $\mathcal{O}(\mathscr{C})$ (hence $\overline{\partial_1(f)}^n \neq 0$). This means exactly that the Kähler differential form $\bar{\omega}$ is torsion.

REMARK 3.6. *In* [16], *the second author proves how to extend this formulation of the Rosenfeld lemma for integral plane curves to the case of arbitrary integral k-varieties.*

3.4. Implication $(1) \Rightarrow (3)$

Let us begin by establishing a technical lemma.

LEMMA 3.7. *Let k be a field of characteristic zero, with fixed algebraic closure $\bar{k}$. Let $\mathscr{C}$ be an integral k-curve. Let $\mathscr{U}$ be an affine open subscheme of $\mathscr{C}$. Let $g \in \mathcal{O}_{\mathscr{C}}(\mathscr{U})$. The following assertions are equivalent:*

  (a) *The section g is zero.*
  (b) *There exists a non-constant morphism of k-schemes φ: $\operatorname{Spec}(\bar{k}[[T]]) \to \mathscr{U}$ such that one has $\varphi^{\sharp}(g) = 0$ in the ring $\bar{k}[[T]]$.*

Let us recall that every morphism of k-schemes φ: $\operatorname{Spec}(\bar{k}[[T]]) \to \mathscr{U}$ corresponds to a morphism of k-algebras $\varphi^{\sharp}$: $\mathcal{O}_{\mathscr{C}}(\mathscr{U}) \to \bar{k}[[T]]$.

PROOF. Note that since $\bar{k}$ is algebraically closed, there exist non-constant morphisms of k-schemes φ: $\operatorname{Spec}(\bar{k}[[T]]) \to \mathscr{U}$. Thus, we only have to show $(b) \Rightarrow (a)$. Assume (b) and $g \neq 0$. Let φ : $\operatorname{Spec}(\bar{k}[[T]]) \to \mathscr{U}$ be a non-constant morphism of k-schemes such that $\varphi^{\sharp}(g) = 0$. Since $\bar{k}[[T]]$ is a domain, the kernel of the morphism $\varphi^{\sharp}$ is a prime ideal containing g. Since $g \neq 0$ and by the dimension assumption, it is a maximal ideal of $\mathcal{O}_{\mathscr{C}}(\mathscr{U})$. Since φ is non-constant, this is a contradiction. $\qquad\square$

Let us prove our equivalence. By assumption, there exists $\gamma \in \mathcal{O}(\mathscr{C}) \setminus \{0\}$ such that $\gamma\bar{\omega} = 0$. Let us denote $(\varphi^{\sharp}(X), \varphi^{\sharp}(Y)) \in k[[t]]^2$ by $\varphi(T)$. This is a non-constant pair of power series such that $f(\varphi(T)) = 0$. Then, we have $\gamma(\varphi(T))\varphi^*\omega = 0$. Since $\gamma \neq 0$, by Lemma 3.7, we deduce that $\gamma(\varphi(T)) \neq 0$; hence, we have $\varphi^*\omega = 0$, since the $\bar{k}[[T]]$-module $\hat{\Omega}^1_{\bar{k}[[T]]/k}$ is free of rank 1 generated by dT.

REMARK 3.8. Assertion (3) of Theorem 1.1 is equivalent to saying that, for every k-algebra A which is a domain and for every morphism of k-scheme $\varphi \in \mathscr{C}(A[[T]])$ we have $\varphi^*\omega = 0$ (see [**2**, Theorem 1.4]). It also can be translated in the context of arc schemes (see Section 6). Indeed, in this framework, this assertion says that ω induces a (non-trivial) nilpotent function on the associated arc scheme by [**12**, Lemma 3.10].

3.5. Implication $(3) \Rightarrow (4)$

Let us denote $(\varphi^{\sharp}(X), \varphi^{\sharp}(Y)) \in k[[t]]^2$ by $\varphi(T)$. This is a non-constant pair of power series such that $f(\varphi(T)) = 0$. By assumption, since the $\bar{k}[[T]]$-module $\hat{\Omega}^1_{\bar{k}[[T]]/k}$ is free of rank 1 generated by dT, we have

$$\omega_1(\varphi(T))\partial_T(\varphi^{\sharp}(X)) + \omega_2(\varphi(T))\partial_T(\varphi^{\sharp}(Y)) = 0.$$

Since $f(\varphi(T)) = 0$ and $\partial_T(\varphi^\sharp(X))$ or $\partial_T(\varphi^\sharp(Y))$ is not zero, it follows from equation (3.2) that

$$(3.9) \qquad \begin{vmatrix} \omega_1(\varphi(T)) & \partial_1(f)(\varphi(T)) \\ \omega_2(\varphi(T)) & \partial_2(f)(\varphi(T)) \end{vmatrix} = 0.$$

By equation (3.9), we deduce that $D_\omega(f)(\varphi(T)) = 0$. It follows that $D_\omega(f) = 0$ by Lemma 3.7 applied to $g := D_\omega(f)$ and φ. This concludes the proof of this equivalence.

3.6. Equivalence (1) $\Leftrightarrow$ (5)

Let $\omega_1, \omega_2 \in k[X, Y]$ such that $\omega = \omega_1 dX + \omega_2 dY$. Assertion (1) means that there exists a polynomial $\gamma \in k[X, Y] \setminus \langle f \rangle$, a polynomial $\alpha \in k[X, Y]$ and a Kähler differential form $\eta \in \Omega^1_{k[X,Y]/k}$ such that $\gamma\omega = f\eta + \alpha df$. Since $\bar{\omega}$, $\partial_1(\bar{f})$ and $\partial_2(\bar{f})$ are not zero, the latter condition is equivalent to the following relation:

$$(3.10) \qquad \begin{vmatrix} \bar{\omega}_1 & \partial_1(\bar{f}) \\ \bar{\omega}_2 & \partial_2(\bar{f}) \end{vmatrix} = 0$$

in the ring $\mathcal{O}(\mathscr{C})$. Furthermore, let us note that

$$(3.11) \qquad \omega \wedge df = \begin{vmatrix} \omega_1 & \partial_1(f) \\ \omega_2 & \partial_2(f) \end{vmatrix} dX \wedge dY.$$

Formulas (3.10) and (3.11) conclude the proof of this equivalence.

4. Singular Locus of Torsion Kähler Differential Forms

4.1. Let k be a field of characteristic zero. Let $\mathscr{V}$ be an integral affine k-variety, with generic point η. In this section, we define the notion of *singular locus* of Kähler differential forms on $\mathscr{V}$. For every Kähler differential form $\omega \in \Omega^1_{\mathscr{V}/k}$, we can consider the subset of $\mathscr{V}$ formed by the points $x \in \mathscr{V}$ such that the image ω_x of ω in $\Omega^1_{\mathscr{V}/k}(\mathscr{V}) \otimes_{\mathcal{O}(\mathscr{V})} \kappa(x)$ vanishes. The intersection of this subset with $\mathrm{Sing}(\mathscr{V})$ is called the *singular locus* of ω and we denote it by $\mathrm{Sing}(\omega)$. In the particular case of hypersurfaces of affine spaces, we introduce a concrete description of this locus, which is more suitable for our purposes. Let $f \in k[X_1, \ldots, X_N]$ be an irreducible polynomial. Let $\mathscr{V} = \mathrm{Spec}(k[X_1, \ldots, X_N]/\langle f \rangle)$. Let $\mathrm{Jac}(f)$ be the *Jacobian ideal* of $\mathscr{V}$ defined to be the ideal of $k[X_1, \ldots, X_N]$ generated by f and the $\partial_i(f)$ for every integer $i \in \{1, \ldots, N\}$. Let $\omega \in \Omega^1_{\mathscr{V}/k}(\mathscr{V})$ be a Kähler differential form and $\tilde{\omega} \in \Omega^1_{k[X_1,\ldots,X_N]/k}$ be a lifting of ω. We set $\tilde{\omega} = \sum_{i=1}^N a_i dX_i$. The closed subscheme of $\mathrm{Sing}(\mathscr{V})$ associated with the ideal $\mathrm{Jac}(f) + \langle a_i \rangle_{i \in \{1, \ldots, N\}}$ of

$k[X_1, \ldots, X_N]$ does not depend on the choice of the lifting of ω. We observe that it coincides with $\mathrm{Sing}(\omega)$.

4.2. Let us state the following application of Theorem 1.1 with respect to singular locus of torsion Kähler differential forms..

PROPOSITION 4.1. *Let k be a field of characteristic zero. Let $f \in k[X, Y]$ be an irreducible polynomial. Let $\mathscr{C} = \mathrm{Spec}(k[X, Y]/\langle f \rangle)$ be the associated integral k-curve. Let $\omega \in \Omega^1_{k[X,Y]/k}$ be a Kähler differential form whose image is torsion in $\Omega^1_{\mathcal{O}(\mathscr{C})/k}$. Then, we have $\mathrm{Sing}(\mathscr{C}) = \mathrm{Sing}(\omega)$.*

REMARK 4.2. In [2], the authors have shown how to generalize this result in arbitrary dimensions.

PROOF. By Theorem 1.1, we observe that it is equivalent to consider the corresponding question for the associated dual derivation D_ω. Then, we apply Lemma 4.3. □

LEMMA 4.3. *Let k be a field of characteristic zero. Let $f \in k[X, Y]$ be a non-constant reduced polynomial and $\mathscr{C}$ be the associated affine plane curve. Let $a, b \in k[X, Y]$ and $D \in \mathrm{Der}_k(\log(\mathscr{C}))$. We set $D = a\partial_1 + b\partial_2$. Let $\mathfrak{m}$ be a maximal ideal of the ring $k[X, Y]$ containing f which corresponds to a singular point of $\mathscr{C} = \mathrm{Spec}(k[X, Y]/\langle f \rangle)$. Then, we have $a, b \in \mathfrak{m}$.*

PROOF. We would like to thank the referee for pointing out the following argument, which is an elementary alternative to our original argument. By assumption, there exist polynomials $\alpha, \beta, \gamma \in k[X, Y]$ such that

$$(4.4) \qquad \alpha\partial_1(f) + \beta\partial_2(f) = \gamma f.$$

Since D preserves $\langle f \rangle$, by differentiating equation (4.4), we observe that D preserves also $\langle f, \partial_1 f, \partial_2 f \rangle$ hence also the associated primes of the latter ideal. Thus, D preserves $\mathfrak{m}$. Since the field $K = k[X, Y]/\mathfrak{m}$ is a separable finite extension of k, the element of $\mathrm{Der}_k(K)$ induced by D is trivial. Thus, we deduce that $D(k[X, Y]) \subset \mathfrak{m}$ which concludes the proof. □

5. A Structure Statement on Derivation Module of Plane Curves

5.1. Let k be a field. We still denote by f an irreducible element of $k[X, Y]$ and by $\mathscr{C}$ the associated integral affine plane curve. By isomorphism (2.1), we have the formula

$$(5.1) \qquad \frac{\mathrm{Der}_k(\log(\mathscr{C}))}{\langle f \rangle \mathrm{Der}_k(k[X, Y])} \cong \mathrm{Der}_k(\mathcal{O}(\mathscr{C})).$$

By Theorem 1.1, we observe that the inverse of the $k[X, Y]$-linear map $\theta \colon \omega \mapsto D_\omega$ induces, by restriction, the following commutative diagram of

$k[X, Y]$-linear maps:

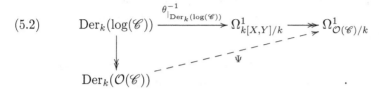

(5.2)

and the image of Ψ is the torsion submodule of the $\mathcal{O}(\mathscr{C})$-module $\Omega^1_{\mathcal{O}(\mathscr{C})/k}$. Then, we easily deduce the following statement from this observation.

COROLLARY 5.3. *Let k be a field of characteristic zero. Let $\mathscr{C}$ be an integral affine plane k-curve defined by the datum of $f \in k[X, Y]$. The kernel of the map Ψ equals the submodule of $\mathrm{Der}_k(\mathcal{O}(\mathscr{C}))$ generated by the image $\bar{J}_f$ of the Jacobian derivation J_f. In particular, the map Ψ induces an isomorphism of $\mathcal{O}(\mathscr{C})$-modules*

$$\mathrm{Der}_k(\mathcal{O}(\mathscr{C}))/\mathcal{O}(\mathscr{C}) \cdot \bar{J}_f \cong \mathrm{Tors}(\Omega^1_{\mathcal{O}(\mathscr{C})/k}) \cong \mathrm{Der}_k(\log(\mathscr{C}))/\langle J_f, f\rangle.$$

5.2. For example, by [**13**], one knows that the rank of the $\mathcal{O}(\mathscr{C})$-module $\mathrm{Der}_k(\log(\mathscr{C}))$ is at most two. Besides, since the Lipman–Zariski conjecture holds for curves ([**10**, Theorem 1]) and $\mathrm{Der}_k(\mathcal{O}(\mathscr{C}))$ is torsion-free, we conclude that it equals 2 if and only if the curve $\mathscr{C}$ is singular. We infer also from Corollary 5.3 that the rank (as $\mathcal{O}(\mathscr{C})$-module) of $\mathrm{Tors}(\Omega^1_{\mathcal{O}(\mathscr{C})/k})$ is at most 2. When the k-curve $\mathscr{C}$ is singular, we also conclude that at least one of the generators of $\mathrm{Der}_k(\mathcal{O}(\mathscr{C}))$ does not belong to the submodule generated by the image of the Jacobian derivation J_f (otherwise, $\mathrm{Der}_k(\mathcal{O}(\mathscr{C}))$ would be generated by $\bar{J}_f$ and would be free). So, the torsion submodule $\mathrm{Tors}(\Omega^1_{\mathcal{O}(\mathscr{C})/k})$ is of rank at least one and at most two. We summarize this observation in the following statement.

COROLLARY 5.4. *Let k be a field of characteristic zero. Let $\mathscr{C}$ be an integral affine plane k-curve, which is assumed to be singular. Then, the torsion submodule $\mathrm{Tors}(\Omega^1_{\mathcal{O}(\mathscr{C})/k})$ of $\Omega^1_{\mathcal{O}(\mathscr{C})/k}$ is of rank 1 or 2.*

5.3. For the convenience of the reader, let us describe the situation, when the k-curve $\mathscr{C}$ is *quasi-homogeneous*, i.e., defined by the datum of an irreducible polynomial $f \in k[X, Y]$ satisfying $f \in \langle \partial_1(f), \partial_2(f)\rangle$. It is well known that the Jacobian derivation J_f and the so-called *Euler derivation* E induce a minimal system of generators of $\mathrm{Der}_k(\mathcal{O}(\mathscr{C}))$. Since f is quasi-homogeneous, there exist $\alpha, \beta \in k$, $q \in k^*$ such that

$$\alpha X \partial_1(f) + \beta Y \partial_2(f) = qf.$$

(This is the so-called *Euler identity*.) The Euler derivation then is defined to be $E = \alpha X \partial_1 + \beta Y \partial_2$ and thus belongs to $\mathrm{Der}_k(\log(\mathscr{C}))$. Let $D \in \mathrm{Der}_k(\log(\mathscr{C}))$. Let $P \in k[X, Y]$ such that $D(f) = Pf$. We consider $\tilde{D} := D - Pq^{-1}E \in \mathrm{Der}_k(\log(\mathscr{C}))$. Thus, we have $\tilde{D}(f) = 0$, which means that the image of $\tilde{D}$ belongs to the submodule of $\mathrm{Der}_k(\mathcal{O}(\mathscr{C}))$ generated by the Jacobian derivation J_f. By Corollary 5.3, we conclude that the torsion submodule $\mathrm{Tors}(\Omega^1_{\mathcal{O}(\mathscr{C})/k})$ is generated by the Kähler differential form $\beta x dy - \alpha y dx$.

6. A Consequence on the Schematic Structure of Arc Schemes Associated with Plane Curves

6.1. Let $\mathscr{V}$ be a k-variety. Let us denote by $R = k[[T]]$ the ring of formal power series and, for every integer $n \in \mathbf{N}$, by $\mathscr{L}_n(\mathscr{V})$ the restriction *à la* Weil of the R_n-scheme $\mathscr{V} \otimes_R R_n$ with respect to the morphism of k-algebras $k \hookrightarrow R_n$, where $R_n = k[[T]]/\langle T^{n+1} \rangle$. This object exists in the category of k-schemes; it is a k-scheme of finite type which we call the *n-jet scheme* of $\mathscr{V}$. In particular, we observe that $\mathscr{L}_1(\mathscr{V}) \cong \mathrm{Spec}(\mathrm{Sym}(\Omega^1_{\mathscr{V}/k})) = T_{\mathscr{V}/k}$. We introduce now the *arc scheme* associated with $\mathscr{V}$ by

$$\mathscr{L}_\infty(\mathscr{V}) := \varprojlim_n (\mathscr{L}_n(\mathscr{V})),$$

which exists in the category of k-schemes. For every integer $n \in \mathbf{N}$, the canonical morphism of k-schemes $\pi_n^\infty : \mathscr{L}_\infty(\mathscr{V}) \to \mathscr{L}_n(\mathscr{V})$ is called the *truncation morphism* of level n. If the k-variety $\mathscr{V}$ is quasi-separated, for every k-algebra A, there exists by [1] a natural bijection

(6.1) $\mathrm{Hom}_k(\mathrm{Spec}(A), \mathscr{L}_\infty(\mathscr{V})) \cong \mathrm{Hom}_R(\mathrm{Spec}(A[[T]]), \mathscr{V}).$

A point $\gamma \in \mathscr{L}_\infty(\mathscr{V})$ is called an *arc* on the variety $\mathscr{V}$. Differential algebra also provides a description of the k-scheme $\mathscr{L}_\infty(\mathscr{V})$ when the k-variety $\mathscr{V}$ is assumed to be affine. Let us assume that $\mathscr{V} = \mathrm{Spec}(k[X_1, \ldots, X_N]/I)$, then we have the formula

$$\mathscr{L}_\infty(\mathscr{V}) \cong \mathrm{Spec}(k\{X, Y\}/[I]).$$

6.2. The following statement (see [6]) deals with the geometric structure of arc/jet schemes associated with plane curves. It interprets Theorem 1.1 as a criterion for smoothness of varieties in terms of the reducedness of associated arc/jet schemes.

THEOREM 6.2. *Let k be a field of characteristic zero. Let $f \in k[X, Y]$ be an irreducible polynomial. If $\mathscr{C} := \mathrm{Spec}(k[X, Y]/\langle f \rangle)$, then the following assertions are equivalent:*

(1) *The k-curve $\mathscr{C}$ is smooth.*
(2) *The associated arc scheme $\mathscr{L}_\infty(\mathscr{C})$ is reduced.*
(3) *For every $n \in \mathbf{N}$, the associated jet scheme $\mathscr{L}_n(\mathscr{C})$ is reduced.*
(4) *For every $n \in \mathbf{N}$, the associated jet scheme $\mathscr{L}_n(\mathscr{C})$ is irreducible.*
(5) *There exists an integer $n \in \mathbf{N}^*$ such that the associated jet scheme $\mathscr{L}_n(\mathscr{C})$ is reduced.*

PROOF. We only justify the non-obvious assertions (1) ⇒ (2). If $\mathscr{C}$ is smooth, then [**15**, Proposition 3.2] proves that $\mathscr{L}_\infty(\mathscr{C})$ is reduced. Assertion (2) ⇒ (1) is implied by the assertion (3) of Theorem 1.1. Assertion (1) ⇒ (3) can be deduced from [**14**, Lemme 3.4.2]. Assertion (3) ⇒ (2) directly follows from the definition of the functor $\mathscr{L}_\infty(\cdot)$. Assertion (4) ⇔ (2) follows from [**11**, Corollary 4.2; **14**, Lemme 3.4.2]. □

REMARK 6.3. In [**16**], the second author has in particular shown how to extend Theorem 6.2 to all the classes of curves which satisfy the Berger conjecture in characteristic zero (see [**7**, p. 135]).

7. A SAGE Code to Compute Nilpotent Kähler Differential Forms of Plane Curves

Let k be a field of characteristic zero. Let $f \in k[X, Y]$ be an irreducible polynomial, and $\mathscr{C} := \mathrm{Spec}(k[X, Y]/\langle f \rangle)$. In this last section, we are interested in the problem of computing torsion Kähler differential forms on $\mathscr{C}$. This problem is well known; we only show here how our point of view provides a way to solve this problem.

7.1. Theorem 1.1 emphasizes the fact that the (non-trivial) elements of weight 1 in $\mathrm{Sym}(\Omega^1_{\mathcal{O}(\mathscr{C})/k})$ correspond to the (non-trivial) elements in $\mathrm{Tors}(\Omega^1_{\mathscr{C}/k})(\mathscr{C})$. Following this basic idea, we propose the following SAGE code, which provides a Gröbner basis of the ideal

$$N_{11} := (\langle f, \Delta(f) \rangle : \partial_1(f)^\infty)$$

of the ring $k[X_0, Y_0, X_1, Y_1]$, and whose entry is the datum of f. Our algorithm is based on the Rosenfeld lemma (see Proposition 3.5), we know that every torsion differential form $\bar{\omega}$ of $\mathscr{C}$ defines an element $P(\omega)$ of weight 1 in this ideal N_{11}. In particular, a system of generators of $\mathrm{Tors}(\Omega^1_{\mathcal{O}(\mathscr{C})/k})$ is given by the image modulo $\langle f, \Delta(f) \rangle$ of the elements of weight 1 in this basis. To realize the computation, we use the Buchberger algorithm to compute a Gröbner basis of our ideal N_{11}, thanks to the classical formula:

$$N_{11} = \langle f, \Delta(f), (1 - \partial_1(f)Z) \rangle k[X_0, X_1, Y_0, Y_1, Z] \cap k[X_0, X_1, Y_0, Y_1].$$

See, e.g., [**4**, Chapter 4, §4].

7.2. Here, one can find a SAGE code of our algorithm.

```
field = QQ

monomial_order_R0 = 'lex'
variables = ['x0','x1','y0','y1']
R0 = PolynomialRing(field,variables,order=monomial_order_R0)
R0.inject_variables()
```

```
def der (f):
#
# adjunction derivation
#
    return (f.derivative(x0)*x1+f.derivative(y0)*y1)

def Display(list):
    for i in [0..len(list)-1]:
        print(list[i])

def Ideal_generators(I):
    Display(I.gens())

#Declare the expression of F

F=x0^3-y0^2

DF=der(F)
SF = F.derivative(x0)

monomial_order = 'lex'
R = PolynomialRing(field,['T']+variables,order=monomial_order)
R.inject_variables()
I_aux=R.ideal([F,DF,1-T*SF])
N_11=I_aux.elimination_ideal([T])
Ideal_generators(N_11)
```

Bibliography

[1] B. Bhatt, Algebraization and Tannaka duality. *Camb. J. Math.*, 4(4):403–461, 2016.

[2] D. Bourqui and J. Sebag, On torsion Kähler differential forms. *J. Pure Appl. Algebra*, 222(8):2229–2243, 2018.

[3] R.-O. Buchweitz and G.-M. Greuel, The Milnor number and deformations of complex curve singularities. *Invent. Math.*, 58(3):241–281, 1980.

[4] D. Cox, J. Little and D. O'Shea, *Ideals, Varieties, and Algorithms*. Undergraduate Texts in Mathematics. Springer-Verlag, New York, 1992. An introduction to computational algebraic geometry and commutative algebra.

[5] E. R. Kolchin, *Differential Algebra and Algebraic Groups*. Pure and Applied Mathematics, Vol. 54. Academic Press, New York, 1973.

[6] K. Kpognon and J. Sebag, Nilpotency in arc scheme of plane curves. *Comm. Algebra*, 45(5):2195–2221, 2017.

[7] E. Kunz, *Kähler Differentials*. Advanced Lectures in Mathematics. Friedr. Vieweg & Sohn, Braunschweig, 1986.

[8] E. Kunz, On the tangent bundle of a scheme. *Univ. Iagel. Acta Math.*, (37):9–24, 1999. Effective methods in algebraic and analytic geometry (Bielsko-Biała, 1997).

[9] E. Kunz and R. Waldi, Über den Derivationenmodul und das Jacobi-Ideal von Kurvensingularitäten. *Math. Z.*, 187(1):105–123, 1984.

[10] J. Lipman, Free derivation modules on algebraic varieties. *Amer. J. Math.*, 87:874–898, 1965.

[11] M. Mustaţă, Jet schemes of locally complete intersection canonical singularities. *Invent. Math.*, 145(3):397–424, 2001. With an appendix by David Eisenbud and Edward Frenkel.

[12] J. Nicaise, and J. Sebag, Greenberg approximation and the geometry of arc spaces. *Comm. Algebra*, 38(11):4077–4096, 2010.

[13] K. Saito, Theory of logarithmic differential forms and logarithmic vector fields. *J. Fac. Sci. Univ. Tokyo Sect. IA Math.*, 27(2):265–291, 1980.

[14] J. Sebag, Intégration motivique sur les schémas formels. *Bull. Soc. Math. France*, 132(1):1–54, 2004.

[15] J. Sebag, Arcs schemes, derivations and Lipman's theorem. *J. Algebra*, 347:173–183, 2011.

[16] J. Sebag, A remark on Berger's conjecture, Kolchin's theorem, and arc schemes. *Arch. Math. (Basel)*, 108(2):145–150, 2017.

[17] J. Sebag, On logarithmic differential operators and equations in the plane. *Illinois J. Math.*, 62(1–4):215–224, 2018.

7

Models of Affine Curves and $\mathbb{G}_a$-actions[*]

Kevin Langlois

Mathematisches Institut, Heinrich Heine Universität,
40225 Düsseldorf, Germany
langlois.kevin18@gmail.com

Using the approach of Barkatou and El Kaoui, we classify certain affine curves over discrete valuation rings having a free additive group action. Our classification generalizes the results of Miyanishi in equicharacteristic zero.

1. Introduction

Let $\mathcal{O}$ be a discrete valuation ring. Choose a uniformizer $t \in \mathcal{O}$, such that $k = \mathcal{O}/(t)$ is the residue field and write K for the fraction field of $\mathcal{O}$. A faithful flat integral affine scheme of finite type over $\mathcal{O}$, is an *affine $\mathcal{O}$-curve* if it has relative dimension 1 and if K is algebraically closed in its function field. Our aim is to classify the models of the affine line (i.e., affine $\mathcal{O}$-curves whose generic fiber is isomorphic to $\mathbb{A}_K^1$). In particular, the arithmetic surface in question inherits a non-trivial action of the additive group scheme $\mathbb{G}_{a,\mathcal{O}}$.

Miyanishi described any affine $\mathcal{O}$-curve with a (free) $\mathbb{G}_{a,\mathcal{O}}$-action, such that the special fiber is integral and under the condition that $\mathcal{O}$ is equicharacteristic zero [**5**, Theorem 4.3]. Barkatou and El Kaoui extended this result in [**1**] for reduced special fibers over an equicharacteristic zero principal ideal domain. Using the approach of [**1**], we obtain the following generalization which is valid in any characteristic.

---

[*]Dedicated to Mikhail Zaidenberg on the occasion of his 70th birthday.

THEOREM 1.1. *Assume that k is perfect. Let C be an affine $\mathcal{O}$-curve with a free $\mathbb{G}_{a,\mathcal{O}}$-action and reduced special fiber. Then, there exist a natural number $n \geq 1$ and polynomials $f_i \in \mathcal{O}[x_1, \ldots, x_i]$ for $1 \leq i \leq n$, such that*

$$C \simeq \operatorname{Spec} \mathcal{O}[x_1, \ldots, x_{n+1}]/(tx_2 - f_1, \ldots, tx_{n+1} - f_n).$$

Moreover, the following properties are fulfilled:

(i) *The ideal $(t, f_1, \ldots, f_n)$ is zero-dimensional and radical, and the reduction modulo t of the $\partial f_i / \partial x_i$ is invertible.*

(ii) *Consider the subalgebra $B = \mathcal{O}[\alpha_1(x), \ldots, \alpha_n(x)] \subseteq K[x]$, where the α_i's are defined as $\alpha_1(x) = x$ and $\alpha_i(x) = t^{-1} f_{i-1}(\alpha_1(x), \ldots, \alpha_{i-1}(x))$ for $2 \leq i \leq n$. Then, under the previous isomorphism, the $\mathbb{G}_{a,\mathcal{O}}$-actions on C are in one-to-one correspondence with the $\mathbb{G}_{a,K}$-actions $x \mapsto x + \sum_{j=1}^{r} c_j \lambda^{p^{s_j}}$ on $\mathbb{A}_K^1 = \operatorname{Spec} K[x]$ that let stable the algebra B, where $s_j \in \mathbb{Z}_{\geq 0}, c_j \in K$, and p is the characteristic exponent of the field K.*

Section 2 sets the notation of the paper, while the proof of Theorem 1.1 is given in Section 3.

2. Basics

A $\mathbb{G}_{a,\mathcal{O}}$-action on the affine $\mathcal{O}$-curve $C = \operatorname{Spec} B$ is equivalent to a sequence

$$\delta^{(i)} : B \to B, \quad i = 0, 1, 2, \ldots$$

of $\mathcal{O}$-linear maps sharing the conditions (see [4]):

(a) the map $\delta^{(0)}$ is the identity;

(b) for any $b \in B$, there is $i \in \mathbb{Z}_{>0}$, such that $\delta^{(j)}(b) = 0$ for any $j \geq i$;

(c) we have the *Leibniz rule*

$$\delta^{(i)}(b_1 \cdot b_2) = \sum_{i_1 + i_2 = i} \delta^{(i_1)}(b_1) \cdot \delta^{(i_2)}(b_2),$$

where $i \in \mathbb{Z}_{\geq 0}$ and $b_1, b_2 \in B$;

(d) for all indices $i, j \in \mathbb{Z}_{\geq 0}$,

$$\delta^{(i)} \circ \delta^{(j)} = \binom{i+j}{i} \delta^{(i+j)}.$$

The sequence $\delta = (\delta^{(i)})$ is called a *locally finite iterative higher derivation* (LFIHD). The kernel $\ker(\delta)$ is the intersection of the linear subspaces $\ker(\delta^{(i)})$ where i runs over $\mathbb{Z}_{>0}$. Since K is algebraically closed in the fraction field of B, we have $\ker(\delta) = \mathcal{O}$ if the action is non-trivial. The $\mathbb{G}_{a,\mathcal{O}}$-action on C is *free* if the ideal generated by $\{\delta^{(i)}(b) \, ; i \in \mathbb{Z}_{>0} \text{ and } b \in B\}$ is B. The *exponential morphism* is

$$\exp(\delta T) : B \to B[T], \quad b \mapsto \sum_{i \in \mathbb{Z}_{\geq 0}} \delta^{(i)}(b) T^i.$$

LEMMA 2.1. *Assume that the* $\mathbb{G}_{a,\mathcal{O}}$*-action on* C *is free. Then, there exist an LFIHD* δ *on* B *corresponding to a free action and* $x \in B$*, such that* $B \otimes_{\mathcal{O}} K = K[x]$ *and*

$$\exp(\delta T)(x) = x + \sum_{i=1}^{m} t^{n_i} T^{e_i}$$

for some natural numbers n_i *and some powers*

$$1 \leq e_1 = p^{r_1} < \cdots < e_m = p^{r_m} = e,$$

where p *is the characteristic exponent of* $\mathcal{O}$*. Moreover, for any* $\kappa \gg 0$*, we may choose* δ*, such that* $\delta^{(e)}$ *is an* $\mathcal{O}$*-derivation on the monomials of* $K[x]$ *of degree less than* κ*. Assume further that the special fiber of* C *is reduced. Then* $B = \mathcal{O}[x]$ *provided that* $\min_{1 \leq i \leq m} n_i = 0$*.*

PROOF. We may assume that $p > 1$. Let δ_1 be the LFIHD defined by the $\mathbb{G}_{a,\mathcal{O}}$-action. Choose $x \in B$, such that $\exp(T\delta_1)(x)$ has positive minimal degree. Then $B \otimes_{\mathcal{O}} K = K[x]$ [**2**, Lemma 2.2(c)]. As the extension δ_K of δ_1 on $K[x]$ corresponds to a $\mathbb{G}_{a,K}$-action on $\mathbb{A}^1_K$, we have $\delta_1^{(j)}(x) \in \ker(\delta_K) \cap B = \mathcal{O}$ for any $j \in \mathbb{Z}_{>0}$. So, if $\delta_1^{(j)}(x) \neq 0$, then $\delta_1^{(j)}(x) = c_j t^{m_j}$ for some $m_j \in \mathbb{Z}_{\geq 0}$ and $c_j \in \mathcal{O}^*$. Consequently, we modify δ_1 by changing the c_j's by 1. Now, write

$$\exp(\delta_1 T)(x) = x + \sum_{i=1}^{m-1} t^{n_i} T^{e_i}, \quad \text{where } m \geq 1.$$

We introduce a new LFIHD δ on $K[x]$ (trivial on K) defined by

$$\exp(\delta T)(x) = \exp(\delta_1 T)(x) + t^n T^e,$$

where e and n satisfy the following conditions. Let $b_1, \ldots, b_s \in B$ such that $B = \mathcal{O}[b_1, \ldots, b_s]$ and consider a relation

$$1 = \sum_{j=1}^{\alpha} c_j \delta_1^{(\beta_j)}(d_j) \quad \text{for } c_j, d_j \in B \text{ and } \beta_j \in \mathbb{Z}_{>0},$$

which is guaranteed from the freeness assumption. Let κ be a constant greater than the degrees in x of the $b_j \in K[x]$ and take e a power of p, verifying

$$e > (\kappa + 2) \max \{e_j, \beta_\ell \,|\, 1 \leq j < m \text{ and } 1 \leq \ell \leq \alpha\}.$$

Finally, let $n \in \mathbb{Z}_{>0}$, such that $t^n b_j \in \mathcal{O}[x]$ for $1 \leq j \leq s$. We claim that δ induces an LFIHD on B with the required properties. Indeed, if $i < e$, then $\delta^{(i)}(b) = \delta_1^{(i)}(b)$ for any $b \in B$. Now, assume that $i \geq e$ and let

$$b_j = \sum_u \lambda_u x^u, \quad \text{where } \lambda_u \in K \text{ and } d = \deg_x(b_j).$$

Let e_δ reach the maximum of e_j's. By a direct induction on $u \le d$, t^n divides $\delta^{(\beta)}(x^u)$ if $\beta > u e_\delta$ (note that $\delta^{(\beta)}(x^u) = u x^{u-1} t^n$ if $\beta = e$). Thus, $\delta^{(i)}(b_j) \in \mathcal{O}[x]$ for any j and δ induces a free $\mathbb{G}_{a,\mathcal{O}}$-action on C. This yields the first claim.

Let us show the second one. The assumption $\min_{1 \le i \le m} n_i = 0$ implies that $\delta^{(\gamma)}(x) = 1$ for some γ and that the residue class $\bar{x}$ of x modulo t is not algebraic over k. Indeed, if $\bar{x}$ would admit an algebraic dependence relation, then, applying the exponential map (from the $\mathbb{G}_{a,k}$-action on $\operatorname{Spec} B/tB$) to this relation, we would get a contradiction. Let $b \in B \setminus tB$. Since $B \subseteq K[x]$, there is a primitive polynomial $s(T) \in \mathcal{O}[T]$, such that $s(x) = t^r b$ for some $r \in \mathbb{Z}_{\ge 0}$. Observe that $s(x) \equiv 0 \bmod tB$ if $r > 0$. So, the previous step implies that $r = 0$ and $b \in \mathcal{O}[x]$. Now, let $c \in tB$. Write $c = t^\ell a$ for some $\ell \in \mathbb{Z}_{>0}$ and $a \in B \setminus tB$. As $a \in \mathcal{O}[x]$, we have $c \in \mathcal{O}[x]$. Thus, $B = \mathcal{O}[x]$, as required. □

3. Proof of the Main Result

Let $C = \operatorname{Spec} B$ be an affine $\mathcal{O}$-curve with a free $\mathbb{G}_{a,\mathcal{O}}$-action. Assume that the special fiber is reduced and that k is perfect. Let δ be an LFIHD on B as in the proof of Lemma 2.1. Set $e := e_m$ and $n := n_m$. Consider $B_1 := \mathcal{O}[x], I_1 := (tB) \cap B_1$ and inductively define the other rings and ideals by

$$B_{i+1} = B_i[t^{-1} I_i] \text{ and } I_{i+1} := (tB) \cap B_i \quad \text{for } i = 1, \ldots, n.$$

The inclusions $B_i \subseteq B_{i+1}$ yield a sequence of $\mathbb{G}_{a,\mathcal{O}}$-equivariant *affine modifications* (cf. [3])

$$\operatorname{Spec} B_{n+1} \to \operatorname{Spec} B_n \to \cdots \to \operatorname{Spec} B_2 \to \mathbb{A}^1_{\mathcal{O}} = \operatorname{Spec} B_1,$$

where $\operatorname{Spec} B_{i+1}$ is the complement of the hypersurface $\mathbb{V}(t)$ in the blow-up of $\operatorname{Spec} B_i$ with center I_i. The following lemma is analogous to [1, Lemma 4.4]. We use the adapted result [1, Lemma 4.3] for the ring $\mathcal{O}$ where k needs to be perfect. Note that in the argument of the proof, we will vary δ and e, while the number n will be constant in the entire paper.

LEMMA 3.1. *There exist* $x_1, \ldots, x_{n+1} \in B$ *and* $f_i \in \mathcal{O}[T_1, \ldots, T_i]$ *for* $1 \le i \le n$, *such that the following conditions hold:*

(i) $x_1 = x$, t *divides* $f_i(x_1, \ldots, x_i)$, *and if*

$$x_{i+1} = t^{-1} f_i(x_1, \ldots, x_i),$$

then t^{n-i+1} *divides* $\delta^{(e)}(x_i)$ *for an appropriate choice of* δ.

(ii) $B_i = \mathcal{O}[x_1, \ldots, x_i]$, $I_i = (t, f_1(x_1), \ldots, f_i(x_1, \ldots, x_i)) \subseteq B_i$, *and the reduction modulo* t *of* $\partial f_i / \partial T_i$ *is invertible.*

(iii) *We have* $C = \operatorname{Spec} B_{n+1}$.

PROOF. We show the existence of f_i's and (ii) by induction on i. We treat the case $i = 1$. Let δ_k be the LFIHD corresponding to the $\mathbb{G}_{a,k}$-action on Spec B/tB and set $R := \ker(\delta_k)$. By our assumption and [2, Lemmata 2.1 and 2.2], the scheme Spec R is zero-dimensional and reduced. Therefore, $B_1/I_1 \subseteq R$. By [1, Lemma 4.3], there is $f_1 \in \mathcal{O}[T_1]$, such that $I_1 = (t, f_1(x)) \subseteq B_1$ and the reduction modulo t of $\partial f_1/\partial T_1$ is a unit. If $x_2 := t^{-1}f_1(x)$, then $B_2 = \mathcal{O}[x_1, x_2]$. Assume that statement (ii) holds for $i < n$. It follows that $B_{i+1} = \mathcal{O}[x_1, \ldots, x_{i+1}]$. Let J be the preimage of I_{i+1} by the morphism $E := \mathcal{O}[T_1, \ldots, T_{i+1}] \to B_{i+1}$, $T_j \mapsto x_j$. Since Spec E/J is reduced and zero-dimensional, by [1, Lemma 4.3], there exist polynomials $g_j \in \mathcal{O}[T_1, \ldots, T_j]$, such that

$$I_{i+1} = (t, g_1(x_1), \ldots, g_{i+1}(x_1, \ldots, x_{i+1})) \subseteq B_{i+1},$$

and the reduction modulo t of $\partial g_j/\partial T_j$'s is invertible. From property (ii) in [1], we may choose $g_1, \ldots, g_i$, such that $I_i = (t, g_1(x_1), \ldots, g_i(x_1, \ldots, x_i))$. Therefore, we take $g_j = f_j$ for any $1 \le j \le i$ (from our induction hypothesis) and let $f_{i+1} = g_{i+1}$, as required.

Choose an LFIHD δ with $e \gg 0$ as in Lemma 2.1, such that $\delta^{(e)}$ acts as an $\mathcal{O}$-derivation on $x_1, \ldots, x_{n+1}$ (seen as polynomials in x). We show (i) by induction on i. For $i = 1$, we have $\delta^{(e)}(x_1) = \delta^{(e)}(x) = t^n$, and for $i = 2$, we get $\delta^{(e)}(f_1(x)) = t^n \partial f_1(x)/\partial x$. As $x_2 := t^{-1}f_1(x)$, it follows that t^{n-1} divides $\delta^{(e)}(x_2)$. Assume that statement (i) holds for $i < n$. By induction hypothesis, t^{n-j+1} divides $\delta^{(e)}(x_j)$ for any $j \le i$. This implies

$$\delta^{(e)}(x_{i+1}) = t^{-1}\left(\sum_{j=1}^{i} \partial f_j/\partial x_j \cdot \delta^{(e)}(x_j)\right),$$

and so, t^{n-i} divides $\delta^{(e)}(x_{i+1})$, proving (i).

(iii) If $n = 0$, then $B = \mathcal{O}[x] = B_1$ (see Lemma 2.1). Thus, we assume $n > 0$. Let $b \in B$, such that $tb \in B_{n+1}$ and write $tb = \sum_{j=1}^{r} a_j x_{n+1}^j$ for some $a_j \in \mathcal{O}[x_1, \ldots, x_n]$. By the reasoning we did before, we may choose δ as in Lemma 2.1, such that $\delta^{(e)}$ acts as an $\mathcal{O}$-derivation on the polynomials $a_j x_{n+1}^j$. We show by induction on r that $a_j \in I_n$ for any j. The case $r = 0$ being obvious, assume that the statement holds true for $r - 1$. Then

$$t\delta^{(e)}(b) = \sum_{j=1}^{r} \delta^{(e)}(a_j)x_{n+1}^j + \delta^{(e)}(x_{n+1}) \cdot \left(\sum_{j=1}^{r} ja_j x_{n+1}^{j-1}\right).$$

By statement (i) and the fact that $n > 0$, t divides $\delta^{(e)}(a_j)$'s. From a direct computation,

$$\delta^{(e)}(x_{n+1}) = t\alpha + \prod_{i=1}^{n} \partial f_i/\partial x_i \quad \text{for some } \alpha \in B.$$

Therefore, t divides $\sum_{j=1}^{r} j a_j x_{n+1}^{j-1}$. Let p be the characteristic exponent of $\mathcal{O}$ and assume that $p > 1$. By induction assumption, $a_j \in I_n$ for any non-zero $j \notin p\mathbb{Z}$. Now, there is $b' \in B$, such that

$$tb' = \sum_{1 \le j \le r, j \notin p\mathbb{Z}} a_j x_{n+1}^j.$$

Letting $b_1 = b - b'$, we may write

$$tb_1 = \sum_{u=1}^{s} a_{up^\ell} (x_{n+1}^{p^\ell})^u \quad \text{for some } s, \ell \ge 1.$$

Here, ℓ is taken so that $a_{up^\ell} \ne 0$ for some $1 \le u \le s$ with $u \notin p\mathbb{Z}$. Now, applying $\delta^{(ep^\ell)}$, we obtain

$$t\delta^{(ep^\ell)}(b_1) = \sum_{u=1}^{s} \delta^{(ep^\ell)}(a_{up^\ell})(x_{n+1}^{p^\ell})^u$$
$$+ (\delta^{(e)}(x_{n+1}))^{p^\ell} \cdot \left(\sum_{u=1}^{s} u a_{up^\ell} (x_{n+1}^{p^\ell})^{u-1} \right).$$

By the Leibniz rule and the fact that $n > 0$, t divides $\delta^{(ep^\ell)})(a_{up^\ell})$'s. Thus,

$$t \text{ divides } \sum_{u=1}^{s} u a_{up^\ell}(x_{n+1}^{p^\ell})^{u-1}.$$

According to our induction hypothesis, $a_{up^\ell} \in I_n$ for any $u \notin p\mathbb{Z}$. Continuing this process, we arrive at $a_j \in I_n$ for any j. Let us show that $B = B_{n+1}$. By the previous step, we have

$$a_j = a_{0,j} t + \sum_{i=1}^{n} a_{i,j} f_i \quad \text{for } a_{i,j} \in B_n \text{ and } j = 1, \ldots, r.$$

Setting $c_i = \sum_{j=1}^{r} a_{i,j} x_{n+1}^j$, we have

$$b = t^{-1} \left(\sum_{j=1}^{r} a_j x_{n+1}^j \right) = \sum_{i=1}^{n} c_i x_i \in B_{n+1}, \quad \text{where } x_0 = 1.$$

Hence, $b \in B_{n+1}$ provided that $t^\varepsilon b \in B_{n+1}$ for some $\varepsilon \in \mathbb{Z}_{\ge 0}$. From the equality $B_{(t)} = K[x] = (B_{n+1})_{(t)}$, we conclude that $B = B_{n+1}$. This completes the proof of the lemma. $\square$

PROOF OF THEOREM 1.1. We follow the proof of [1, Theorem 3.1]. Consider the surjective morphism $\psi : E := \mathcal{O}[T_1, \ldots, T_{n+1}] \to B_{n+1} = B$, $T_i \mapsto x_i$, the ideal $I = (tT_2 - f_1, \ldots, tT_{n+1} - f_n)$ and let $J = \bigcup_i (I : t^i E)$.

We show that $J = \ker \psi$. Note that $J \subseteq \ker \psi$ is clear. Let $b \in \ker \psi$. By performing several Euclidean divisions, we get an equality

$$t^\ell b = \sum_{j=1}^{r} \gamma_j (tT_{j+1} - f_j) + \beta, \text{ where } \beta \in \mathcal{O}[T_1], \ell \in \mathbb{Z}_{\geq 0}, \text{ and } \gamma_j \in E.$$

Since $\psi(t^\ell b) = 0$ and $x_1 = x$ is transcendental over K, we have $\beta = 0$. Thus, $\ker \psi = J$. It remains to be proved that $I = J$. From [1, Lemma 4.5] (which is valid in our context), we only need to have $J \subseteq tE + I = (t, f_1, \ldots, f_n)$. Let $b \in J$. Then $b = \sum_{j=0} a_j T_{n+1}^j$ for some $a_j \in \mathcal{O}[T_1, \ldots, T_n]$. Since $\psi(b) = 0$, the argument of the proof of Lemma 3.1(iii) implies that $\psi(a_j) \in I_n$ for any j. Thus, $b \in tE + I$, establishing the theorem. $\square$

Acknowledgments

This research was conducted in the framework of the research training group *GRK 2240: Algebro-geometric Methods in Algebra, Arithmetic and Topology*, which is funded by the DFG. This research is supported by ERCEA Consolidator Grant 615655 — NMST and also by the Basque Government through the BERC 2014–2017 program and by the Spanish Ministry of Economy and Competitiveness MINECO: BCAM Severo Ochoa excellence accreditation SEV-2013-0323.

Bibliography

[1] M. A. Barkatou and M. El Kahoui, Locally nilpotent derivations with a PID ring of constants. *Proc. Amer. Math. Soc.*, 140(1):119–128, 2012.

[2] A. J. Crachiola and L. G. Makar-Limanov, On the rigidity of small domains. *J. Algebra*, 284(1):1–12, 2005.

[3] S. Kaliman and M. Zaidenberg, Affine modifications and affine hypersurfaces with a very transitive automorphism group. *Transform. Groups*, 4(1):53–95, 1999.

[4] M. Miyanishi, A remark on an iterative infinite higher derivation. *J. Math. Kyoto Univ.*, 8:411–415, 1968.

[5] M. Miyanishi, Additive group scheme actions on integral schemes defined over discrete valuation rings. *J. Algebra*, 322(9):3331–3344, 2009.

8

Théorèmes de Structure sur les Espaces d'Arcs

Alexis Bouthier

Institut de Mathématiques de Jussieu, 4 place Jussieu,
75005 Paris, France
alexis.bouthier@imj-prg.fr

The goal of this chapter is to survey different results on the structure of arc spaces. The main input is the use of a new family of morphisms, the Weierstrass morphisms and a new family of schemes, the type (S) schemes, which are the main pieces to understand the geometry of the singularities of arc spaces.

1. Introduction

Le but de cet article est de présenter des résultats récents qui s'attachent à étudier les singularités de l'espace d'arcs en tant que schéma non-noethérien. Le premier résultat dans cette direction a été établi dans les années 2000 par Grinberg–Kazhdan et Drinfeld. Il donne un résultat de structure sur les voisinages formels des espaces d'arcs et montre qu'ils se décomposent comme des produits d'un k-schéma de type fini avec un espace affine de dimension infinie.

Pour plusieurs applications en théorie des représentations et au programme de Langlands, il est important de pouvoir développer une théorie des faisceaux sur de tels espaces. Une première étape est donc d'obtenir un énoncé qui globalise Drinfeld–Grinberg–Kazhdan et de voir dans quel mesure l'espace d'arcs peut se décomposer «globalement» comme un produit.

2. Préliminaires

2.1. Division de Weierstrass et approximation d'Artin

DÉFINITION 2.1. Soit A un anneau local commutatif, $\mathfrak{m}$ son idéal maximal, k son corps résiduel. On dit qu'il vérifie la division de Weierstrass, si pour tout $f \in A[[t]]$ tel qu'en réduction $\bar{f} = t^n u(t)$ avec $u(t) \in k[[t]]^{\times}$, le module $A[[t]]/(f)$ est un A-module libre de type fini de base $(1, \ldots, t^{n-1})$.

PROPOSITION 2.2. *Si A satisfait la division de Weierstrass, soit $f \in A[[t]]$ comme ci-dessus et $g \in A[[t]]$, alors il existe un unique couple $(b, r) \in A[[t]] \times A[t]$ avec $\deg(r) < n$, tel que:*

$$g = bf + r.$$

De plus, il existe un unique polynôme unitaire $q \in A[t]$ de degré n tel que:

$$f = qv,$$

avec $v \in A[[t]]^{\times}$.

Le théorème fondamental est le suivant [**7**, §3, no. 9, Proposition 6]:

THÉORÈME 2.3. *Si A est local complet, alors il satisfait la division de Weierstrass.*

En revanche, on ne peut espérer obtenir un théorème de Weierstrass pour un anneau local hensélien. Il suffit de considérer A_n l'anneau local des germes de fonctions holomorphes de $\mathbb{C}^n$ dans un voisinage de zéro. Sa complétion est donnée par $R_n := \mathbb{C}[[t_1, \ldots, t_n]]$. Si $f \in A_n[[t]]$, on peut l'écrire:

$$f = qu$$

dans $R_n[[t]]$. En revanche, si f n'est pas convergente dans $\mathbb{C}^{n+1}$, il n'y a pas de raison que $q \in A_n[t]$.

On va avoir besoin de cet énoncé dans un cadre légèrement plus général. Soit k un corps. On considère la catégorie Inf_k des épaississements de k. Elle consiste en les k-algèbre locales A, d'idéal maximal $\mathfrak{m}$, de corps résiduel k, tel que $\mathfrak{m}^m = 0$ pour un certain $m \in \mathbb{N}$. Les morphismes sont les morphismes de k-algèbres.

Remarques:

(1) Dans la suite, on appelle les objets $A \in Inf_k$ des anneaux quasi-artiniens. Cela tient au fait que si A est noethérien et est dans Inf_k alors il est artinien.

(2) Il résulte du théorème 2.3 que si A est quasi-artinien, alors il satisfait la division de Weierstrass.

On rappelle également le théorème fondamental suivant qui permet d'algébriser des isomorphismes formels [**2**]:

THÉORÈME 2.4 (Théorème d'approximation d'Artin). *Soient Z et Z' des k-schémas de type fini tels que l'on a un isomorphisme des anneaux locaux complétés:*

$$\mathcal{O}^{\wedge}_{Z,z} \simeq \mathcal{O}^{\wedge}_{Z',z'}$$

alors ils admettent un voisinage étale commun U:

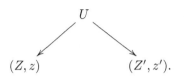

$$(Z', z').$$

Remarques:

(1) Cet énoncé admet d'autres variantes, pour les séries convergentes ou algébriques, mais nous n'aurons pas besoin de tels énoncés et seule cette version du théorème d'approximation d'Artin nous sera utile. On renvoie à [**18**] pour un panorama général.

(2) L'énoncé n'est pas connu sous des hypothèses non-noethériennes.

2.2. Complétion d'anneaux

Soit A un anneau, $I \subset A$ un idéal, on définit la complétion I-adique de A:

$$\hat{A} := \varprojlim A/I^n.$$

Plus généralement pour un A-module M, on définit:

$$\hat{M} := \varprojlim M/I^n.$$

Le A-module M est dit I-adiquement complet si $M \to \hat{M}$ est un isomorphisme. Un des exemples qui apparaîtra dans ce texte est la complétion t-adique:

$$A[[t]] = \varprojlim A[t]/(t^n).$$

La complétion se comporte de façon radicalement différente si A est noethérien ou pas. Si A est noethérien alors d'après [**29**, Tag. 0BNH], la flèche:

$$\mathrm{Spec}(\hat{A}) \to \mathrm{Spec}(A)$$

est plate et pour tout A-module de type fini M, on a:

$$\hat{M} \simeq M \otimes_A \hat{A}$$

et $\hat{M}$ est I-adiquement complet. Si A n'est pas noethérien, la complétion I-adique n'est plus plate. On a le résultat suivant [**29**, Tag. 0AL8]:

PROPOSITION 2.5. *Soit A un anneau dénombrable, alors $A[[t]]$ est plat sur A si et seulement si A est cohérent.*

On obtient donc que si A est dénombrable et non-cohérent, $A[[t]]$ n'est *jamais* un $A[t]$-module plat. Parmi les anneaux non-noethériens, on peut opérer une nouvelle dichotomie, suivant que l'on complète par rapport à un idéal I de type fini ou un idéal quelconque.

Si l'on a une paire (A, I) avec I de type fini, d'après [**29**, Tag. 00M9], pour tout A-module M de type fini, $\hat{M}$ est I-adiquement complet. Un autre avantage de compléter par un idéal de type fini I est que si $\bar{A} := A/I$, alors la flèche canonique :

$$(2.6) \qquad A[[t]] \otimes_A \bar{A} \to \bar{A}[[t]]$$

est un isomorphisme. Enfin, dans certains cas non-noethériens avec I de type fini, Kato-Fujiwara-Gabber [**14**] montrent des énoncés de platitude pour la complétion I-adique.

En revanche, si I n'est plus de type fini, alors (2.6) est faux (cf. [**24**, Lem. 2.1]) et $\hat{A}$ n'est même plus I-adiquement complet; l'exemple classique est $A = k[x_i]_{i \in \mathbb{N}}$ et $I = (x_i)_{i \in \mathbb{N}}$. Alors la complétion I-adique de A s'identifie à $k[[x_i]]_{i \in \mathbb{N}}$ qui n'est pas I-adiquement complète. On renvoie à [**14**, Ex. 2.1.2] ou Yekutieli [**30**, Ex. 1.8] pour plus de détails. On montre même [**16**] que $\hat{A}$ n'est même pas $\hat{I}$-adiquement complet.

Le point de départ de ce travail est la situation la plus «pathologique» où l'on s'intéresse à des complétions I-adiques avec I qui n'est pas de type fini. Le théorème de structure 6.1 rend la situation un peu meilleure en réduisant à des complétions d'anneaux non-noethériens pour I de type fini.

3. Espace D'arcs

3.1. Définition

Soit k un corps. Soit X un k-schéma de type fini. Pour $n \in \mathbb{N}$, le foncteur $R \mapsto X(R[t]/(t^{n+1}))$ est représentable par un k-schéma de type fini $\mathcal{L}_n X$. On considère alors l'espace d'arcs:

$$\mathcal{L}X = \varprojlim_{n \in \mathbb{N}} \mathcal{L}_n X.$$

En général, c'est un schéma non-noethérien. On a le théorème suivant dû à Bhatt [**5**]:

THÉORÈME 3.1. *Le schéma $\mathcal{L}X$ représente le foncteur sur les k-algèbres:*

$$R \mapsto X(R[[t]]).$$

Si on part de $X = \mathbb{A}^1$, on a $\mathcal{L}X = \operatorname{Spec}(k[a_i]_{i \in \mathbb{N}})$. Plus généralement si X est une hypersurface affine définie par $f(x_1, \ldots, x_l) = 0$. Alors $\mathcal{L}X$ est le sous-schéma fermé de $\mathcal{L}\mathbb{A}^l$ défini par:

$$f\left(\sum a_i^{(1)} t^i, \ldots, \sum a_i^{(l)} t^i\right) = 0.$$

Si $Y \to X$ est étale, alors on a, par critère infinitésimal de relèvement:

$$\mathcal{L}Y \simeq \mathcal{L}X \times_X Y.$$

En particulier, si U est ouvert dans X, $\mathcal{L}U$ est un ouvert quasi-compact dans $\mathcal{L}X$. Si F est fermé dans X, alors $\mathcal{L}F$ est fermé dans $\mathcal{L}X$ mais l'immersion fermée n'est pas de type fini. Pour un k-schéma X réduit, on s'intéresse à:

$$\mathcal{L}^\bullet X = \mathcal{L}X - \mathcal{L}X_{\text{sing}}$$

avec X_{sing} le lieu singulier. C'est un ouvert non-quasi compact de $\mathcal{L}X$. En particulier, on a:

$$\mathcal{L}^\bullet X(k) = X(k[[t]]) \cap X_{\text{reg}}(k((t))).$$

Plus généralement, étant donné un fermé $F \to X$ d'ouvert complémentaire U, on peut considérer:

$$\mathcal{L}_U X := \mathcal{L}X - \mathcal{L}F.$$

En prenant $U = \mathbb{G}_m$ et $X = \mathbb{A}^1$, on a:

$$\mathcal{L}_U X(k) = k[[t]] \cap k((t))^*.$$

En revanche, cette égalité est fausse sur des anneaux tests plus généraux, si R est intègre, on a seulement l'inclusion:

$$R[[t]] \cap R((t))^* = \coprod_{n \geq 0} t^n R[[t]]^* \subset \mathcal{L}_U X(k) = R[[t]] \cap K((t))^*.$$

où $K := \text{Frac}(R)$. En particulier, le terme de gauche est disconnexe, tandis que celui de droite est connexe, il recolle les différentes strates. De plus, il est ouvert dans $\mathcal{L}\mathbb{A}^1$, ce qui n'est pas le cas du membre de gauche.

On s'intéresse à la structure de l'espace d'arcs et de ses tronqués. Si X est lisse, alors pour tout $n \in \mathbb{N}$, les applications:

$$\mathcal{L}_{n+1}X \to \mathcal{L}_n X \text{ et } \mathcal{L}X \to \mathcal{L}_n X$$

sont formellement lisses, plates et surjectives, toujours par le critère de relèvement infinitésimal.

Si X est singulier, rien de tout cela ne vaut. De plus, la topologie est radicalement différente. On commence par le théorème de Mustaţă [23]:

THÉORÈME 3.2. *Soit X un k-schéma localement intersection complète avec $\text{car}(k) = 0$, alors pour tout $n \in \mathbb{N}$, $\mathcal{L}_n X$ est irréductible si et seulement si X a des singularités rationnelles.*

En revanche, on a le théorème suivant dû à Kolchin [20] :

THÉORÈME 3.3. *Si X est irréductible et $\text{car}(k) = 0$, alors $\mathcal{L}X$ est irréductible.*

On a un théorème de structure formelle pour $\mathcal{L}X$ qui dit que l'espace d'arcs bien qu'infini, se comporte mieux ([12; 15]):

THÉORÈME 3.4 (Drinfeld–Grinberg–Kazhdan). *Soit X un k-schéma de type fini, $\gamma(t) \in \mathcal{L}^\bullet X(k)$ tel que $\gamma(0)$ ne soit pas un point isolé de X, alors on a un isomorphisme de voisinage formels:*

$$\mathcal{L}X_\gamma^\wedge \simeq (Y \times \mathbb{A}^{\mathbb{N}})^\wedge_{(y,0)},$$

avec Y un k-schéma de type fini et $y \in Y(k)$.

Remarques:

(1) Par un théorème de Bourqui–Sebag [8], ce théorème est optimal, à savoir qu'il est faux pour les arcs $\gamma \in \mathcal{L}X_{\mathrm{sing}}$.

(2) La preuve se fait en deux temps; tout d'abord, pour un k-schéma S et $x_0 \in S(k[[t]])$, l'anneau local complété $\mathcal{O}^\wedge_{\mathcal{L}S,x_0}$ pro-représente le foncteur des déformations:

$$F : Inf_k \to \mathrm{Ens}$$

avec $F(A) := \{\tilde{x} \in S(A[[t]]) \mid \tilde{x} = x_0 \bmod \mathfrak{m}\}$. Puis ensuite, il suffit donc de démontrer que le foncteur des déformations sur la catégorie Inf_k sont les mêmes. A cette étape, le théorème de division de Weierstrass apparaît de manière cruciale.

(3) On voit dans la proposition 3.5 à quel point la partie de dimension finie est unique.

Exemples: Si $X := \{(x,y,z) \mid xy = z^2\}$ et l'on prend les arcs $\gamma_1 = (t,0,0)$ et $\gamma_2(t) = (t^2,0,0)$. Alors, on a $Y_1 = \{v \in \mathbb{C} \mid v^2 = 0\}$ et $Y_2 = \{(a,b,v,w) \in \mathbb{C}^4 \mid aw^2 = v^2, 2wv = bw^2\}$ et $y = (0)$.

Expliquons le deuxième exemple. Soit A quasi-artinien, $\mathfrak{m}$ son idéal maximal. On regarde un point $(x(t), y(t), z(t)) \in \mathcal{L}X(A)$ qui relève γ_1. Comme A satisfait la division de Weierstrass, on a:

$$x(t) = q(t)u(t),$$

où $u(t)$ est une unité avec $q(t) = a + bt + t^2$ et $a, b \in \mathfrak{m}$. On fait alors la division de Weierstrass pour $z(t) = v + wt + q(t)h(t)$. Nous avons alors:

$$q(t)u(t)y(t) = z(t)^2 = v^2 + 2wt + w^2t^2 + q(t)h(t)[2(v + wt) + q(t)h(t)],$$

en particulier $v^2 + 2vwt + w^2t^2 \in (q(t))$. Comme A est quasi-artinien, on a $A[[t]]/(q(t)) \simeq A[t]/(q(t))$. Cela force $aw^2 = v^2$ et $bw^2 = 2wv$. On en déduit une flèche:

$$\phi_2 : Y_2 \times \mathcal{L}\mathbb{G}_m \times \mathcal{L}\mathbb{A}^1 \to \mathcal{L}X$$

donnée par

$$(a,b,v,w,x(t),z(t)) \mapsto (q(t)x(t), y(t), v + wt + q(t)z(t)).$$

avec $q(t) = a + bt + t^2$.

Remarque: Les modèles formels que l'on obtient sont arbitraires. Drinfeld donne l'exemple de:

$$yx_{n+1} + f(x_1, \ldots, x_n) = 0,$$

et de l'arc $\gamma = (0, \ldots, 0, t)$. Le modèle obtenu est $Y := \{(x_1, \ldots, x_n) \mid f(x_1, \ldots, x_n) = 0\}$. En effet, pour A quasi-artinien, on écrit $x_{n+1}(t) = (t - \alpha)u(t) \in A[[t]]$, avec $u(t)$ inversible. Si $x_1(t), \ldots, x_n(t)$ sont donnés alors il existe un unique $y(t)$ si et seulement si:

$$f(x_1(\alpha), \ldots, x_n(\alpha)) = 0.$$

3.2. Définition d'une fonction et unicité du modèle

Notons $\mathcal{O} := k[[t]]$ et $F = k((t))$. Une des applications du théorème de Drinfeld–Grinberg–Kazhdan (DGK) est de pouvoir associer à $\mathcal{L}X$ une fonction canonique, dans le cas où le corps k est fini. Dans de nombreux cas, cette fonction, a une interprétation en termes de théorie de la représentation. Plus précisément, dans [**10**], si l'on considère la situation où X est un monoïde, cette fonction est reliée aux facteurs L non-ramifiés. La proposition suivante s'obtient à l'aide de [**10**, Proposition 1.2]. Elle a également été obtenue indépendamment par Gabber [**9**, §7]:

PROPOSITION 3.5. *Si* $(Y \times \mathbb{A}^{\mathbb{N}})^{\wedge}_{(y,0)} \simeq (Y' \times \mathbb{A}^{\mathbb{N}})^{\wedge}_{(y',0)}$ *avec* Y *et* Y' *des* k-schémas de type fini, $y \in Y(k)$, $y' \in Y'(k)$, alors il existe m et l tels que:

$$(Y \times \mathbb{A}^m)_{(y,0)} \simeq (Y' \times \mathbb{A}^l)_{(y',0)}.$$

En particulier, en utilisant l'approximation d'Artin, il existe un voisinage lisse commun U:

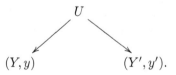

Remarques:

(1) Dans la proposition précédente, on peut même prendre m ou l égal à zéro, par un résultat d'élimination [**17**, Proposition 2.6] qui apparaît également dans [**9**, §7]. On voit donc que dans l'énoncé de DGK, le voisinage formel du facteur de dimension finie est unique à multiplication par un facteur $\mathbb{A}^n$ près.

(2) Il est important d'avoir Y et Y' de type fini, dans la section 5.2, l'espace des arcs non-dégénérés de $\{xy = 0\}$ fournit un contre-exemple, comme il n'est pas localement pour la topologie étale isomorphe à $\mathcal{L}\mathbb{A}^1 \coprod \mathcal{L}\mathbb{A}^1$.

En particulier, en utilisant le théorème d'approximation d'Artin et si k est un corps fini, on obtient une fonction:

$$f_{IC_{\mathcal{L}^\bullet X}} : \begin{array}{ccc} X(\mathcal{O}) \cap X_{\mathrm{reg}}(F) & \to & \overline{\mathbb{Q}}_\ell \\ \gamma(t) & \mapsto & \mathrm{Tr}(\mathrm{Fr}_y, IC_{Y,y}[-\dim(Y_y)]) \end{array}$$

qui est bien définie.

Remarques:

(1) Un des corollaires du théorème d'approximation d'Artin est que le complexe d'intersection d'un k-schéma de type fini, ne dépend que de ses voisinages formels.

(2) On obtient ainsi un invariant de l'espace d'arcs $\mathcal{L}X$, il s'agit de voir comment on le calcule. Pour cela, on va construire un modèle global qui marche pour tous les arcs non-dégénérés.

3.3. Canonicité du modèle de dimension finie

Dans l'énoncé de Drinfeld–Grinberg–Kazhdan, apparaît un espace affine de dimension infinie assez délicat à manipuler. On voudrait une procédure pour extraire la partie de dimension finie. Dans un certain nombre de cas, cela est possible. Un autre inconvénient de l'énoncé DGK est qu'il produit un modèle explicite Y donné par un certain nombre d'équations [**12**], mais *a priori* le modèle change en fonction de l'arc considéré et il n'est pas canonique. Le but de cette section est donc également d'apporter une réponse à ce problème.

3.3.1. *Le cas monoïde*

Soit G connexe réductif déployé sur k. Soit X affine, intègre, normal qui contient une orbite ouverte dense X^0, sur laquelle G agit simplement transitivement, de telle sorte que $X^0 \simeq G$. En particulier, d'après [**11**, 6.2.C], X admet une structure de monoïde. On considère le champ d'Artin $[X/G]$ et l'espace d'arcs du champ quotient $\mathcal{L}[X/G]$, défini de la manière suivante:

$$\mathcal{L}[X/G] \simeq \varprojlim_{n \geq 0} \mathcal{L}_n[X/G].$$

Chaque $\mathcal{L}_n[X/G]$ est un champ d'Artin, il admet un atlas lisse par $\mathcal{L}_n X$ qui est un $\mathcal{L}_n G$-torseur et les flèches de transition sont affines. On a une flèche de projection:

$$\mathcal{L}X \to \mathcal{L}[X/G]$$

qui est un $\mathcal{L}G$-torseur. En particulier, on ne change pas les singularités en passant aux arcs G-équivariants. Soit le champ des sections $\mathrm{Hom}(\mathbb{P}^1, [X/G])$. C'est également un champ d'Artin d'après Olsson [**25**] et on considère le sous-champ ouvert $\mathrm{Hom}^0(\mathbb{P}^1, [X/G])$ des sections qui tombent génériquement dans X^0. Il classifie les paires (E, ϕ) où E est un G-torseur sur $\mathbb{P}^1$ et $\phi \in H^0(\mathbb{P}^1, X \wedge^G E)$.

C'est un espace algébrique, car la condition de généricité tue les automorphismes et on peut montrer même que c'est un schéma qui est une union disjointe dénombrable de schémas projectifs de type fini à l'aide de de la grassmannienne affine [**10**, Proposition 4.2]. On considère l'espace des arcs non-dégénérés:

$$\mathcal{L}^\bullet[X/G] = \mathcal{L}_{X^0}[X/G].$$

Fixons une place $v \in \mathbb{P}^1$. On a une flèche de restriction au voisinage formel en v:

$$\mathrm{res}_v : M_G := \mathrm{Hom}^0(\mathbb{P}^1, [X/G]) \to \mathcal{L}^\bullet[X/G].$$

On a l'énoncé suivant [**10**, Propositions 2.1, 2.2]:

PROPOSITION 3.6. *Soit $\gamma \in \mathcal{L}^\bullet[X/G](k)$, alors il existe $m \in \mathrm{Hom}^0(\mathbb{P}^1, [X/G])(k)$ tel que:*

$$\mathrm{res}_v(m) = \gamma$$

et tel que res_v induit un isomorphisme sur les complétés formels.

Remarque: Dans [**10**], au lieu de regarder les arcs G-équivariants, on considère le $\mathcal{L}G$-torseur $\tilde{M}_G$ où l'on ajoute une trivialisation en v, l'énoncé est plus agréable en considérant $\mathcal{L}[X/G]$.

L'énoncé clé est le théorème de Beauville–Laszlo [**3**]:

THÉORÈME 3.7. *Soit A un anneau, C une courbe projective lisse connexe sur k, $v \in C(k)$. Alors il y a une bijection fonctorielle entre les classes d'isomorphismes suivantes:*

$$G - torseurs \ sur \ C_A \Longleftrightarrow les \ triplets \ (E_{C-v}, E_v, \beta)$$

où E est un G-torseur sur $(C-v)_A$, E_v un G-torseur sur le voisinage formel en v, $D_A \simeq A[[t]]$ (après choix d'une uniformisante) et β un isomorphisme sur $D_A^\bullet \simeq A((t))$.

Remarque: Si A est noethérien, cela se déduit presque de la descente fidèlement plate (il faut construire la relation de cocycle), en général non.

Expliquons en particulier, comment dans l'énoncé 3.6, on globalise γ en m:

Soit $\gamma \in \mathcal{L}^\bullet[X/G](k)$, alors $\gamma \in X(\mathcal{O}) \cap G(F)/G(\mathcal{O})$, cela correspond à un G-torseur sur $\mathcal{O}$ et une section $\phi \in H^0(\mathrm{Spec}(\mathcal{O}), E \wedge^G X)$ qui induit une trivialisation sur $\mathrm{Spec}(F)$. On recolle alors avec le torseur trivial sur $\mathbb{P}^1 - v$, muni de la section unité et on obtient une paire $m := (E, \phi) \in M_G(k)$.

Remarques:

(1) En particulier, en remplaçant un disque formel par une courbe projective lisse, on obtient un objet de dimension finie qui a bien les mêmes singularités que le modèle local.

(2) Toutefois, l'inconvénient de remplacer le disque formel par une courbe est qu'on ajoute des singularités en d'autres points de $\mathbb{P}^1$, en particulier, seule une partie de M_G a les mêmes singularités que $\mathcal{L}[X/G]$. En général, cette partie a l'inconvénient d'être constructible dans M_G.

Exemple: Prenons $X = \mathbb{A}^1, G = \mathbb{G}_m$ et l'arc $\gamma = (E_0, t^n) \in \mathcal{L}[\mathbb{A}^1/\mathbb{G}_m](k)$. Le modèle global est dans ce cas:

$$\coprod_{d \in \mathbb{N}} \mathbb{P}^d,$$

qui classifie les diviseurs effectifs de degré d. On globalise l'arc $\gamma(t)$ en le diviseur effectif $D = n[v]$. La flèche de restriction au voisinage formel en v induit un isomorphisme formel:

$$\mathrm{Spf}(k[[t_1, \ldots, t_n]]) \simeq \mathcal{L}[\mathbb{A}^1/\mathbb{G}_m]_\gamma^\wedge.$$

Remarque: Soit IC_{M_G} le complexe d'intersection du modèle global, alors en vertu de la proposition 3.6, pour calculer $f_{\mathrm{IC}_{\mathcal{L}X}}$, il suffit de calculer $f_{\mathrm{IC}_{M_G}}$. On calcule dans [**10**, Propositions 3.8, and 4.3] ce complexe d'intersection global.

3.3.2. Remarques sur le cas général

Dans une prepublication récente [**13**], Drinfeld explique les éléments nécessaires pour adapter la discussion précédente à un cas plus général. Plus précisément, on dit qu'un k-schéma X est localement intersection complète spécial s'il est défini dans $k[x_1, \ldots, x_m, y_1, \ldots, y_n]$ par des équations $p_1 = \cdots = p_n = 0$ et que le lieu singulier est défini par le diviseur:

$$\det\left(\frac{\partial p}{\partial y}\right) = 0.$$

Drinfeld construit alors dans le cas lci spécial, un groupoïde lisse Γ dit de Newton, qui agit simplement transitivement sur X_{reg} et forme le champ quotient $[X/\Gamma]$. Dans la discussion précédente, le rôle de Γ est joué par G. L'avantage de ce champ quotient (et de $[X/G]$ dans la section précédente) est qu'il contient un point comme ouvert dense. Drinfeld parle de «pointy stacks». On peut de même que ci-dessus considérer le champ:

$$\mathrm{Hom}^0(\mathbb{P}^1, [X/\Gamma]),$$

des sections qui sont génériquement dans l'ouvert lisse. On s'attend à ce que ce sous-champ soit un modèle canonique de la même manière que dans la proposition 3.6. Drinfeld formule une conjecture sur les champs de morphismes de «champs pointus» [**13**, 4.2.3] partant d'une courbe quelconque sur k. Dans notre situation, on sait déjà que l'espace considéré est un champ d'Artin, comme on part d'une courbe projective et la construction de Γ

implique qu'il n'y a pas d'automorphismes, donc d'après [**29**, Tag. 03YR], on obtient un espace algébrique localement de type fini.

3.4. Recollement de modèles formels

Comme on l'a vu dans la section précédente, l'inconvénient du modèle global est qu'il ajoute des singularités en d'autres points. On aimerait donc recoller les différents modèles de dimension finie purement localement. On va voir que cela ne peut être fait avec des k-schémas de type fini, ainsi que le montre l'exemple suivant. Le théorème 6.1 montre que cela est en revanche possible en considérant certains types de schémas non-noethériens.

Exemple: On considère le cas du cône quadratique $X := \{(x, y, z) \mid xy = z^2\}$. Si l'on prend les arcs de valuation un et deux, $\gamma_1(t) = (t, 0, 0)$ et $\gamma_2(t) = (t^2, 0, 0)$, les modèles formels Y_1 et Y_2 sont donnés par $\{v^2 = 0\}$ et $\{(a, b, v, w) \mid aw^2 = v^2,\ bw^2 = 2wv\}$. On a alors un morphisme:

$$\phi : Y_2 \times \mathcal{L}\mathbb{G}_m \times \mathcal{L}\mathbb{A}^1 \to \mathcal{L}X,$$

donné par:

$$((a, b, v, w), u(t), \xi(t)) \mapsto (qu(t), y(t), v + wt + q\xi(t)),$$

avec $y(t)$ défini uniquement par $(x(t), z(t))$ et $q = a + bt + t^2$. Le morphisme ϕ induit un isomorphisme formel sur les points qui s'envoient sur $(t^2 u(t), y, z)$ avec $u(t)$ une série inversible. En revanche, cela n'est plus vrai aux autres points où les singularités des deux côtés sont différentes. En effet, le quadruplet $(1, 0, 0, 0)$ est un point singulier de Y_2 mais s'envoie sur un point lisse de $\mathcal{L}X$.

4. Morphismes Pro-lisses

On veut obtenir le théorème de Drinfeld–Grinberg–Kazhdan dans un certain voisinage. On est confronté à plusieurs problèmes:

(1) Le théorème de Weierstrass ne marche pas pour des anneaux henséliens.
(2) Le théorème d'approximation d'Artin n'est pas connu pour des anneaux non-noethériens.
(3) La complétion $\mathcal{O}_{S,x}^\wedge$ n'est pas plate si S est non-noethérien.
(4) La flèche

$$\begin{array}{rccc} \phi & : & \mathbb{A}^1 \times \mathcal{L}\mathbb{G}_m & \to & \mathcal{L}\mathbb{A}^1 \\ & & (a, x(t)) & \mapsto & (t - a)x(t) \end{array}$$

est un isomorphisme formel en $(0, 1)$ mais n'est pas formellement lisse. En particulier, la topologie naturelle pour ce type de schémas n'est aucune des topologies usuelles; étales, lisses ou pro-étales.

La preuve de DGK fait apparaître comme outil fondamental le théorème de Weierstrass. Il s'agit donc de faire un changement de base par une

certaine classe de morphismes de telle sorte qu'on aura tautologiquement
une décomposition de Weierstrass après changement de base. On veut également que cette nouvelle classe de morphismes respecte les singularités de
$\mathcal{L}X$. L'ingrédient fondamental est donc l'introduction des morphismes pro-
lisses et le fait que les opérations (cf. Propositions 4.7 et 4.9) du type
«Weierstrass» sont pro-lisses.

Soit Sch_k la catégorie des k-schémas quasi-compacts quasi-séparés. Dans
[**4**], on introduit les notions suivantes:

DÉFINITION 4.1. Soit $f : Z \to S$ un morphisme de k-schémas.

(1) f est pro-lisse (resp. pro-étale) s'il s'écrit comme une limite projective filtrante de morphismes lisses (resp. étales) avec des morphismes de transition affines.

(2) f est placide s'il s'écrit comme une limite projective filtrante
$Y \simeq \varprojlim Z_i$ où les Z_i sont de présentation finie sur S et où les
morphismes de transition sont affines, lisses et surjectifs.

(3) Un k-schéma S est pro-lisse (resp. placide) si $S \to \mathrm{Spec}(k)$ est
pro-lisse (resp. placide).

Remarques:

• Il est crucial de remarquer que pour un morphisme pro-lisse, les
morphismes de transition sont arbitraires. En revanche, pour un
morphisme pro-étale, les morphismes de transition sont automa-
tiquement étales. Les morphismes pro-lisses et placides sont stables par changement de base arbitraire et par composition.

• Un exemple typique de morphisme placide est le morphisme:

$$S \times \mathbb{A}^I \to S$$

où I est un ensemble et $\mathbb{A}^I := \mathrm{Spec}(k[x_i]_{i \in I})$ ou plus généralement
l'espace d'arcs d'un k-schéma lisse de type fini. Le théorème 4.3
explicite quels types de morphismes sont pro-lisses.

Les morphismes pro-lisses apparaissent déjà dans les travaux de Popescu
([**27**; **28**]). Le théorème 4.4 décrit la structure de ceux-ci dans le cas
noethérien.

PROPOSITION 4.2. *Si $f : Z \to S$ est pro-lisse entre k-schémas, alors Z
est plat sur S et le complexe cotangent $L_{Z/S}$ est concentré en degré zéro et
plat sur Z.*

DÉMONSTRATION. Cela vient directement du fait que le complexe cotangent commute aux limites inductives filtrantes d'algèbres [**21**, II.1.2.3.4] et
qu'une limite inductive filtrante d'algèbres plates est plate. □

Remarque: La réciproque est malheureusement fausse d'après un
contre-exemple de Gabber, rédigé par Bhatt [**6**]. De plus, l'exemple rédigé
par Bhatt pourrait amener à croire qu'il s'agit d'une pathologie typique de
la caractéristique p. Néanmoins, Gabber nous a communiqué que le résultat

reste faux en caractéristique zéro. En remplaçant la perfection de l'algèbre B_i de [**6**] par l'algèbre où l'on ajoute les racines carrés, on obtient à nouveau une algèbre non-réduite dont le complexe cotangent est plat et placé en degré zéro.

La première classe de morphismes pro-lisses intéressants nous est donnée par le théorème de Popescu ([**26–28**]):

THÉORÈME 4.3. *Soit $f : \operatorname{Spec}(B) \to \operatorname{Spec}(A)$ un morphisme régulier entre schémas noethériens, alors f est pro-lisse. En particulier, si Y est un k-schéma excellent et $y \in Y$, le morphisme de complétion:*

$$\operatorname{Spec}(\hat{\mathcal{O}}_{Y,y}) \to \operatorname{Spec}(\mathcal{O}_{Y,y})$$

est pro-lisse.

Remarques:

(1) On rappelle qu'un morphisme régulier est un morphisme plat à fibres géométriquement régulières et que si A est un anneau excellent, une des propriétés est que pour tout $\mathfrak{p} \in \operatorname{Spec}(A)$, le morphisme de complétion $A_{\mathfrak{p}} \to \hat{A}_{\mathfrak{p}}$ est régulier.

(2) On rappelle que si A est excellent, il en est de même de toute A-algèbre de type fini. De plus, les exemples typiques d'anneaux excellents qui apparaîtront ici sont donnés par les corps et les anneaux locaux complets noethériens. On renvoie à [**29**, Tag. 07QS] pour plus de détails.

(3) De ce théorème, on remarque qu'en particulier, un morphisme pro-lisse n'est pas nécessairement formellement lisse.

A l'aide du théorème de Popescu, on peut clarifier la relation entre régulier et pro-lisse dans le cas noethérien:

THÉORÈME 4.4. *Soit $f : \tilde{Y} \to Y$ un morphisme entre schémas affines noethériens, on a les équivalences suivantes:*

(1) *f est régulier.*

(2) *f est pro-lisse.*

(3) *f est plat et le complexe cotangent $L_{\tilde{Y}/Y}$ est concentré en degré zéro et plat sur $\tilde{Y}$.*

DÉMONSTRATION. (1) $\Rightarrow$ (2) est l'objet du théorème de Popescu, (2) $\Rightarrow$ (3) se déduit de la proposition 4.2 et (3) $\Rightarrow$ (1) se déduit d'un théorème d'André [**1**, Théoréme 30, p. 331]. $\square$

DÉFINITION 4.5. Soit $f : Z \to S$ entre k-schémas. On dit que :

(1) f est Weierstrass, si f est pro-lisse et pour tout k-point $x \in Z(k)$ et $y = f(x)$, le morphisme au niveau des complétés formels:

$$\hat{f}_x : \hat{Z}_x \to \hat{S}_y$$

est formellement lisse.

(2) f est un isomorphisme formel si pour tout k-point $x \in Z(k)$, $\hat{f}_x$ est un isomorphisme.

(3) f est strictement Weierstrass, si f est pro-lisse et est un isomorphisme formel.

Remarque: Les morphismes de Weierstrass, strictement Weierstrass et les isomorphismes formels sont stables par changement de base arbitraire et composition.

La proposition clé est la suivante [**4**, Proposition 1.15]:

PROPOSITION 4.6. *Soit $f : Z \to S$ un morphisme entre k-schémas placides affines, on suppose qu'en tout k-point $x \in Z(k)$, $\hat{f}_x$ est formellement lisse, alors f est de Weierstrass.*

La terminologie de morphisme de Weierstrass se justifie par ce qui suit. Pour tout entier $d \geq 1$, on considère le k-schéma $\mathcal{Q}_d$ qui classifie les polynômes unitaires de degré d et A_d le k-schéma qui classifie les polynômes de degré au plus $d - 1$. On définit l'ouvert:

$$\mathcal{L}\mathbb{A}^{\leq d} := ev_d^{-1}(\mathcal{L}_d\mathbb{A}^1 - \{0\}),$$

avec $ev_d : \mathcal{L}\mathbb{A}^1 \to \mathcal{L}_d\mathbb{A}^1$. On considère le morphisme:

$$\alpha_d : \mathcal{Q}_d \times \mathcal{L}\mathbb{G}_m \to \mathcal{L}\mathbb{A}^{\leq d}$$

donné par $(q, u) \mapsto qu$. On a alors [**4**, Proposition 1.16]:

PROPOSITION 4.7. *Le morphisme α_d est Weierstrass, surjectif et est strictement Weierstrass en les points de la forme $(t^d, u(t))$.*

Le morphisme α_d admet un compagnon:

(4.8) $$\beta_d : \mathcal{Q}_d \times A_d \times \mathcal{L}\mathbb{A}^1 \to \mathcal{Q}_d \times \mathcal{L}\mathbb{A}^1,$$

donné par $(q, v, \xi) \mapsto (q, v + q\xi)$.

PROPOSITION 4.9. *Le morphisme β_d est Weierstrass, surjectif et est strictement Weierstrass en les points de la forme $(t^d, u(t))$.*

DÉMONSTRATION. cf. [**4**, Proposition 1.17]. □

5. Sur Certains Espaces Non-noethériens

5.1. Schémas de type (S)

On construit une famille de schémas non-noethériens qui interviennent de manière fondamentale dans la construction de nos atlas pro-lisses.

Son importance a été dégagée dans [**24**]. Soit $d \geq 1$, on considère le carré cartésien:

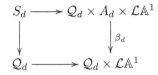

où la flèche horizontale du bas est donnée par $q \mapsto (q,0)$. Le schéma S_d est pro-lisse sur $\mathcal{Q}_d$ et est non-noethérien car β_d n'est pas de type fini. Il admet une section:

$$(5.1) \qquad\qquad \sigma : \mathcal{Q}_d \to S_d$$

donné par $q \mapsto (q,0,0)$. Décrivons-le de manière explicite pour $d = 1$. Le schéma $\mathcal{Q}_1 \times A_1 \times \mathcal{L}\mathbb{A}^1$ classifie les triplets (q,v,ξ) avec $q = t + a$ le polynôme universel de degré un et $\mathcal{Q}_1 := \mathrm{Spec}(k[a])$. On écrit $\mathcal{L}\mathbb{A}^1 := \mathrm{Spec}([\xi_0, \xi_1, \ldots])$. Les équations pour $S_1 = \mathrm{Spec}(A)$ sont alors données par:

$$(5.2) \qquad\qquad v + a\xi_0 = 0$$

$$(5.3) \qquad\qquad \xi_0 + a\xi_1 = 0$$

$$(5.4) \qquad\qquad \xi_1 + a\xi_2 = 0$$

$$(5.5) \qquad\qquad \cdots$$

avec $\xi = (\xi)_{i \in \mathbb{N}}$. En particulier, la fibre au-dessus de S_1 pour $a \neq 0$ est isomorphe à $\mathbb{A}^1$ et la fibre au-dessus de 0 est nulle. On obtient ainsi que S_1 est un schéma non-noethérien, stratifié par des schémas noethériens.

Il résulte de ces équations que pour tout n, a^n divise tous les ξ_i et v. On a donc:

$$\mathfrak{p}_0 = (a) = (a, v, \xi_i),$$

et $A/\mathfrak{p}_0 = k$, donc $\mathfrak{p}_0$ est maximal. De plus, on trouve que la complétion formelle en $\mathfrak{p}_0$ du schéma S_1 est isomorphe à $k[[a]]$. On peut décrire S_1 comme un pro-schéma:

$$S_1 \simeq \varprojlim_{n \in \mathbb{N}} \mathrm{Spec}(A_n)$$

où $A = \varinjlim A_n = \varinjlim_{n \in \mathbb{N}} k[q, x_n]$ avec les morphismes de transition $k[q, x_n] \to k[q, x_{n+1}]$ qui envoient:

$$q \mapsto q \text{ et } x_n \mapsto q x_{n+1}.$$

En tant qu'ensemble, S_1 correspond à $\mathbb{A}^2$ où l'on enlève l'axe vertical et on rajoute l'origine.

DÉFINITION 5.6. Les schémas de type (S) sont les schémas affines de présentation finie sur un produit fini $S_d \times_{\mathcal{Q}_d} \cdots \times_{\mathcal{Q}_d} S_d$. En particulier, ils admettent un morphisme canonique:

$$\psi : T \to \mathcal{Q}_d.$$

Le schéma $\mathcal{Q}_d$ admet une stratification naturelle en sous-schémas localement fermés:

$$\mathcal{Q}_d = \coprod_{0 \leq r \leq d} H_r,$$

avec pour $r \leq d-1$, $H_r := \{q \in \mathcal{Q}_d \mid t^r | q, a_r \in \mathbb{G}_m\} \simeq \mathbb{G}_m \times \mathbb{A}^{d-r-1}$ et $H_d := \{q = t^d\}$. Il résulte de la définition de S_d en tant que produit fibré que:

$$S_d \times_{\mathcal{Q}_d} H_0$$

est un k-schéma de type fini, fibré au-dessus de $\mathbb{G}_m \times \mathbb{A}^{d-1}$ en $\mathbb{A}^d$.

PROPOSITION 5.7. *Soit T un schéma de type (S), $\psi : T \to \mathcal{Q}_d$, alors pour tout $1 \leq r \leq d$, $\psi^{-1}(H_r)$ est un k-schéma de type fini. En particulier, T admet une stratification finie constructible par des k-schémas de type fini. On obtient ainsi que l'espace topologique de T est noethérien et est de dimension finie.*

DÉMONSTRATION. Soit T de type (S), on commence par le cas $T = S_d$. On a une flèche canonique:

$$S_d \to \mathcal{Q}_d$$

Pour $1 \leq r \leq d$, on considère le fermé $F_r \subset \mathcal{Q}_d$ donné par:

$$a_0 = \cdots = a_{r-1} = 0.$$

On rappelle que S_d est donné par une infinité d'équations:

$$(v_0 + \cdots + v_{d-1}t^{d-1}) + (a_0 + a_1 t + \cdots + a_{d-1}t^{d-1} + t^d)\xi = 0,$$

avec $\xi = \sum \xi_i t^i$. En particulier, l'image inverse dans S_d de F_r est donné par les équations:

$$v_0 = \cdots = v_{r-1} = 0, t^r(v_r(t) + q_r\xi) = 0,$$

avec $v_r(t) = v_r + \cdots + v_{d-1}t^{d-r}$, $q_r = a_r + \cdots + a_{d-1}t^{d-1-r} + t^{d-r}$ et $\xi = \sum \xi_i t^i$. On obtient ainsi un isomorphisme canonique:

$$S_d^{(r)} \simeq S_{d-r}$$

et un diagramme:

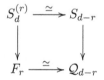

En particulier, par récurrence, au-dessus de l'ouvert H_r de F_r, S_d est un k-schéma de type fini. En particulier, pour tout produit fini $S_d \times_{\mathcal{Q}_d} \cdots \times_{\mathcal{Q}_d} S_d$, on obtient au-dessus de la stratification H_r de $\mathcal{Q}_d$, des k-schémas de type fini. Il en est donc de même pour tout schéma de type (S) comme ils sont de présentation finie sur de tels produits. □

5.2. Anneaux décents

DÉFINITION 5.8. Soit A un anneau commutatif, il est décent si:

$$\mathrm{nil}_\infty(A) := \bigcap_{n \geq 0} \bigcap_{\mathfrak{p} \in \mathrm{Spec}(A)} \mathfrak{p}^{[n]} = \{0\}.$$

où $\mathfrak{p}^{[n]} := \mathrm{Ker}(A \to A_\mathfrak{p}/\mathfrak{p}^n A_\mathfrak{p})$ est la n-ième puissance symbolique de $\mathfrak{p}$. On a $\mathfrak{p} = \mathfrak{p}^{[1]}$ et $\mathfrak{p}^n \subset \mathfrak{p}^{[n]}$, mais l'inégalité est stricte en général ([**29**, Tag. 05G9]), même pour des k-algèbres de type fini.

Remarques:

(1) Si A est réduit, il est décent, car $\mathrm{nil}(A) = \bigcap_{\mathfrak{p} \in \mathrm{Spec}(A)} \mathfrak{p} = \{0\}$.

(2) Si A est local noethérien, il est décent par le théorème d'intersection de Krull ([**29**, Tag. 00IP]). On obtient ainsi qu'un anneau noethérien est décent car d'après [**29**, Tag. 00L9], on a une injection dans un produit fini:

$$A \to \prod_{\mathfrak{p} \in \mathrm{Ass}(A)} A_\mathfrak{p},$$

où $\mathrm{Ass}(A)$ désigne l'ensemble, fini car A est noethérien, des idéaux premiers minimaux.

(3) $A = \bigcup_{n \geq 0} k[[t^{1/n}]]$ et $B = A/tA$, alors l'idéal maximal $\mathfrak{m}$ de B vérifie $\mathfrak{m} = \mathfrak{m}^2$ et $\mathfrak{m} \subset \mathrm{nil}_\infty(A)$.

On a $x \in \mathrm{nil}_\infty(A)$ si pour tout $\mathfrak{p} \in \mathrm{Spec}(A)$, x est dans le noyau de:

$$A \to A_\mathfrak{p}/\mathfrak{p}^n A_\mathfrak{p}$$

pour tout n. On obtient que:

$$x \notin \mathrm{nil}_\infty(A) \iff \exists\, \nu : A \to R \text{ tel que } \nu(x) \neq 0 \text{ et } R \in \mathit{Inf}_k.$$

On s'intéresse à un autre phénomène qui apparaît dans les espaces d'arcs. Soit $X := \{(x, y) \in \mathbb{A}^2 \mid xy = 0\}$, on considère la flèche:

$$\phi : \mathcal{L}\mathbb{A}^1 \coprod \mathcal{L}\mathbb{A}^1 \to \mathcal{L}X$$

qui envoie $(x(t), y(t))$ sur $(x(t), 0)$ ou $(0, y(t))$, que l'on restreint en enlevant 0.

LEMME 5.9. *Le morphisme ϕ induit un isomorphisme sur les anneaux décents, mais pas en général.*

DÉMONSTRATION. Si $A \in Inf_k$, comme $x(t)$ est non-dégénéré, $x(t) \in A[[t]] \cap A((t))^\times$, car sa réduction sur le corps résiduel est non-nulle. En particulier, la multiplication par x sur $A((t))$ est injective et on a une injection $A[[t]] \to A((t))$. En général, si A est décent et x non-dégénéré, soit $y \in A[[t]]$ tel que $xy = 0$, alors pour tout anneau local $R \in Inf_k$, on a $\nu(y) = 0$ avec $\nu : A \to R$, donc $y = 0$ comme A est décent. $\square$

Ce n'est pas un isomorphisme car il suffit de prendre l'exemple universel:
$$(t + a)\left(\sum \xi_i t^i\right) = 0.$$
La notion d'anneau décent se recolle et pour tout schéma Z on a une notion de schéma décent Z_{dec} où l'on conserve le même espace topologique et où le faisceau des fonctions est $\mathcal{O}_Z/\mathrm{nil}_\infty(\mathcal{O}_Z)$.

DÉFINITION 5.10. Une immersion fermée $i : S_0 \to S$ est dite décente si le faisceau d'idéaux $\mathcal{I}$ qui la définit est contenu dans $\mathrm{nil}_\infty(\mathcal{O}_S)$.

Remarques:
(1) Les immersions décentes sont stables par changement de base.
(2) De plus, toute immersion décente est un isomorphisme formel.
(3) En revanche, si l'on prend l'immersion nilpotente $S_{red} \to S$, elle n'est pas en général un isomorphisme formel.

6. Énoncés Principaux

On considère un k-schéma de type fini X réduit et pur. On dispose de l'ouvert $\mathcal{L}^\bullet X := \mathcal{L}X - \mathcal{L}X_{sing}$, comme c'est un schéma qui n'est pas quasi-compact, on va considérer des ouverts plus petits. Pour tout entier $d \in \mathbb{N}$, on peut alors considérer l'ouvert
$$\mathcal{L}X^{\leq d} := ev_d^{-1}(\mathcal{L}_d X - \mathcal{L}_d X_{sing})$$
avec $ev_d : \mathcal{L}X \to \mathcal{L}_d X$. C'est un schéma quasi-compact quasi-séparé, comme image inverse d'un schéma de type fini par un morphisme affine. Le résultat de structure est le suivant [4, Theorem 1.25]:

THÉORÈME 6.1. *Soit $d \in \mathbb{N}$, alors il existe un morphisme de schémas $f : Z \to \mathcal{L}X^{\leq d}$ avec Z quasi-compact quasi-séparé tel que:*
(1) *f est Weierstrass, affine et surjectif.*
(2) *$Z_{dec} \simeq T \times \mathbb{A}^{\mathbb{N}}$ avec T un k-schéma de type (S).*
On appelle un tel Z un atlas formel.

De plus, si l'on stratifie l'espace d'arcs, il admet une vraie structure produit [4, Théorème 1.29]:

Théorème 6.2. *Le schéma* $\mathcal{L}X^{\leq d}$ *admet une stratification finie constructible par des schémas isomorphes à* $Y \times \mathbb{A}^{\mathbb{N}}$ *pour un k-schéma de type fini* Y.

Remarque: Le théorème 6.2 est en fait un corollaire immédiat de 6.1, en utilisant la proposition 5.7.

Un résultat analogue pour des familles d'espaces d'arcs au-dessus de $\mathbb{A}^1$ a été obtenu par Hauser–Woblistin [**19**]. Ils montrent un résultat plus précis, à savoir que $\mathcal{L}X^{=d}$ est déjà un produit.

Cela peut se déduire également du théorème 6.1. En effet, le schéma T de type (S) a un morphisme canonique vers $\mathcal{Q}_d$ et ce morphisme contrôle précisément l'ordre de la singularité.

En particulier, l'assertion que $\mathcal{L}X^{=d}$ est un produit de la forme $H \times \mathbb{A}^{\mathbb{N}}$ se ramène à voir que $T^{=d} := \psi^{-1}(H_d)$ est un k-schéma de type fini et cela est précisément 5.7.

Pour déduire complétement l'énoncé, il faut également prendre en compte l'immersion décente. En reprenant le cours de la preuve de [**4**], on constate que l'immersion décente est un isomorphisme strates par strates.

Il ne s'agit pas de donner ici une preuve complète de ces deux résultats. Nous les montrons dans le cas du cône quadratique, qui fait déjà apparaître quasiment toutes les étapes de la preuve dans le cas général. Soit donc $X := \{(x,y,z) \in \mathbb{A}^3 \mid xy = z^2\}$. On a la projection sur la première coordonnée:

$$\mathcal{L}X \to \mathcal{L}\mathbb{A}^1$$

et on considère l'ouvert $\mathcal{L}X^{\leq 2} := (p_1)^{-1}(\mathcal{L}\mathbb{A}^{\leq 2})$. Montrons le théorème 6.1 pour cet ouvert. Soit $M := \mathcal{Q}_2 \times A_2 \times \mathcal{L}(\mathbb{G}_m \times \mathbb{A}^2)$, on commence par former le carré cartésien:

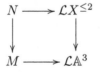

où la flèche horizontale est donnée par:

$$(q, \bar{z}, u, y, \xi) \mapsto (qu, y, \bar{z} + q\xi).$$

Le schéma N classifie les quintuplets $(q, \bar{z}, u, \xi)$ tels que:

$$quy = \bar{z}^2 + qH(q, \bar{z}, \xi),$$

que l'on récrit en:

(6.3) $$(\bar{z}^2 - A_0 q) + tqA_1(t) = 0,$$

avec $A(t) = uy - H(q, \bar{z}, \xi)$ et $A(t) = A_0 + tA_1(t)$. On vérifie alors que N s'insère dans le carré cartésien:

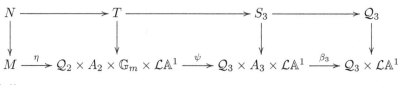

où l'on a:

$$\eta(q, \bar{z}, u, \xi) \quad := \quad (q, \bar{z}, u_0, A(t)),$$

$$\psi(q, \bar{z}, u_0, A(t)) \quad := \quad (tq, \bar{z}^2 - qA_0, A_1(t)).$$

avec $u = u_0 + tu_1$. On termine la preuve par le lemme suivant:

LEMME 6.4. *On a un isomorphisme $N \simeq T \times \mathbb{A}^{\mathbb{N}}$.*

DÉMONSTRATION. Par changement de base, il suffit de voir que c'est déjà le cas pour M et la flèche η. On a:

$$u = u_0 + tu_1,$$

$$A(t) = uy - H(q, \bar{z}, \xi) \Longrightarrow y = \frac{A + H}{u}.$$

En particulier, la donnée de u_1 et de ξ déterminent uniquement y et u, ce qu'on voulait. $\square$

Enfin, pour le théorème 6.2 dans ce cas, on a:

$$\mathcal{L}X^{=2} := \{(t^2 u(t), y, z) \in \mathcal{L}X, u \in \mathcal{L}\mathbb{G}_m\}.$$

On vérifie immédiatement que:

$$\mathcal{L}X^{=2} = D_2 \times \mathcal{L}(\mathbb{G}_m \times \mathbb{A}^1) = (D_2 \times \mathbb{G}_m) \times \mathcal{L}\mathbb{A}^2$$

avec $D_2 := \{(z_0, z_1) \in \mathbb{A}^2 \mid z_0^2 = 0, z_0 z_1 = 0\}$. De même, on a:

$$\mathcal{L}X^{=1} = (D_1 \times \mathbb{G}_m) \times \mathcal{L}\mathbb{A}^2$$

avec $D_1 = \{z_0^2 = 0\}$ et $\mathcal{L}X^{=0}$ est l'espace d'arcs d'un schéma lisse.

Remarque: Pour le cas du cône quadratique, cela ne fait pas apparaître d'immersion décente. En revanche, pour le cas général, l'exemple de l'espace d'arcs de $xy = 0$ montre que c'est indispensable.

7. Vers une Théorie des Faisceaux

Dans cette section, on explique comment le théorème 6.1 permet d'obtenir une «bonne» théorie des faisceaux constructibles sur les espaces d'arcs. Nous donnons ici quelques éléments de réponse, qui font partie d'un travail en cours de l'auteur avec D. Kazhdan.

Dans la suite, on choisit un entier n premier à la caractéristique de k. Dans un premier temps, une des raisons pour lesquelles la catégorie des morphismes pro-lisses est qu'on dispose d'un énoncé de changement de base [**22**, Exp. XIV, 2.5.3].

PROPOSITION 7.1. *On considère un carré cartésien de k-schémas qcqs:*

$$\begin{array}{ccc} T' & \xrightarrow{g} & T \\ {\scriptstyle f}\downarrow & & \downarrow{\scriptstyle f} \\ S' & \xrightarrow{g} & S \end{array}$$

avec g pro-lisse, alors pour tout faisceau étale K de $\mathbb{Z}/n$-modules, le morphisme de changement de base:

$$g^* Rf_* K \to Rf_* g^* K$$

est un isomorphisme.

Remarque: Dans loc.cit., l'énoncé est dans le cas noethérien, avec pro-lisse remplacé par régulier, mais la preuve vaut telle quelle dans ce contexte plus général et se ramène par des techniques de passage à la limite au changement de base lisse usuel.

On considère alors la catégorie des faisceaux constructibles $D_c^b(\mathcal{L}X^{\leq d}, \mathbb{Z}/n)$. Il s'agit de voir qu'elle est stable par le formalisme des six opérations. Le point clé pour obtenir un tel formalisme est un énoncé de finitude pour j_* où $j : U \to \mathcal{L}X^{\leq d}$ est une immersion ouverte quasi-compacte. En combinant le théorème 6.1 et l'énoncé de changement de base 7.1, on est donc ramené à montrer un énoncé de constructibilité pour un ouvert quasi-compact U de $T \times \mathbb{A}^{\mathbb{N}}$.

Maintenant, par descente noethérienne [**29**, Tag. 095M], on sait que pour une paire (K, U) avec U quasi-compact et $K \in D_c^b(U, \mathbb{Z}/n)$, on sait qu'il provient d'une paire (K_0, U_0) où $K_0 \in D_c^b(U_0, \mathbb{Z}/n)$ et $U_0 \hookrightarrow T \times \mathbb{A}^m$ pour un certain m. Ainsi, il s'agit donc de comprendre ce qu'il se passe pour les schémas de type (S).

Le premier gain est que contrairement à la situation pour les espaces d'arcs où l'on considérait des complétions par rapport à des idéaux qui ne sont pas de type fini. Une fois que l'on passe aux schémas de type (S), on complète par rapport à un idéal de **type fini**. On a vu que les schémas de type (S) sont des schémas non-noethériens dont les anneaux locaux formels sont noethériens. Pour démontrer l'énoncé de constructibilité, il suffit de comprendre du point de vue cohomologique le passage de l'hensélianisé au complété formel. C'est ici qu'apparaît de manière fondamentale le théorème suivant dû à Gabber [**22**, Exp. XX, 4.4] :

THÉORÈME 7.2. *Soit A un anneau commutatif, $I \subset A$ un idéal de type fini. On considère un morphisme $(A, I) \to (A', I')$ de paires henséliennes avec $I' = IA'$, de telle sorte que l'on a un isomorphisme au niveau des complétions I et I'-adiques:*

$$A^\wedge \simeq A'^\wedge.$$

$X = \mathrm{Spec}(A)$, $X' = \mathrm{Spec}(A')$, $U := \mathrm{Spec}(A) - V(I)$, $U' := \mathrm{Spec}(A') - V(I')$, $\pi : X' \to X$, $j : U \to X$, $j' : U' \to X'$, *alors pour tout faiseau de torsion K sur U, le morphisme de changement de base:*

$$\pi^* R^m j_* K \to R^m j'_* \pi^* K$$

est un isomorphisme pour tout m.

Remarque: Il est important de noter que l'on ne fait pas d'hypothèses sur A.

Voyons, comment on applique ce résultat dans le cas concret de S_1. Pour rappel, on a vu dans la section 5.1:

$$S_1 = \mathrm{Spec}(A)$$

avec $A = \varinjlim k[q, x_l]$ avec comme morphismes de transition $x_l \mapsto q x_{l+1}$. On a un morphisme pro-lisse surjectif:

$$p : S_1 \to \mathbb{A}^1.$$

On a $p^{-1}(\mathbb{G}_m) = \mathbb{G}_m \times \mathbb{A}^1$. Notons, $j : U := \mathbb{G}_m \times \mathbb{A}^1 \to S_1$. Son complémentaire F est de présentation finie dans S_1 et on a:

$$F = \mathrm{Spec}(A/(q)) \simeq \mathrm{Spec}(k).$$

L'idéal $\mathfrak{p}_0 = (q)$ est premier et la complétion en cet idéal de A s'identifie à $k[[q]]$.

PROPOSITION 7.3. *Soit K un faisceau constructible de $\mathbb{Z}/n$-modules sur U, alors $Rj_* K$ est constructible.*

DÉMONSTRATION. Il suffit de calculer la fibre au point 0 de $Rj_* K$. Notons $\tilde{A}$ l'hensélisé en 0 de S_1, $\tilde{j} : \tilde{U} := U \times_{S_1} \mathrm{Spec}(\tilde{A}) \to \mathrm{Spec}(A)$. Par changement de base à $\tilde{A}$, il suffit de prouver la constructibilité de $R\tilde{j}_* K$ avec K un faisceau sur $\tilde{U}$. On considère le carré cartésien:

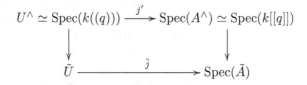

Comme l'idéal $\mathfrak{p}_0$ est de type fini, on peut appliquer le changement de base de Gabber 7.2 et on est ramené à prouver la constructibilité pour l'immersion ouverte:

$$\mathrm{Spec}(k((q))) \to \mathrm{Spec}(k[[q]]),$$

ce qui fait à nouveau l'objet d'un théorème de Gabber [22, Exp. 0, Théorème 1]. □

Le cas général s'obtient de manière analogue et nous reviendrons là-dessus dans un futur travail.

Bibliographie

[1] M. André, *Homologie des Algèbres Commutatives*. Grundlehren der Math. Wiss., Vol. 206. Springer, Berlin, 1974.

[2] M. Artin, Algebraic approximation of structures over complete local rings. *Publ. Math. IHES*, 36:23–58, 1969.

[3] A. Beauville Y. Laszlo, Un lemme de descente. *C. R. Acad. Sci. Paris*, 320:335–340, 1995.

[4] A. Bouthier and D. Kazhdan, Faisceaux pervers sur les espaces d'arcs. Preprint, 2015, arXiv:1509.02203 (v5).

[5] B. Bhatt, Algebraization and Tannaka duality. *Cambridge J. Math.*, 4(4):403–461, 2016.

[6] B. Bhatt, An imperfect ring with a trivial cotangent complex. http://www-personal.umich.edu/~bhattb/math/trivial-cc.pdf.

[7] N. Bourbaki, *Algèbre Commutative*, Chapter VII. Hermann, Paris, 1972.

[8] D. Bourqui and J. Sebag, Drinfeld–Grinberg–Kazhdan's theorem is false for singular arcs. *J. de l'IMJ*, 16(4):879–885, 2017.

[9] D. Bourqui and J. Sebag, The Drinfeld–Grinberg–Kazhdan theorem for formal schemes and singularity theory. *Confluentes Math.*, 1:29–64, 2017.

[10] A. Bouthier, B. C. Ngô and Y. Sakellaridis, On the formal arc space of a reductive monoid (with Erratum). *Amer. J. Math.*, 138:81–109, 2016.

[11] M. Brion and S. Kumar, *Frobenius Splitting Methods in Geometry and Representation Theory*. Progress in Mathematics, Birkhäuser, Boston, 2005.

[12] V. Drinfeld, On the Grinberg–Kazhdan formal arc theorem. Preprint, 2002, arXiv:math-AG/0203263.

[13] V. Drinfeld, Grinberg–Kazhdan theorem and Newton groupoids. Preprint, 2002, arXiv:math-AG/0203263.

[14] K. Fujiwara, O. Gabber and F. Kato, On Hausdorff completions of commutative rings in rigid geometry. *J. Algebra*, 332:293–321, 2011.

[15] M. Grinberg and D. Kazhdan, Versal deformations of formal arcs. *Geom. Funct. Anal.*, 10(3):543–555, 2000.

[16] M. Haiech, On noncomplete completions. In *A paraître aux Proc. Conf. Arc Schemes and Singularity Theory*.

[17] H. Hauser and C. Chiu, On the formal neighbourhood of degenerated arcs. https://homepage.univie.ac.at/herwig.hauser/.

[18] H. Hauser and G. Rond, Artin approximation. https://homepage.univie.ac.at/herwig.hauser/.

[19] H. Hauser and S. Woblistin, On the structure of varieties of power series in one variable. Preprint, https://homepage.univie.ac.at/herwig.hauser/.

[20] E. Kolchin, *Differential Algebra and Algebraic Groups*. Academic Press, New York, 1973.

[21] L. Illusie, *Complexe Cotangent et Déformations I et II*. Lecture Notes in Mathematics, Vols. 239–283. Springer, Berlin, 1972.

[22] L. Illusie, Y. Laszlo and F. Orgogozo, Travaux de Gabber sur l'uniformisation locale et la cohomologie étale des schémas quasi-excellents. *Astérisque*, 363–364: 2014.

[23] M. Mustata, avec un appendice de D. Eisenbud, E. Frenkel, Jet schemes of locally complete intersections canonical singularities. *Inventiones*, 145:397–424, 2001.

[24] B.C. Ngô, Weierstrass preparation theorem and singularities in the space of non-degenerated arcs. Preprint, 2017, arXiv:1706.05926.

[25] M. Olsson, Hom-stacks and restriction of scalars. *Duke Math. J.*, 134:139–164, 2006.

[26] D. Popescu, General Néron desingularization. *Nagoya Math. J.*, 100:97–126, 1985.

[27] D. Popescu, General Néron desingularization and approximation. *Nagoya Math. J.*, 104:85–115, 1986.

[28] D. Popescu, Letter to the editor: General Néron desingularization and approximation. *Nagoya Math. J.*, 118:45–53, 1990.

[29] Stack Project, http://stacks.math.columbia.edu/.

[30] A. Yekutieli, On flatness and completions of infinitely generated modules over noetherian rings. *Comm. Algebra*, 39(11):4221–4245, 2011.

9

Partition Identities and Application to Infinite-Dimensional Gröbner Basis and Vice Versa

Pooneh Afsharijoo* and Hussein Mourtada[†]

*Equipe Géométrie et Dynamique, Institut Mathématique de Jussieu-Paris
Rive Gauche, Université Paris Diderot, Bâtiment Sophie Germain,
case 7012, 75205 Paris Cedex 13, France
*pooneh.afsharijoo@imj-prg.fr
†hussein.mourtada@imj-prg.fr

In the first part of this chapter, we consider a Gröbner basis of the differential ideal $[x_1^2]$ with respect to the weighted lexicographical monomial order and show that its computation is related to an identity involving the partitions that appear in the first Rogers–Ramanujan identity. We then prove that a Gröbner basis of this ideal is not differentially finite in contrary to the case of the weighted reverse lexicographical order. In the second part, we give a simple and direct proof of a theorem of Nguyen Duc Tam about the Gröbner basis of the differential ideal $[x_1 y_1]$; we then obtain identities involving partitions with two colors.

1. Introduction

An integer partition of a positive integer number n is a decreasing sequence of positive integers

$$\lambda = (\lambda_1 \geq \lambda_2 \geq \cdots \geq \lambda_l),$$

such that $\lambda_1 + \lambda_2 + \cdots + \lambda_l = n$. The λ_i's are called the parts of λ and l is its size. The book [A] is a classic in the theory of integer partitions. A famous identity related to partitions and which plays an important role in this chapter is the **First Rogers–Ramanujan Identity**:

The number of partitions of n with neither consecutive parts, nor equal parts is equal to the number of partitions of n whose parts are congruent to 1 or 4 modulo 5.

In [3] (see also [4]), we obtained this identity by considering the space of arcs of the double point $X = \mathrm{Spec}(\mathbf{K}[x]/x^2)$ centered at the origin and denoted by X_∞^0. The coordinate ring A of X_∞^0 is naturally graded and we associate with it its Hilbert–Poincaré series that we call the Arc Hilbert–Poincaré series; we denote it by $\mathrm{AHP}_{X,0}(t)$. Note that this is an invariant of singularities of algebraic varieties [9, 3]. We prove in [3] that $\mathrm{AHP}_{X,0}(t)$ is equal to the generating sequence of the number of partitions appearing in the Rogers–Ramanujan identities. The proof uses a Gröbner basis computation associated with a monomial ordering (reverse lexicographical). Note that we have

$$A = \frac{\mathbf{K}[x_i, i \in \mathbf{Z}_{>0}]}{[x_1^2]},$$

where $[x_1^2]$ is the differential ideal generated by x_1^2 and its iterated derivative by the derivation D which is determined by $D(x_i) = x_{i+1}$. So,

$$[x_1^2] = (x_1^2, 2x_1x_2, 2x_1x_3 + 2x_2^2, \ldots).$$

The grading of A is determined by giving to x_i the weight i.

In Section 1, we consider a different monomial ordering (lexicographical) and we find that the Gröbner basis computation with respect to this monomial ordering is related with another identity involving the partitions that appear in the first Rogers–Ramanujan identity. We then use this other member in the Rogers–Ramanujan identity to carry the computations of the Gröbner basis in small degrees (usual degree) but in all weights. This leads us to prove, in contrast to the case of the reverse lexicographical ordering, that a Gröbner basis of our ideal with respect to the lexicographical ordering is not differentially finite.

In Section 2, we will give a direct and simpler proof of a theorem by Nguyen Duc Tam [11], where he computes a Gröbner basis of the arc space of $X = \mathrm{Spec}(\mathbf{K}[x, y]/(xy))$ or simply of the differential ideal generated by xy. We then use this theorem to obtain identities of partitions with two colors.

2. Hilbert Series and Integer Partitions

In this section, we will begin by considering the Hilbert–Poincaré series of some graded algebras that are inspired by Gröbner basis computations; we then interpret these series as generating sequences of partitions having special properties. At the end of the section, we will give more explanations on how we have found these graded algebras.

We will denote by $\mathbf{K}$ an algebraically closed field of characteristic 0. Recall that for a graded $\mathbf{K}$-algebras, $A = \oplus_{i \in \mathbf{N}} A_i$ (we assume that

$\dim_{\mathbf{K}}(A_i) < \infty$) the Hilbert–Poincaré series of A, that we denote by $\mathrm{HP}(A)$, is by definition

$$\mathrm{HP}(A) = \sum_{i \in \mathbf{N}} \dim_{\mathbf{K}}(A_i) q^i,$$

where q is a variable. For more details about Hilbert–Poincaré series, see the appendix in [3] and the references there.

Let $n, l \geq 1$ be integer numbers. We consider the graded algebra $\mathbf{K}[x_l, x_{l+1}, \ldots]$, where we give x_i the weight i, for $i \geq l$. This grading induces a grading on the $\mathbf{K}$-algebra $\frac{\mathbf{K}[x_l, x_{l+1}, \ldots]}{(x_{i_1} \cdots x_{i_n}, \; i_j \geq l)}$.

We consider the Hilbert–Poincaré series

$$H_n^l = \mathrm{HP}\left(\frac{\mathbf{K}[x_l, x_{l+1}, \ldots]}{(x_{i_1} \cdots x_{i_n}, \; i_j \geq l)}\right).$$

LEMMA 2.1. We have

$$H_n^l = q^l H_{n-1}^l + H_n^{l+1}.$$

PROOF. Using Corollary 6.2 in [3], we have

$$H_n^l = \mathrm{HP}\left(\frac{\mathbf{K}[x_l, x_{l+1}, \ldots]}{(x_{i_1} \cdots x_{i_n}, \; i_j \geq l)}\right)$$

$$= \mathrm{HP}\left(\frac{\mathbf{K}[x_l, x_{l+1}, \ldots]}{(x_l, x_{i_1} \cdots x_{i_n}, \; i_j \geq l)}\right) + q^l \mathrm{HP}\left(\frac{\mathbf{K}[x_l, x_{l+1}, \ldots]}{((x_{i_1} \cdots x_{i_n}, \; i_j \geq l) : x_l)}\right)$$

$\square$

The last term is exactly $H_n^{l+1} + q^l H_{n-1}^l$.

The above lemma allows us to determine H_n^l.

PROPOSITION 2.2. *We have*

$$H_n^l = 1 + \frac{q^l}{1-q} + \frac{q^{2l}}{(1-q)(1-q^2)} + \cdots + \frac{q^{(n-1)l}}{(1-q)(1-q^2)\cdots(1-q^{n-1})}.$$

PROOF. The proof is by induction on the integer n. Note that for $n = 1$, $H_1^l = \mathrm{HP}(\mathbf{K}) = 1$. For $n = 2$, the weighted-homogeneous components of

$$\frac{\mathbf{K}[x_l, x_{l+1}, \ldots]}{(x_{i_1} x_{i_2}, \; i_j \geq l)}$$

are generated by 1 in degree 0 and x_i in degree i for $i \geq l$; for $i = 1, \ldots, l-1$, the weighted-homogeneous components of degree i is the null vector space. Let us assume that the formula is true for $H_j^l, j \leq n - 1$ and prove it for H_n^l. Using Lemma 2.1 repetitively, we obtain

$$H_n^l = q^l H_{n-1}^l + H_n^{l+1} = q^l H_{n-1}^l + q^{l+1} H_{n-1}^{l+1} + H_n^{l+2}$$

$$= \cdots = q^l H_{n-1}^l + q^{l+1} H_{n-1}^{l+1} + \cdots + q^m H_{n-1}^m + H_n^{m+1}.$$

But we have

$$\lim_{m \to \infty} H_n^m = 1,$$

where the limit is considered for the q-adic topology in $\mathbf{C}[[q]]$; hence, we can write

$$H_n^l = 1 + q^l H_{n-1}^l + q^{l+1} H_{n-1}^{l+1} + \cdots + q^m H_{n-1}^m + q^{m+1} H_{n-1}^{m+1} + \cdots$$

$$= 1 + \sum_{m \geq l} q^m H_{n-1}^m.$$

By the induction hypothesis, we obtain

$$H_n^l = 1 + q^l + \frac{q^{2l}}{1-q} + \frac{q^{3l}}{(1-q)(1-q^2)} + \cdots$$

$$+ \frac{q^{(n-1)l}}{(1-q)(1-q^2)\cdots(1-q^{n-2})}$$

$$+ q^{l+1} + \frac{q^{2(l+1)}}{1-q} + \frac{q^{3(l+1)}}{(1-q)(1-q^2)} + \cdots$$

$$+ \frac{q^{(n-1)(l+1)}}{(1-q)(1-q^2)\cdots(1-q^{n-2})} + \cdots$$

(summing by columns)

$$= 1 + (q^l + q^{l+1} + \cdots) + \left(\frac{q^{2l}}{1-q} + \frac{q^{2(l+1)}}{1-q} + \cdots \right)$$

$$+ \cdots + \left(\frac{q^{(n-1)l}}{(1-q)(1-q^2)\cdots(1-q^{n-2})} \right.$$

$$+ \left. \frac{q^{(n-1)(l+1)}}{(1-q)(1-q^2)\cdots(1-q^{n-2})} + \cdots \right).$$

The formula for H_n^l follows from the formula

$$q^{jl} + q^{j(l+1)} + q^{j(l+2)} + \cdots = \frac{q^{jl}}{1-q^j}, \ j = 1, \ldots, n-1. \qquad \square$$

COROLLARY 2.3. *Let $l, n \geq 1$ be integers. The generating series of the integer partitions with parts greater or equal to l and size (number of parts) less or equal to $n-1$ is*

$$1 + \frac{q^l}{1-q} + \frac{q^{2l}}{(1-q)(1-q^2)} + \cdots + \frac{q^{(n-1)l}}{(1-q)(1-q^2)\cdots(1-q^{n-1})}.$$

PROOF. This is just the combinatorial interpretation of Proposition 2.2: A basis of the ith weighted-homogeneous component of $\frac{\mathbf{K}[x_l, x_{l+1}, \ldots]}{(x_{i_1} \cdots x_{i_n}, \ i_j \geq l)}$ is given by the monomials $x_{h_1} \cdots x_{h_r}$ such that $h_1 + \cdots + h_r = i, h_j \geq l$ and $x_{h_1} \cdots x_{h_r} \notin (x_{i_1} \cdots x_{i_n}, \ i_j \geq l)$; this corresponds to the partitions of i with parts greater or equal to l and size (number of parts) less or equal to $n - 1$. $\qquad \square$

We now look at the Hilbert series that makes the link to the first Rogers–Ramanujan identity. For $j \geq 1$, let

$$(2.4) \qquad \mathrm{HP}\left(\frac{\mathbf{K}[x_j, x_{j+1}, \ldots]}{(x_{i_1} \cdots x_{i_k} x_k, \ i_1 \geq i_2 \geq \cdots \geq i_k \geq k \geq j)} \right).$$

LEMMA 2.5. We have

$$H_j = q^j H_j^j + H_{j+1}.$$

PROOF. Using Corollary 6.2 in [3], we have

$$H_j = \mathrm{HP}\left(\frac{\mathbf{K}[x_j, x_{j+1}, \ldots]}{(x_{i_1} \cdots x_{i_k} x_k, \ i_1 \geq i_2 \geq \cdots \geq i_k \geq k \geq j)} \right)$$

$$= \mathrm{HP}\left(\frac{\mathbf{K}[x_j, x_{j+1}, \ldots]}{(x_j, x_{i_1} \cdots x_{i_k} x_k, \ i_1 \geq i_2 \geq \cdots \geq i_k \geq k \geq j)} \right)$$

$$+ q^j \mathrm{HP}\left(\frac{\mathbf{K}[x_j, x_{j+1}, \ldots]}{((x_{i_1} \cdots x_{i_k} x_k, \ i_1 \geq i_2 \geq \ldots \geq i_k \geq k \geq j) : x_j)} \right).$$

The last term is exactly $H_{j+1} + q^j H_j^j$. $\qquad \square$

THEOREM 2.6. *We have*

$$H_1 = 1 + \frac{q}{1-q} + \frac{q^4}{(1-q)(1-q^2)} + \frac{q^9}{(1-q)(1-q^2)(1-q^3)} + \cdots$$

$$= 1 + \sum_{n \geq 1} \frac{q^{n^2}}{(1-q)(1-q^2) \cdots (1-q^n)}.$$

PROOF. Applying Lemma 2.5 for $j = 1$, we obtain that $H_1 = qH_1^1 + H_2$; applying the same lemma for $j = 2$ we have that $H_1 = qH_1^1 + q^2 H_2^2 + H_3$; Applying repetitively and in the same way Lemma 2.5, we obtain that for $m \geq 2$,

$$H_1 = qH_1^1 + q^2 H_2^2 + \cdots + q^m H_m^m + H_{m+1}.$$

Noticing that

$$\lim_{m \to \infty} H_m = 1$$

(where the limit is considered for the q-adic topology in $\mathbf{C}[[q]]$, $\mathbf{C}$ being the field of complex numbers) we can write

$$H_1 = 1 + qH_1^1 + q^2 H_2^2 + \cdots + q^m H_m^m + \cdots .$$

Using Proposition 2.2, we obtain that H_1 is equal to

$$1 \qquad +q$$

$$+q^2 \qquad +\frac{q^4}{1-q}$$

$$+q^3 \qquad +\frac{q^6}{1-q} \qquad + \frac{q^9}{(1-q)(1-q^2)}$$

$$+q^4 \qquad +\frac{q^8}{1-q} \qquad + \frac{q^{12}}{(1-q)(1-q^2)} \qquad + \cdots$$

Summing by columns we obtain the result. □

An equivalent statement of the theorem is the following.

THEOREM 2.7. *Let $n \geq 1$ be a positive integer. The number of partitions of n with size less than or equal to the smallest part is equal to the number of partitions of n without consecutive nor equal parts.*

PROOF. This follows from the known fact (see [2, 3]) that the series obtained in Theorem 2.6 is also the generating sequence of the partitions without consecutive or equal parts. □

It was pointed to us by Jan Schepers that the theorem [13] is mentioned in [13] but without a clear reference. We have rediscovered this theorem from Gröbner basis computations (see the explanations after Theorem 2.8); our point of view allows us to obtain the following family of identities (indexed by an integer number k), which are generalizations of Theorem 2.7.

THEOREM 2.8. *Let $n \geq k$ be a positive integer. The number of partitions of n with parts larger or equal to k and size less than or equal to (the smallest part minus $k-1$) is equal to the number of partitions of n with parts larger or equal to k and without consecutive nor equal parts.*

PROOF. We denote by F_k the Hilbert series of

$$\frac{\mathbf{K}[x_k, x_{k+1}, \ldots]}{(x_{i_1} \cdots x_{i_{j-k+1}} x_j, \ i_1 \geq i_2 \geq \cdots \geq i_{j-k+1} \geq j \geq k)}.$$

It follows from similar computations to those in the proof of Theorem 2.6 that

$$F_k = 1 + q^k H_1^k + q^{k+1} H_2^{k+1} + \cdots + q^{k+m} H_{m+1}^{k+m} + \cdots .$$

By Proposition 2.2, we have

$$F_k = 1 \quad +q^k$$

$$+q^{k+1} \quad +\frac{q^{2k+2}}{1-q}$$

$$+q^{k+2} \quad +\frac{q^{2k+4}}{1-q} \quad +\frac{q^{3k+6}}{(1-q)(1-q^2)}$$

$$+q^{k+3} \quad +\frac{q^{2k+6}}{1-q} \quad +\frac{q^{3k+9}}{(1-q)(1-q^2)} \quad + \cdots.$$

Summing by columns, we obtain that

$$F_k = 1 + \sum_{n \geq 1} \frac{q^{n(n+k-1)}}{(1-q)(1-q^2)\cdots(1-q^n)}.$$

This implies that

$$F_{k+1} + q^k F_{k+2}$$

$$= \quad 1 \qquad +\frac{q^{k+1}}{1-q} \quad +\frac{q^{2k+4}}{(1-q)(1-q^2)} \quad +\frac{q^{3k+9}}{(1-q)(1-q^2)(1-q^3)} \quad + \cdots$$

$$+q^k \quad +\frac{q^{2k+2}}{1-q} \quad +\frac{q^{3k+6}}{(1-q)(1-q^2)} \quad + \cdots.$$

Again summing by columns, we obtain that

$$F_{k+1} + q^k F_{k+2} = 1 + \sum_{n \geq 1} \frac{q^{n(n+k-1)}}{(1-q)(1-q^2)\cdots(1-q^n)} = F_k.$$

It follows from Proposition 5.8 in [**3**] (with a shift in the indices) that for $k \geq 1$, there exist $A_j, B_j \in \mathbf{C}[[q]]$ such that

$$F_k = A_{k+i}F_{k+i} + B_{k+i+1}F_{k+i+1},$$

$A_{k+1} = 1, B_{k+2} = q^k$, and for all $i \geq 2$, we have

$$A_{k+i} = A_{k+i-1} + B_{k+i},$$

$$B_{k+i+1} = q^{k+i-1}A_{k+i-1}.$$

Denote now by H'_k the Hilbert series of

$$\frac{\mathbf{K}[x_k, x_{k+1}, \ldots]}{(x_i^2, x_{i+1}^2, i \geq k)}.$$

By equation (7) in [**3**], H'_k satisfies the same recursion formula

$$H'_k = A_{k+i}H'_{k+i} + B_{k+i+1}H'_{k+i+1}.$$

Since $\lim B_j = 0$ and $\lim F_j = \lim H'_j = 1$ (in the (q)-adic topology), we have

$$F_k = \lim \left(A_{k+i} F_{k+i} + B_{k+i+1} F_{k+i+1} \right) = \lim A_{k+i}$$

$$= \lim \left(A_{k+i} H'_{k+i} + B_{k+i+1} H'_{k+i+1} \right) = H'_k.$$

Noticing that F_k is the generating series of the number of partitions with parts larger or equal to k and size less than or equal to the smallest part minus $k-1$ and that H'_k is the generating series of the number of partitions with parts larger or equal to k and without consecutive nor equal parts, we obtain the result in the theorem. $\square$

Theorem 2.7 is inspired from a Gröbner basis computation of the differential ideal $[x_1^2]$: By [3], the initial ideal of $[x_1^2]$ with respect to the reverse lexicographical ordering is $(x_i^2, x_i x_{i+1}, i \geq 1)$, while we can guess that its initial ideal with respect to the lexicographical ordering is $((x_{i_1} \cdots x_{i_k} x_k, i_1 \geq i_2 \geq \cdots \geq i_k \geq k \geq 1)$ (we make use of this guess in Section 2). Hence, the Hilbert series of the quotient rings of $\mathbf{K}[x_1, x_2, \ldots]$ by these ideals are equal. The Hilbert series of the quotient by $(x_i^2, x_i x_{i+1}, i \geq 1)$ is the generating series of the partitions without consecutive nor equal parts; The Hilbert series of the quotient by $(x_{i_1} \cdots x_{i_k} x_k, i_1 \geq i_2 \geq \cdots \geq i_k \geq k \geq 1)$ is the generating series of the partitions with size less or equal to the smallest part. Theorem 2.8 can be guessed in the same way by considering the ideal $[x_k^2]$ in $\mathbf{K}[x_k, x_{k+1}, \ldots]$.

3. The Lex Gröbner Basis of $[x_1^2]$

Again, let $\mathbf{K}$ be a field of characteristic 0, and consider the graded ring $\mathbf{K}[x_1, x_2, \ldots]$, where the weight of x_i is i. So, the weight of the monomial $x^\alpha := x_{i_1}^{\alpha_1} x_{i_2}^{\alpha_2} \cdots x_{i_n}^{\alpha_n}$ is equal to $\sum_{j=1}^n i_j \alpha_j$, and its (usual) degree is equal to n.

Let $f_2 = x_1^2$ and for $i \geq 3, f_i = D^{i-2}(f_2) := D(f_{i-1})$, where D is the derivation determined by $D(x_i) = x_{i+1}$; then $I = [f_2] := (f_2, f_3, \ldots) \subset \mathbf{K}[x_1, x_2, \ldots]$, is the defining ideal (up to isomorphism) of the space of arcs centered at the origin of $X = \mathrm{Spec}(\mathbf{K}[x]/(x^2))$. The ideal I is a differential ideal, i.e., we have $D(I) \subset I$. We are interested in the possibility that I has a differentially finite Gröbner basis with respect to a monomial ordering; see the following definition.

DEFINITION 3.1. Let $J \subset \mathbf{K}[x_1, x_2, \ldots]$ be a differential ideal with respect to D (i.e., $D(J) \subset J$). Let "$<$" be a total monomial order defined on $\mathbf{K}[x_1, x_2, \ldots]$. We say that J has a differentially finite Gröbner basis with respect to "$<$", if there exists a finite number of polynomials $h_1, \ldots, h_r \in \mathbf{K}[x_1, x_2, \ldots]$ such that $J = [h_1, \ldots, h_r]$ and the initial ideal $\mathrm{In}_<(J)$ of J with respect to "$<$" satisfies

$$\mathrm{In}_<(J) = (\mathrm{In}_<(D^i(h_j)), j = 1, \ldots, r; i \geq 0),$$

where D^i denotes the ith iterated derivative and D^0 is the identity.

Note that the notation $[h_1, \ldots, h_r]$ in the definition denotes the differential ideal generated by the h_i, $i = 1, \ldots, r$ and by all their iterated derivatives. Note that there might be different notions of differential Gröbner basis, see [**5, 12**] and their bibliography.

In this section, we prove that no Gröbner basis of I with respect to the weighted lexicographical order is differentially finite. Note that, contrary to this case, it follows for [**3**] that in the case of the weighted reverse lexicographical order, I has a differentially finite Gröbner basis.

We denote the n-th derivative of a polynomial f_i by $f_i^{(n)}$, so we have

$$f_n = f_2^{(n-2)} = (x_1^2)^{(n-2)} = \sum_{i=0}^{n-2} \binom{n-2}{i} x_1^{(i)} x_1^{(n-2-i)}$$

$$= \sum_{i=0}^{n-2} \binom{n-2}{i} x_{1+i} x_{n-i-1}.$$

Denote the leading term of f_n with respect to the weighted lexicographical order by $\mathrm{LT}(f_n)$. So, $\mathrm{LT}(f_n) = 2x_1 x_{n-1}$ for all $n \geq 2$.

Recall that the S-polynomial of $f, g \in \mathbf{K}[x_1, x_2, \ldots]$ is by definition

$$S(f, g) := \frac{x^\gamma}{\mathrm{LT}(f)} f - \frac{x^\gamma}{\mathrm{LT}(g)} g,$$

where x^γ is the least common multiple of the leading monomials of f and g. A possible reference about S-polynomials and Gröbner basis is [**8**].

A direct computation of the S-polynomial of f_3 and f_4 gives

$$S(f_3, f_4) = x_2^3.$$

We set $F_{x_2^3} := S(f_3, f_4)$. For $k > 2$, we recursively define

$$F_{x_2 x_k^2} := S(F_{x_2 x_{k-1}^2}^{(2)}, S(f_{k+1}, f_{k+2})).$$

We then have the following lemma.

LEMMA 3.2. *With respect to the weighted lexicographic order, for $k > 2$, the leading monomial of $F_{x_2 x_k^2}$ is $x_2 x_k^2$.*

PROOF. The proof is by induction on the integer k. Note that for $k = 3$, we have $F_{x_2^3}^{(1)} = 3x_2^2 x_3$, $F_{x_2^3}^{(2)} = 3x_2^2 x_4 + 6x_2 x_3^2$ and $S(f_4, f_5) = x_2^2 x_4 - 3x_2 x_3^2$. So, we have

$$S(F_{x_2^3}^{(2)}, S(f_4, f_5)) = 5x_2 x_3^2 := F_{x_2 x_3^2}.$$

For $k = 4$, we have $F^{(1)}_{x_2 x_3^2} = 10 x_2 x_3 x_4 + 5 x_3^3$, $F^{(2)}_{x_2 x_3^2} = 10 x_2 x_3 x_5 + 10 x_2 x_4^2 + 25 x_3^2 x_4$ and $S(f_5, f_6) = 3 x_2 x_3 x_5 - 4 x_2 x_4^2 - 3 x_3^2 x_4$. So,

$$S(F^{(2)}_{x_2 x_3^2}, S(f_5, f_6)) = \frac{7}{3} x_2 x_4^2 + \frac{7}{2} x_3^2 x_4 := F_{x_2 x_4^2}.$$

Now, assume that the claim holds for $k - 1 \geq 4$. This means that for $k - 1 \geq 4$, we have

$$F_{x_2 x_{k-1}^2} := S(F^{(2)}_{x_2 x_{k-2}^2}, S(f_k, f_{k+1})).$$

Since the leading monomial of $F_{x_2 x_{k-1}^2}$ is $x_2 x_{k-1}^2$, we can assume that $F_{x_2 x_{k-1}^2} = a x_2 x_{k-1}^2 + g_3$ for some rational number a and some polynomial g_3 with the monomials of the form $x_{i_1} x_{i_2} x_{i_3}$ such that $3 \leq i_1 \leq i_2 \leq i_3$. So, on the one hand, the second derivative of $F_{x_2 x_{k-1}^2}$ will be as follows:

$$F^{(2)}_{x_2 x_{k-1}^2} = 2a x_2 x_{k-1} x_{k+1} + 2a x_2 x_k^2 + h_3,$$

where $h_3 = 4a x_3 x_{k-1} x_k + a x_4 x_{k-1}^2 + g_3^{(2)}$. On the other hand, we have

$$S(f_{k+1}, f_{k+2}) = S\left(\sum_{i=0}^{k-1} \binom{k-1}{i} x_{1+i} x_{k-i}, \sum_{i=0}^{k} \binom{k}{i} x_{1+i} x_{k+1-i} \right)$$

$$= \frac{1}{2} \sum_{i=0}^{k-1} \binom{k-1}{i} x_{1+i} x_{k-i} x_{k+1} - \frac{1}{2} \sum_{i=0}^{k} \binom{k}{i} x_{1+i} x_{k+1-i} x_k$$

$$= \frac{1}{2} \sum_{i=1}^{k-2} \binom{k-1}{i} x_{1+i} x_{k-i} x_{k+1} - \frac{1}{2} \sum_{i=1}^{k-1} \binom{k}{i} x_{1+i} x_{k+1-i} x_k.$$

So, by the above equation, we obtain that $\mathrm{LT}(S(f_{k+1}, f_{k+2})) = (k-1) x_2 x_{k-1} x_{k+1}$. Now, we can compute $S(F^{(2)}_{x_2 x_{k-1}^2}, S(f_{k+1}, f_{k+2}))$.

$$S(F^{(2)}_{x_2 x_{k-1}^2}, S(f_{k+1}, f_{k+2}))$$

$$= \frac{1}{2a}(2a x_2 x_{k-1} x_{k+1} + 2a x_2 x_k^2 + h_3)$$

$$- \frac{1}{(k-1)} \left(\frac{1}{2} \sum_{i=1}^{k-2} \binom{k-1}{i} x_{1+i} x_{k-i} x_{k+1} - \frac{1}{2} \sum_{i=1}^{k-1} \binom{k}{i} x_{1+i} x_{k+1-i} x_k \right)$$

$$= x_2 x_k^2 + \frac{1}{2a} h_3 - \frac{1}{(k-1)} \left(\frac{1}{2} \sum_{i=2}^{k-3} \binom{k-1}{i} x_{1+i} x_{k-i} x_{k+1} \right.$$

$$\left. - \frac{1}{2} \sum_{i=1}^{k-1} \binom{k}{i} x_{1+i} x_{k+1-i} x_k \right)$$

$$= x_2 x_k^2 + \frac{1}{2a} h_3 + \frac{k}{k-1} x_2 x_k^2 - \frac{1}{(k-1)} \left(\frac{1}{2} \sum_{i=2}^{k-3} \binom{k-1}{i} x_{1+i} x_{k-i} x_{k+1} \right.$$

$$\left. - \frac{1}{2} \sum_{i=2}^{k-2} \binom{k}{i} x_{1+i} x_{k+1-i} x_k \right)$$

$$= \frac{2k-1}{k-1} x_2 x_k^2 + \frac{1}{2a} h_3 - \frac{1}{(k-1)} \left(\frac{1}{2} \sum_{i=2}^{k-3} \binom{k-1}{i} x_{1+i} x_{k-i} x_{k+1} \right.$$

$$\left. - \frac{1}{2} \sum_{i=2}^{k-2} \binom{k}{i} x_{1+i} x_{k+1-i} x_k \right).$$

In the first sum, $2 \le i \le k-3$ and in the second one $2 \le i \le k-2$. So, each monomial that appears in $S(F^{(2)}_{x_2 x_{k-1}^2}, S(f_{k+1}, f_{k+2}))$ is of the form $x_3 x_{i_1} x_{i_2} x_{i_3}$ such that $3 \le i_1 \le i_2 \le i_3$, except $x_2 x_k^2$ and hence

$$\mathrm{LT}(S(F^{(2)}_{x_2 x_{k-1}^2}, S(f_{k+1}, f_{k+2}))) = \frac{2k-1}{k-1} x_2 x_k^2. \qquad \square$$

THEOREM 3.3. *A Gröbner basis of the ideal I, with respect to the weighted lexicographic order, is not differentially finite.*

PROOF. For proving this fact, we will use the idea of Buchberger's algorithm (mainly that any cancellation of initial monomials comes from an S-polynomial [6]) to construct a part of a Gröbner basis of the ideal I with respect to the weighted lexicographic order, which is differentially infinite.

By Lemma 3.2, we have that for every integer $n \ge 3$, the initial monomial of the polynomial $F_{x_2 x_n^2}$ is included in the initial ideal of I.

Let $G = \{f_i, F_{x_2 x_n^2}, F^{(m)}_{x_2 x_n^2} \mid i \ge 2, m \ge 1, n \ge 3\}$. By Buchberger's algorithm, G may be a part of a Gröbner basis of the ideal I but it is not a Gröbner basis of I because

$$S(F_{x_2 x_3^2}, F^{(1)}_{x_2 x_3^2}) = S(F_{x_2 x_3^2}, F_{x_2 x_3 x_4}) = 5 x_3^4.$$

But the monomial x_3^4 which is a member of the ideal I is not divisible by the leading terms of any element of G.

Note that the (usual) degree of the S-polynomial of two polynomials is at least equal to the maximum of degrees of these two polynomials. On the other hand, the derivative of a polynomial has the same degree as itself. Hence, the degree of the f_i's is equal to two, and other elements of G have degrees strictly bigger than 2.

This means that the monomials of degree 2 that appear as the leading terms of elements of Gröbner basis are of the form $x_1 x_i$ for $i \ge 1$.

So, we do not have any polynomial in a Gröbner basis whose leading monomial is $x_2 x_n$ for some $n \ge 2$, and so a polynomial having the same initial monomial of $F_{x_2 x_n^2}$ should be included in this Gröbner basis for each integer $n \geq 3$. Since the initial monomial of the polynomial $F_{x_2 x_n^2}$ is not the initial of the derivative of any other element of G, the Gröbner basis of the ideal I with respect to the weighted lexicographic order will not be differentially finite: it should contain polynomials whose initial monomials are the initials of $F_{x_2 x_n^2}, n \geq 3$ and one of these initial monomials is the derivative of another initial monomial of an element in G. $\qquad \square$

4. Two Color Partitions and the Node

Let $S := \mathbf{K}[x_1, x_2, \ldots, y_1, y_2, \ldots]$ be the graded polynomial ring where x_i, y_i have the weight i for every $i \geq 1$; the order of appearance of the variables is important since we will use below a reverse lexicographical ordering. We consider the derivation on S defined by $D(x_i) = x_{i+1}$ and $D(y_i) = y_{i+1}$. Let $f_2 = x_1 y_1$, and let

$$I = [f_2] = (x_1 y_1, x_2 y_1 + x_1 y_2, \ldots)$$

be the ideal generated by xy and its iterated derivatives $f_i, i \geq 3$ by D: for $i \geq 3, f_i = D(f_{i-1})$. Note that the scheme defined by I is the space of arcs centered at the origin of the node $X = \{xy = 0\} \subset \mathbf{A}^2$.

In this section, we are interested in determining a Gröbner basis of I with respect to the weighted reverse lexicographical order and then to apply this result to integer partitions. Note that the following Gröbner basis was found by Nguyen Duc Tam [11]; he has a beautiful but very long and difficult proof that this is actually a Gröbner basis. Below, we give a simpler and very short proof.

We will begin by defining elements of I, and we will show later that these elements give the Gröbner basis cited above.

DEFINITION 4.1. ([11]). For $1 \leq i_1 \leq i_2 \leq \cdots \leq i_k$ and for $k \geq 2$, we set

$$G_{i_1, i_2+1, i_3+2, \ldots, i_k+k-1} := \det \begin{bmatrix} x_{i_1-k+2} & x_{i_1-k+3} & \cdots & x_{i_1} & f_{i_1+1} \\ x_{i_2-k+3} & x_{i_2-k+4} & \cdots & x_{i_2+1} & f_{i_2+2} \\ x_{i_3-k+4} & x_{i_3-k+5} & \cdots & x_{i_3+2} & f_{i_3+3} \\ \vdots & \vdots & \ddots & \vdots & \vdots \\ x_{i_k+1} & x_{i_k+2} & \cdots & x_{i_k+k-1} & f_{i_k+k} \end{bmatrix},$$

where det stands for determinant.

Expanding the determinant with respect to the last column, we see that these are elements of I. A direct computation using the definition of the f_i gives the following:

LEMMA 4.2. [**11**]. *The leading term of* $G_{i_1,i_2+1,i_3+2,\ldots,i_k+k-1}$ *with respect to weighted reverse lexicographic order is* $x_{i_1}x_{i_2}x_{i_3}\ldots x_{i_k}y_k$.

We denote by $\mathbb{G}$ the set whose elements are the $G_{i_1,i_2+1,i_3+2,\ldots,i_k+k-1}$ and the f_i. It follows from Lemma 4.2 that the ideal generated by the initials of the elements of $\mathbb{G}$ is

$$J := (x_{i_1}x_{i_2}\ldots x_{i_k}y_k \mid i_j, k \geq 1).$$

First, we are interested in computing the Hilbert–Poincaré series of S/J. For this purpose, we introduce for $n \geq 1$ the Hilbert–Poincaré series

$$\mathrm{HP}_n = \mathrm{HP}\left(\frac{\mathbf{K}[x_i, y_j \mid i \geq 1, j \geq n]}{(x_{i_1}x_{i_2}\ldots x_{i_k}y_k \mid i_j \geq 1, k \geq n)}\right).$$

So $\mathrm{HP}(S/J) = \mathrm{HP}_1$. We will use the following form of H_n^1 from Section 1.

LEMMA 4.3. *For any* $n \geq 2$, *we have*

$$H_n^1 = \frac{1}{(1-q)(1-q^2)\ldots(1-q^{n-1})}.$$

PROOF. By Proposition 2.2, we have

$$H_n^1 = 1 + \frac{q}{1-q} + \frac{q^2}{(1-q)(1-q^2)} + \cdots + \frac{q^{n-1}}{(1-q)\ldots(1-q^{n-1})}.$$

We prove the expression in the lemma by induction on the integer n. For $n = 2$,

$$H_2^1 = 1 + \frac{q}{1-q} = \frac{1}{1-q}.$$

Assume that $H_n^1 = \frac{1}{(1-q)\ldots(1-q^{n-1})}$. Now, we have

$$H_{n+1}^1 = 1 + \frac{q}{1-q} + \cdots + \frac{q^{n-1}}{(1-q)\ldots(1-q^{n-1})} + \frac{q^n}{(1-q)\ldots(1-q^n)}$$

$$= H_n^1 + \frac{q^n}{(1-q)\ldots(1-q^n)}.$$

By induction hypothesis, we obtain

$$H_{n+1}^1 = \frac{1}{(1-q)\ldots(1-q^{n-1})} + \frac{q^n}{(1-q)\ldots(1-q^n)} = \frac{1}{(1-q)\ldots(1-q^n)}.$$

$\square$

LEMMA 4.4.

$$\mathrm{HP}_n = \mathrm{HP}_{n+1} + q^n \prod_{i \geq 1} \frac{1}{1-q^i}.$$

PROOF. Using Corollary 6.2 in [**3**], we have

$$\mathrm{HP}_n = \mathrm{HP}\left(\frac{\mathbf{K}[x_i, y_j \mid i \geq 1, j \geq n]}{(x_{i_1}x_{i_2}\ldots x_{i_k}y_k \mid i_j \geq 1, k \geq n)}\right)$$

$$= \text{HP}\left(\frac{\mathbf{K}[x_i, y_j | i \geq 1, j \geq n+1]}{(x_{i_1} x_{i_2} \dots x_{i_k} y_k | \ i_j \geq 1, k \geq n)}\right)$$

$$+ q^n HP\left(\frac{\mathbf{K}[x_i, y_j | i \geq 1, j \geq n]}{(x_{i_1} x_{i_2} \dots x_{i_n} | \ i_j \geq 1)}\right)$$

$$= \text{HP}_{n+1} + q^n \text{HP}\left(\frac{\mathbf{K}[x_1, x_2, \dots]}{(x_{i_1} x_{i_2} \dots x_{i_n} | \ i_j \geq 1)}\right) \ \text{HP}\left(\mathbf{K}[y_n, y_{n+1}, \dots]\right)$$

$$= \text{HP}_{n+1} + q^n H_n^1 \prod_{i \geq n} \frac{1}{1 - q^i};$$

by Lemma 4.3, this is equal to

$$\text{HP}_{n+1} + \frac{q^n}{(1 - q) \cdots (1 - q^{n-1})} \prod_{i \geq n} \frac{1}{1 - q^i}$$

$$= \text{HP}_{n+1} + q^n \prod_{i \geq 1} \frac{1}{1 - q^i}. \qquad \square$$

PROPOSITION 4.5. *We have*

$$\text{HP}(S/J) = \text{HP}_1 = \frac{1}{1 - q} \prod_{i \geq 1} \frac{1}{1 - q^i}.$$

PROOF. Using Lemma 4.4 repetitively we obtain that, for $m \geq 2$,

$$\text{HP}_1 = q \prod_{i \geq 1} \frac{1}{1 - q^i} + q^2 \prod_{i \geq 1} \frac{1}{1 - q^i} + \cdots q^m \prod_{i \geq 1} \frac{1}{1 - q^i} + \text{HP}_{m+1}.$$

On the other hand,

$$\lim_{m \to \infty} \text{HP}_m = \prod_{i \geq 1} \frac{1}{1 - q^i},$$

where the limit is considered for the q-adic topology in $\mathbf{C}[[q]]$; so, we have

$$\text{HP}_1 = \prod_{i \geq 1} \frac{1}{1 - q^i} + q \prod_{i \geq 1} \frac{1}{1 - q^i} + q^2 \prod_{i \geq 1} \frac{1}{1 - q^i} + \cdots$$

$$= (1 + q + q^2 + \cdots) \prod_{i \geq 1} \frac{1}{1 - q^i}$$

$$= \frac{1}{1 - q} \prod_{i \geq 1} \frac{1}{1 - q^i}.$$

[−2.1pc]

$\square$

We are now ready to prove the following theorem.

THEOREM 4.6. ([11]). *We have that* $\mathbb{G}$ *is a Gröbner basis of* I.

PROOF. Let $\mathrm{In}(I)$ be the initial ideal of I with respect to the weighted reverse lexicographical order. Since all the elements of $\mathbb{G}$ are also in I, we have that $J \subset \mathrm{In}(I)$; to prove that $\mathbb{G}$ is a Gröbner basis of I, we need to prove that $J = \mathrm{In}(I)$.

Noticing that $(f_2, f_3, \ldots)$ is a regular sequence [7] (note that this is rarely the case [10]) and that f_i is of weight i, we deduce that

$$\mathrm{HP}(S/I) = \frac{1}{1-q} \prod_{i \geq 1} \frac{1}{1-q^i},$$

which is equal by Proposition 4.5 to $\mathrm{HP}(S/J)$. But since we have a flat deformation with generic fiber S/I and special fiber $S/\mathrm{In}(I)$, we have that $\mathrm{HP}(S/I) = \mathrm{HP}(S/\mathrm{In}(I))$, hence $\mathrm{HP}(S/\mathrm{In}(I)) = \mathrm{HP}(S/J)$. We deduce that the homogeneous components of the same weight of $S/(\mathrm{In}(I)$ and S/J have the same (finite) dimension, and since we have an inclusion in one sense because $J \subset \mathrm{In}(I)$, they are equal. Hence, $J = \mathrm{In}(I)$. $\quad\square$

We will interpret the above results in terms of two colors partitions: consider that we have two copies of each positive integer number m, one is blue and the other is red; we denote these copies by m_b and m_r. We define an order between the colored integers by $m_b > m_r$ (so that we do not count in a partition $m_b + m_r$ and $m_r + m_b$ as different); if $m > k$, we say $m_c > k_{c'}$ for $c, c' \in \{b, r\}$.

An integer partition of a positive integer number n is a decreasing sequence (with respect to the order that we have just defined) of positive integers of one color or another

$$\lambda = (\lambda_{1,c_1} \geq \lambda_{2,c_2} \geq \cdots \geq \lambda_{l,c_l}),$$

where $c_i \in \{b, r\}$ and such that $\lambda_{1,c_1} + \lambda_{2,c_2} + \cdots + \lambda_{l,c_l} = n$. For example, the two colors integer partitions of 2 are:

$$2_b$$
$$2_r$$
$$1_b + 1_b$$
$$1_r + 1_r$$
$$1_b + 1_r.$$

Colored partitions have already appeared in the work of Andrews and Agarwal [1].

On the one hand, we can interpret the series

$$\frac{1}{1-q} \prod_{i \geq 1} \frac{1}{1-q^i}$$

as the generating series of the partitions with two colors of 1 and only the red color of any other positive integer. So, the partitions of 2 of this type are all the partitions appearing in the above example except the first one.

On the other hand, the monomials in S/J of weight n are in bijection with the partitions with two colors of n whose number of blue parts is strictly less than its smallest red part (if this latter exists). In the above example of partitions of 2, all the partitions except the last one are of this type. The Hilbert–Poincaré series $HP(S/J)$ is then the generating sequence of this type of partitions. Hence, Proposition 4.5 gives the following theorem.

THEOREM 4.7. *The number of partitions of n with two colors of 1 and only the red color of any other positive integer is equal to the number of partitions with two colors of n whose number of blue parts is strictly less than its smallest red part (if this latter exists).*

Playing the same game with the ideal $[x_j y_j]$ instead of $[x_1 y_1]$, we can prove the following generalization of Theorem 4.7.

THEOREM 4.8. *Let j be a positive integer number. The number of partitions of n with two colors of $1, \ldots, 2j - 1$ and only the red color of any other positive integer is equal to the number of partitions with two colors of n whose number of blue parts is strictly less than its smallest red part (if this latter exists) minus $(j - 1)$.*

We recover Theorem 4.7 by putting $j = 1$.

Acknowledgments

We are thankful to Marc Chardin and Bernard Teissier for several discussions during the preparation of this work and to Jan Schepers for the remarks and suggestions he made on an earlier version of this chapter. We also thank the referee for his careful reading, his corrections and suggestions.

Bibliography

[1] A.K. Agarwal, Padmavathamma, and M.V. Subbarao, *Partition Theory*. Atma Ram and Sons, Chandigarh, 2005.

[2] G.E. Andrews, *The Theory of Partitions*, Cambridge Mathematical Library. Cambridge University Press, Cambridge, 1998. Reprint of the 1976 original.

[3] C. Bruschek, H. Mourtada and J. Schepers, Arc spaces and Rogers–Ramanujan identities, *Ramanujan J.* 30(1):9–38, 2013.

[4] C. Bruschek, H. Mourtada, and J. Schepers, Arc spaces and Rogers-Ramanujan identities. In *Discrete Mathematics and Theoretical Computer Science Proceedings, FPSAC*, 2011, pp. 211–220.

[5] G. Carra Ferro, A survey on differential Gröbner bases. In *Gröbner Bases in Symbolic Analysis*, Radon Ser. Comput. Appl. Math., 2, Walter de Gruyter, Berlin, 2007, pp. 77–108.

[6] D. Cox, J. Little, and D. O'Shea, *Donal Ideals, Varieties, and Algorithms. An Introduction to Computational Algebraic Geometry and Commutative Algebra*, 4th edn. Undergraduate Texts in Mathematics. Springer, Cham, 2015.

[7] R. Goward and K. Smith, The jet scheme of a monomial scheme, *Comm. Algebra* 34(5):1591–1598, 2006.

[8] G.M. Greuel and G. Pfister, *A Singular Introduction to Commutative Algebra. With Contributions by O. Bachmann, C. Lossen and H. Schonemann*. Springer-Verlag, Berlin, 2002.

[9] H. Mourtada, Jet schemes of rational double point surface singularities. In *Valuation Theory in Interaction*, EMS Ser. Congr. Rep., Eur. Math. Soc., Sept. 2014, pp. 373–388.

[10] H. Mourtada, Jet schemes of complex plane branches and equisingularity. *Ann. Ins. Fourier*, 61(6):2313–2336, 2011.

[11] T. D. Nguyen Duc Tam, Combinatorics of jet schemes and its applications, PhD Thesis, University of Tokyo.

[12] F. Ollivier, Standard bases of differential ideals. In *Applied Algebra, Algebraic Algorithms and Error-Correcting Codes*. Lecture Notes in Computer Science, vol. 508, Springer, Berlin, 1991, pp. 304–321.

[13] On-line Encyclopedia of Integer Sequences, entry A003114.

10

The Algebraic Answer to the Nash Problem for Normal Surfaces According to de Fernex and Docampo[*]

Monique Lejeune-Jalabert

Laboratoire de Mathématiques, UMR 8100 CNRS
45 Avenue des Etats-Unis, Université de Versailles Saint,
Quentin F78035, Versailles, France
Monique.Lejeune-Jalabert@uvsq.fr

We give a detailed proof of the bijectivity of the Nash map for normal surface singularities in characteristic 0; this means that the number of irreducible components of the space of arcs on the surface centered at a singular point coincides with the number of irreducible exceptional curves above this point on its minimal resolution of singularities. This proof is extracted from the results of de Fernex Docampo concerning the Nash map for singularities in any dimension.

1. Introduction

In *"Arc Structure of Singularities"* [10] written in 1968, Nash initiated the study of the space of arcs of a singular algebraic variety X defined over the complex numbers. From the existence of a resolution of singularities for varieties over a field of characteristic zero proved by Hironaka in 1964, he deduced that the set of arcs on X originating in the singular locus Sing X of X has a finite number of irreducible components. Moreover, to each one of them, he associated an irreducible component of the exceptional locus Ex f on any resolution of singularities $f\colon Y \to X$ inducing an isomorphism over $X \setminus \operatorname{Sing} X$, an "essential component" in this terminology, since "it must appear" in the inverse image of Sing X on Y for every resolution. This "correspondence" is now called the Nash map and understanding its image is the Nash problem.

---

[*]The author and the editors are grateful to Mercedes Haiech for translating and redacting the work and putting it in Latex.

After giving the definition of the Nash map, we shall mention some recent progress on the Nash problem in Section 2.

The aim of these notes is to present a detailed algebraic proof of the surjectivity of the Nash map for normal surfaces defined over an uncountable algebraically closed field of characteristic zero. This may be rephrased by saying that the irreducible components of the space of arcs originating at an isolated singular point of such a surface are in one-to-one correspondence with the irreducible exceptional curves on its minimal desingularization.

This was first proved using topological methods by Fernandez de Bobadilla and Pe Pereira in [1] and follows as a particular case of the results of de Fernex and Docampo in [4] which are proved in a purely algebraic way. An overview of this proof has already appeared in the survey paper by de Fernex [3].

2. Arcs, the Nash Map and the Nash Problem

Let k be a field. Given a variety X over k (i.e., a reduced separated k-scheme of finite type) and a field extension K of k, a K-arc on X is a k-morphism:

$$\operatorname{Spec} K[[t]] \to X.$$

The image of the closed point of $\operatorname{Spec} K[[t]]$ in X is called the center of the arc. There exists a k-scheme X_∞, called the space of arcs of X, whose K-rational points are the K-arcs on X. In general, X_∞ is not a noetherian scheme. For every k-algebra A, there is a natural isomorphism

$$\operatorname{Hom}_k(\operatorname{Spec} A, X_\infty) \simeq \operatorname{Hom}_k(\operatorname{Spec} A[[t]], X).$$

Given $P \in X_\infty$, not necessarily a closed point of X_∞, let $k(P)$ be its residue field. We denote by $h_P \colon \operatorname{Spec}(k(P)[[t]]) \to X$ the $k(P)$-arc on X given by this isomorphism. Let $j_0 \colon X_\infty \to X$ be the morphism mapping P to the center of h_P.

We set $X_\infty^{\operatorname{Sing}} = j_0^{-1}(\operatorname{Sing} X)$.

Assume that there exists a resolution of singularities $f \colon X \to Y$ of X inducing an isomorphism over $X \setminus \operatorname{Sing} X$ and let E be an irreducible component of the exceptional locus $\operatorname{Ex} f$ of f (the component E need not be of codimension 1 in Y). Let $Y_\infty^E := (j_0^Y)^{-1}(E)$, where $j_0^Y \colon Y_\infty \to Y$ is the natural map, and let $N_E := \overline{f_\infty(Y_\infty^E)}$, where $f_\infty \colon Y_\infty \to X_\infty$ is the map induced by f. In view of the properness of f, we have

$$X_\infty^{\operatorname{Sing}} = \left(\bigcup_{\text{irreducible components } E \text{ of } \operatorname{Ex} f} N_E \right) \cup (\operatorname{Sing} X_\infty).$$

Actually, if char $k = 0$, no irreducible components of $X_\infty^{\operatorname{Sing}}$ is contained in $(\operatorname{Sing} X)_\infty$ (see [5]).

THEOREM 2.1 (Nash [10]). *If k is an algebraically closed field of characteristic zero, then $X_\infty^{\operatorname{Sing}}$ has a finite number of irreducible components. Each* one of them coincides with some N_E, where E is an irreducible component of the exceptional locus $\mathrm{Ex}\, f$ for a resolution of singularities $f\colon Y \to X$. Furthermore, E is the center on Y of a divisorial valuation ν of $k(X)$, whose center on any resolution $\tilde{f}\colon \tilde{Y} \to X$ is again an irreducible component of $\mathrm{Ex}\, \tilde{f}$.

DEFINITION 2.2. A divisorial valuation ω of $k(X)$, whose center on any resolution of singularities $f\colon X \to Y$ inducing an isomorphism over $X \setminus \mathrm{Sing}\, X$ is an irreducible component of the exceptional locus $\mathrm{Ex}\, f$, is said to be an essential valuation over X and its center on Y an essential component on Y.

Essential components on various resolutions of singularities of X are identified if they are the centers of the same essential valuation over X.

Assume from now on that k is an algebraically closed field of characteristic zero.

The Nash map is defined as

$$\mathcal{N}_X \colon \left\{ \begin{array}{c} \text{irreducible} \\ \text{components of } X^{\mathrm{Sing}}_\infty \end{array} \right\} \to \left\{ \begin{array}{c} \text{essential components} \\ \text{on a resolution} \\ \text{of singularities } f\colon Y \to X \end{array} \right\}.$$

Normal surfaces have isolated singularities and minimal resolution of singularities (any desingularization of a surface is obtained after a finite number of point blowing-ups from its minimal one). So, the set of essential components on the minimal resolution is the set of its irreducible exceptional curves.

The *Nash problem* was to decide whether $\mathcal{N}_X$ is surjective. For normal surfaces, this amounts to deciding whether the number of irreducible components of X^{Sing}_∞ coincides with the number of irreducible components of $\mathrm{Ex}\, f$, where f is the minimal resolution of singularities of X.

Kollár and Ishii [5], de Fernex [2], Johnson and Kollár [6] have given examples of varieties of dim ≥ 3, for which the Nash map is not surjective ; Bobadilla and Pereira [1] have given a topological proof of the surjectivity of the Nash map for surfaces.

Recently, de Fernex and Docampo [4] have characterized a subset of the image of $\mathcal{N}_X$ (the so-called terminal valuations). For surfaces, this is equivalent to the surjectivity of $\mathcal{N}_X$. Their discussion is algebraic and inspired by the minimal model program.

3. Arcs and Wedges

As usual, the proof of de Fernex and Docampo [4] is based on lifting wedges. For a field extension K of k, a K-wedge on X is a k-morphism $\mathrm{Spec}\, K[[\xi, t]] \to X$. For algebraic varieties $V \subsetneq W$, the curve selection lemma tells us that, for any $O \in V$, there exists an arc h on W whose center is O and which sends the generic point of $\mathrm{Spec}\, k[[t]]$ in $W \setminus V$. The same does not hold if we work with non-noetherian schemes, in particular X_∞.

First, note that the arc space $(X_\infty)_\infty$ of X_∞ is nothing but the space of wedges of X. Indeed,

$$\mathrm{Hom}_k(\mathrm{Spec}\, A[[\xi]], X_\infty) \simeq \mathrm{Hom}(\mathrm{Spec}\, A[[\xi, t]], X).$$

For a K-wedge $\phi \colon \mathrm{Spec}\, K[[\xi, t]] \to X$, its special (resp. generic) arc is defined to be the image of the closed point (resp. the generic point) of $\mathrm{Spec}\, K[[\xi]]$ by $h_\phi \colon \mathrm{Spec}\, K[[\xi]] \to X_\infty$. For simplicity, assume from now on that X is a normal surface. Let $f : Y \to X$ be its minimal resolution of singularities. It follows from a theorem of Reguera in [11, Theorem 5.1, p. 128].

PROPOSITION 3.1. *An irreducible component E of $\mathrm{Ex}\, f$ belongs to $\mathrm{Im}(\mathcal{N}_X)$ if and only if for any field extension K of the residue field $k(P_E)$ of the generic point P_E of N_E, any K-wedge ϕ whose special arc is h_{P_E} and whose generic arc belongs to X_∞^{Sing} lifts to Y.*

In [7, Corollary 2.15, p. 10], Reguera and Lejeune-Jalabert have proved the following.

PROPOSITION 3.2. *Assume k to be uncountable. For any closed point Q in an irreducible component E of $\mathrm{Ex}\, f$, let $N^+(Q)$ be the image by $f_\infty \colon Y_\infty \to X_\infty$ of the set of arcs on Y whose center is Q and which intersect transversally E at Q. If the set of closed points Q in E such that every k-wedge, whose special arc lies in $N^+(Q)$ and whose generic arc belongs to X_∞^{Sing}, lifts to Y, has a non-empty intersection with $\cap_{n \in \mathbf{N}} U_n$, for every countable family of dense open subsets U_n of E, then $E \in \mathrm{Im}(\mathcal{N}_X)$.*

4. Lifting Wedges

In view of Propositions 3.1 and 3.2, the surjectivity of the Nash map for a normal surface defined over an uncountable algebraically closed field of characteristic zero is an immediate consequence of the following theorem.

THEOREM 4.1. *Let k be an uncountable algebraically closed field of characteristic zero. Let X be a normal algebraic surface defined over k and $f : Y \to X$ be its minimal resolution of singularities. Let O be a k-rational singular point of X. Let $\phi \colon S = \mathrm{Spec}\, k[[\xi, t]] \to (X, O)$ be a wedge, such that the strict transform of the image Γ of the t-axis in S by ϕ intersects transversally $\mathrm{Ex}\, f = f^{-1}(O)$ at a non-singular point of $\mathrm{Ex}\, f$. Then ϕ lifts to Y.*

PROOF. The proof proceeds by contradiction. Assume that the statement is false. Then, there exists a commutative diagram

$$
\begin{array}{ccc}
Z & \xrightarrow{\;\varphi\;} & Y \\
{\scriptstyle g}\downarrow & & \downarrow{\scriptstyle f} \\
S & \xrightarrow[\;\phi\;]{} & X
\end{array}
$$

such that g is a composition of point blowing-ups. We assume that g is the minimal number of point blowing-ups to perform which by assumption is ≥ 1. The minimality of the resolution of singularities is characterized by the fact that in $\operatorname{Ex} f$, there are no $E \simeq \mathbb{P}^1$, such that $E^2 = -1$. Actually, this is equivalent to the fact that the canonical divisor K_Y of Y is f-nef (numerically effective) $(\mathcal{O}_Y(K_Y) = \Lambda^2 \Omega^1_{Y/k})$, i.e., $K_Y.E_i \geq 0$ for any irreducible component E_i in $\operatorname{Ex} f$. This is an immediate consequence of the adjunction formula:

$$
\omega_{E_i} \simeq \omega_Y \otimes_{\mathcal{O}_Y} \mathcal{O}_Y(E_i) \otimes_{\mathcal{O}_Y} \mathcal{O}_{E_i},
$$

where ω_{E_i} (resp. ω_Y) is the canonical sheaf on E_i (resp. Y), which implies

$$
2g_{E_i} - 2 = K_Y \cdot E_i + E_i^2,
$$

so $K_Y \cdot E_i = 2g_{E_i} - 2 - E_i^2$; since $g_{E_i} \in \mathbf{Z}_{\geq 0}$ and $-E_i^2 \in \mathbf{Z}_{>0}$, we have $K_Y \cdot E_i < 0$ if and only if $g_{E_i} = 0$ and $E_i^2 = -1$. Now, let K_Z (resp. K_S) be the canonical divisor on Z (resp. S). Since S is an algebroïd and not algebraic scheme, one has to introduce "special differentials" $\Omega'_{Z/k}$ and $\Omega'_{S/k}$. For details on their definition, have a look at [4, Sections 4.2 and 4.3]. We also consider the relative canonical divisors:

$$
K_{Z/Y} := K_Z - \varphi^* K_Y
$$

and

$$
K_{Z/S} := K_Z - g^* K_S \simeq K_Z
$$

since K_S is linearly equivalent to 0 because $S = \operatorname{Spec} k[[\xi, t]]$. $K_Y, K_Z, K_{Z/Y}$ and $K_{Z/S}$ are Cartier divisors.

Now, the irreducible exceptional curves $G_i \in \operatorname{Ex} g$ are of two types. Their image by φ may be either a point or a curve. Note that since ϕ is assumed not to lift to Y, at least the image of one of the G_i is a curve. A new commutative diagram is obtained by contracting all the G_i, such that

$\varphi(G_i)$ is a point

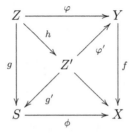

i.e., for $G_i \in \operatorname{Ex} g$, we have $\varphi(G_i)$ is a point if and only if $h(G_i)$ is a point, i.e., $G_i \in \operatorname{Ex} h$. The surface Z' is a normal surface with rational singularities (called sandwiched) and h is its minimal desingularization, since g is the minimal sequence of point blowing-ups to do to get the commutative diagram $f \circ \varphi = \phi \circ g$. Recall that Z' is $\mathbf{Q}$-factorial (i.e., every Weil divisor on Z' has a multiple who is Cartier) since Z' has only rational singularities [8, Proposition 17.1].

de Fernex and Docampo consider the following Weil divisors on Z':

$$K_{Z'} := h_* K_Z, \quad K_{Z'/Y} := h_* K_{Z/Y}, \quad K_{Z'/S} := K_{Z'} - g'^* K_S$$

and on Z,

$$K_{Z/Z'} := K_Z - h^* K_{Z'}.$$

There are two ways of thinking about $h^* K_{Z'}$, either by using the $\mathbf{Q}$-factoriality of Z' or the definition by Mumford of the total transform of a divisor on a normal surface to a desingularization (see [9, p. 17]). Finally, they introduce the decomposition

$$K_{Z'/Y} := K_{Z'/Y}^{\mathrm{exc}} + K_{Z'/Y}^{\mathrm{hor}},$$

where every component in $K_{Z'/Y}^{\mathrm{exc}}$ is g'-exceptional and none of the components of $K_{Z'/Y}^{\mathrm{hor}}$ is. The contradiction will follow from several inequalities.

LEMMA 4.2. $K_{Z/S} \leq h^* K_{Z'/S}$.

PROOF. (i) We first prove that $K_{Z/Z'} \leq 0$. Indeed, since h is a desingularization of Z', we have $K_{Z/Z'} \in \oplus_{F_i \in \operatorname{Ex} h} \mathbf{Z} F_i$. Note that every $F_i \simeq \mathbb{P}^1$. Since the intersection matrices of the exceptional curves over the singular points of Z' are negative definite, it is enough to prove that $K_{Z/Z'} \cdot F_i \geq 0$ for any $F_i \in \operatorname{Ex} h$ (see [8, §18, p. 238]). Now, $K_{Z/Z'} \cdot F_i = (K_Z - h^* K_{Z'}) \cdot F_i = K_Z \cdot F_i = 2g_{F_i} - 2 - F_i^2$ (by the adjunction formula as above); h being the minimal desingularization of Z', and $F_i \simeq \mathbb{P}^1$, we have $F_i^2 \leq -2$, so $K_{Z/Z'} \cdot F_i \geq 0$.

(ii) This implies that $K_{Z/S} \leq h^* K_{Z'/S}$. Indeed, we have

$$
\begin{aligned}
K_{Z/S} &= K_Z - g^* K_S \\
&= K_{Z/Z'} + h^* K_{Z'} - h^* g'^* K_S \\
&= K_{Z/Z'} + h^*(K_{Z'} - g'^* K_S) \\
&= K_{Z/Z'} + h^* K_{Z'/S} \\
&\leq h^* K_{Z'/S}.
\end{aligned}
$$

$\square$

LEMMA 4.3. $h^*(K_{Z'/S}) \leq h^*(K_{Z'/Y}^{\mathrm{exc}})$.

PROOF. We first note that

$$
(4.4) \qquad K_{Z'/S} - K_{Z'/Y}^{\mathrm{exc}} \simeq \varphi'^* K_Y + K_{Z'/Y}^{\mathrm{hor}}
$$

(where $\simeq$ stands for linearly equivalent). Indeed, we have

$$
\begin{aligned}
K_{Z'/S} - K_{Z'/Y}^{\mathrm{exc}} &\simeq K_{Z'} - K_{Z'/Y}^{\mathrm{exc}} \\
&= K_{Z'} - K_{Z'/Y} + K_{Z'/Y}^{\mathrm{hor}} \\
&= h_*(K_Z - K_{Z/Y}) + K_{Z'/Y}^{\mathrm{hor}} \\
&= h_* \varphi^* K_Y + K_{Z'/Y}^{\mathrm{hor}} \\
&= h_* h^* \varphi'^* K_Y + K_{Z'/Y}^{\mathrm{hor}} \\
&= \varphi'^* K_Y + K_{Z'/Y}^{\mathrm{hor}}.
\end{aligned}
$$

To get the last equality, we have used that for a Cartier divisor D on Z', we have $h_* h^* D = D$, since $h^* D = D' + F$, where D' is the proper transform of D and $F \in \mathrm{Ex}\, h$. Since $h_*(F) = 0$ and $h_*(D') = D$, we thus get $h_* h^* D = D$.

Since the intersection matrix of the exceptional curves of g is negative definite, in order to prove that $h^* K_{Z'/S} \leq h^* K_{Z'/Y}^{\mathrm{exc}}$, it is enough to prove that $(h^* K_{Z'/S} - h^* K_{Z'/Y}^{\mathrm{exc}}) \cdot G_i \geq 0$ for any G_i irreducible in $\mathrm{Ex}\, g$. If $h(G_i)$ is a point (hence singular) of Z', i.e., if $G_i \in \mathrm{Ex}\, h$, this intersection number is 0 (by the definition of Mumford [9]). If $h(G_i)$ is a curve G_i', i.e., if $h(G_i) \in \mathrm{Ex}\, g'$, it follows from (4.4) that

$$
(h^* K_{Z'/S} - h^* K_{Z'/Y}^{\mathrm{exc}}) \cdot G_i = \varphi^* K_Y \cdot G_i + h^* K_{Z'/Y}^{\mathrm{hor}} \cdot G_i = \varphi^* K_Y \cdot G_i
$$

$$
+ K_{Z'/Y}^{\mathrm{hor}} \cdot G_i'
$$

(again by Mumford).

Now recall that, by definition, none of the components of $K_{Z'/Y}^{\mathrm{hor}}$ are in $\mathrm{Ex}\, g'$. Hence, $K_{Z'/Y}^{\mathrm{hor}} \cdot G_i' \geq 0$. We have observed that, since f is the minimal desingularization of X, K_Y is f-nef. This implies that $\varphi^* K_Y$ is g-nef. Indeed, despite the non-properness of φ, since the exceptional curves in $\mathrm{Ex}\, g$ are projective ($\simeq \mathbb{P}^1$), we have for $G_i \in \mathrm{Ex}\, g$,

$$
\varphi_*(G_i \cdot \varphi^* K_Y) = \varphi_*(G_i) \cdot K_Y \geq 0,
$$

since $\varphi(G_i) \subset \mathrm{Ex}\, f$. $\square$

So far, we have not used that char$(k) = 0$ and the assumption on the arc $\Gamma \in X_\infty^0$, image of the t-axis in S by ϕ. Let G (resp. E) be the irreducible component in Ex g (resp. Ex f), intersecting the proper transform $\tilde{T}$ of the t-axis in S on Z (resp. $\tilde{\Gamma}$ of Γ on Y). Recall that $\tilde{\Gamma}$ intersects E transversally at a regular point of E which does not lie on any other component of Ex f.

LEMMA 4.5. *Let C be any irreducible component of Ex g' containing $h(G)$. Then $\varphi'(C) = E$ and ord$_C(K_{Z'/Y}) = $ ord$_C(\varphi'^*E) - 1$.*

PROOF. First, note that if $\dim h(G) = 1$ or equivalently $\dim \varphi(G) = 1$, we have $C = h(G)$ and that if $h(G)$ is a point, such a C exists because ϕ is assumed not to lift to Y.

Now, $\varphi'(C)$ is an irreducible curve in Ex f, since $f(\varphi'(C)) = \phi \circ g'(C) = 0$. Since $h(G) \subset C$, we have $\varphi(G) \subset \varphi'(C)$. By assumption, the intersection point of the proper transform of the t-axis on Z with Ex g lies on G, so the intersection point of the proper transform of Γ on Y lies on $\varphi(G)$. But E is the only irreducible component of Ex f on which this point lies. Hence, $\varphi'(C) = E$.

We now choose a closed point z on C which is regular on C and Z' and which does not belong to any other irreducible component of Ex g' than C and to the proper transform on Z' of $\phi^{-1}(0)$ if $\dim \phi^{-1}(0) = 1$. Since $\varphi'(C) = E$, then $\varphi'^{-1}(E)$ coincides with C in a neighborhood of z.

Let $y = \varphi'(z)$. We identify $\widehat{\mathcal{O}_{Y,y}}$ with $k[[y_1, y_2]]$ in such a way that $(E, y) = \operatorname{div} y_1$ and $\widehat{\mathcal{O}_{Z',z}}$ with $k[[u, v]]$ in such a way that $(C, z) = \operatorname{div} u$. Let $\varphi'^\#: \widehat{\mathcal{O}_{Y,y}} \to \widehat{\mathcal{O}_{Z',z}}$ denote the map induced by φ' and set $\varphi'^\#(y_1) = \varphi_1(u, v)$ and $\varphi'^\#(y_2) = \varphi_2(u, v)$. We thus have $\varphi_1(u, v) = u^n U(u, v)$ with $n \geq 1$ and U a unit and $\varphi_2(u, v) = v^q V(v) + u R(u, v)$ with $q \geq 1$ and V a unit. Hence,

$$\operatorname{Jac} \varphi'^\# = \frac{\partial \varphi_1}{\partial u} \frac{\partial \varphi_2}{\partial v} - \frac{\partial \varphi_1}{\partial v} \frac{\partial \varphi_2}{\partial u} = n u^{n-1} v^{q-1} U \left(qV + v \frac{dV}{dv} \right) \mod (u^n).$$

Since $qV + v\frac{dV}{dv}$ is a unit in $k[[v]]$ and char$(k) = 0$, we have

$$\operatorname{ord}_u \operatorname{Jac} \varphi'^\# = n - 1.$$

Remembering that φ'^*E is $\operatorname{div} \varphi'^\#(y_1) = \operatorname{div} \varphi_1(u, v) = nE$ in a neighborhood of z, we get that ord$_C K_{Z'/Y} = \operatorname{ord}_u \operatorname{Jac} \varphi'^\# = \operatorname{ord}_C \varphi'^*E - 1$. $\square$

COROLLARY 4.6. *If $h(G)$ is a point, $\varphi^*E - K_{Z'/Y}^{\mathrm{exc}}$ is effective and nontrivial in a neighborhood of this point. In addition, we have ord$_G \; \varphi^*E > $ ord$_G \; h^*K_{Z'/Y}^{\mathrm{exc}}$.*

PROOF. Since E is an effective Cartier divisor on Y, φ'^*E is an effective Cartier divisor on Z' (note that the proper transform on Z' of some curves $\mathcal{C}_i$ on S such that $\phi(\mathcal{C}_i) = 0$ may appear among its components). Hence, the only possible negative component of $\varphi'^*E - K_{Z'/Y}^{\mathrm{exc}}$ are components of $K_{Z'/Y}^{\mathrm{exc}}$, and so they belong to Ex g'. But, in view of the above lemma, for any irreducible C in $\mathrm{Ex}\, g'$ containing $h(G)$, we have $\mathrm{ord}_C(\varphi'^*E - K^{\mathrm{exc}}_{Z'/Y}) = 1$. Hence, $\varphi'^*E - K^{\mathrm{exc}}_{Z'/Y}$ is effective and non-trivial in a neighborhood of $h(G)$ on Z'.

Now, in view of Mumford [9] p. 17, Prop (ii), $h^*(\varphi'^*E - K^{\mathrm{exc}}_{Z'/Y}) = \varphi^*E - h^*K^{\mathrm{exc}}_{Z'/Y}$ is again effective on a neighborhood of $h^{-1}(h(G))$; moreover, for any irreducible component G_i of $\mathrm{Ex}\, h$, such that $h(G_i) = h(G)$, its coefficient in $\varphi^*E - h^*K^{\mathrm{exc}}_{Z'/Y}$ is strictly positive, in particular for G, we get that $\mathrm{ord}_G\, \varphi^*E > \mathrm{ord}_G\, h^*K^{\mathrm{exc}}_{Z/Y}$.   □

End of the proof of the theorem. First, assume that $h(G)$ is a point: Lemmas 4.2, 4.3 and Corollary 4.6 give us the following inequalities:

$$\mathrm{ord}_G\, K_{Z/S} \leq \mathrm{ord}_G\, h^*K_{Z'/S} \leq \mathrm{ord}_G\, h^*K^{\mathrm{exc}}_{Z'/Y} < \mathrm{ord}_G\, \varphi^*E.$$

Now, we have $1 \leq \mathrm{ord}_G\, K_{Z/S}$, and $\mathrm{ord}_G\, \varphi^*E$ remains to be computed. We have that $\varphi(G) = \varphi'(h(G))$ is a point y in E. Since Y is non-singular, we have an isomorphism $\widehat{\mathcal{O}_{Y,y}} \simeq k[[y_1, y_2]]$ such that $(E, y) = \mathrm{div}\, y_1$. Let $z \in G$ be the intersection point of G with the proper transform $\tilde{T}$ of the t-axis in S. Recall that the proper transform of the arc $\Gamma \in X^0_\infty$ which is $\varphi(\tilde{T})$ is non-singular and intersects transversally E at $\varphi(z) \in \varphi(G) = y$. We have an isomorphism $\widehat{\mathcal{O}_{Z,z}} \simeq k[[u, v]]$ such that $(G, z) = \mathrm{div}\, u$.

Let $\varphi^\#\colon \widehat{\mathcal{O}_{Y,y}} \to \widehat{\mathcal{O}_{Z,z}}$, and set $\varphi^\#(y_1) = \varphi_1(u, v)$, $\varphi^\#(y_2) = \varphi_2(u, v)$. We have $\mathrm{ord}_G\, \varphi^*E = \mathrm{ord}_u\, \varphi_1(u, v)$. The morphism $\varphi : Z \to Y$ induces an isomorphism $\tilde{T} \to \tilde{\Gamma}$, so $\mathrm{ord}_u\, \varphi_1(u, 0) = 1$. Since $\varphi(G) = y$, we have that $\varphi_1(0, v) = 0$. Hence, we have $\varphi_1(u, v) = \varphi_1(u, 0) + vR(u, v)$, so $\varphi_1(0, v) = vR(0, v) = 0$. Therefore, $R(u, v) = u^s\tilde{R}(u, v)$, with $s \geq 1$, and $\varphi_1(u, v) = uU(u) + u^s v\tilde{R}(u, v) = u[U(u) + u^{s-1}v\tilde{R}(u, v)]$ with $U(u)$ a unit in $k[[u]]$. Since $U(u) + u^{s-1}v\tilde{R}(u, v)$ is a unit in $k[[u, v]]$, this implies that $\mathrm{ord}_u\varphi_1(u, v) = 1$ and finally $1 < 1$.

We have thus got the contradiction.

Now, assume that $h(G)$ is a curve, so that $h(G) \in \mathrm{Ex}\, g'$ and $\varphi(G) = E$ by Lemma 4.5. Again, Lemmas 4.2 and 4.3 give us the following inequalities:

$$\mathrm{ord}_G\, K_{Z/S} \leq \mathrm{ord}_G\, h^*K_{Z'/S} \leq \mathrm{ord}_G\, h^*K^{\mathrm{exc}}_{Z'/Y}$$

and we have $1 \leq \mathrm{ord}_G\, K_{Z/S}$. Here, since $h(G) \in \mathrm{Ex}\, g'$, it is a component of $K^{\mathrm{exc}}_{Z'/Y}$ and h is an isomorphism on an open dense subset of G. So $\mathrm{ord}_G\, h^*K^{\mathrm{exc}}_{Z'/Y} = \mathrm{ord}_G\, K_{Z/Y}$.

As above, let $z \in G$ be the intersection point of G with the proper transform $\tilde{T}$ of the t-axis in S, and $\varphi_{|\tilde{T}}$ induces an isomorphism from $\tilde{T}$ to the proper transform $\tilde{\Gamma}$ of the arc $\Gamma \in X^0_\infty$ on Y. Let $y := \varphi(z)$. We have an isomorphism $\widehat{\mathcal{O}_{Z,z}} \simeq k[[u, v]]$ such that $(G, z) = \mathrm{div}\, u$ and an isomorphism $\widehat{\mathcal{O}_{Y,y}} \simeq k[[y_1, y_2]]$, such that $(E, y) = \mathrm{div}\, y_1$. Let $\varphi^\#\colon \widehat{\mathcal{O}_{Y,y}} \to \widehat{\mathcal{O}_{Z,z}}$ denote the induced map by φ, and set $\varphi^\#(y_1) = \varphi_1(u, v)$ and $\varphi^\#(y_2) = \varphi_2(u, v)$. Since $G \subset \varphi^{-1}(E)$, we have that $\mathrm{ord}_u\, \varphi_1(u, v) \geq 1$. Since $\varphi_{|\tilde{T}}\colon \tilde{T} \to \tilde{\Gamma}$ is an isomorphism, we have that $\operatorname{ord}_u \varphi_1(u,0) = 1$. Thus, $\operatorname{ord}_u \varphi_1(u,v) = 1$, and $\varphi_1(u,v) = uU(u,v)$ with $\operatorname{ord}_u U(u,v) = 0$. Since $\varphi(G) = E$, the map $\varphi_{|G} \colon G \to E$ gives rise to an injective map $\widehat{\mathcal{O}_{E,y}} \simeq k[[y_2]] \to \widehat{\mathcal{O}_{G,z}} \simeq k[[v]]$, so $1 \le \operatorname{ord}_v \varphi_2(0,v) = s < +\infty$, and $\varphi_2(u,v) = v^s V(v) + uW(u,v)$, where $V(v)$ is a unit in $k[[v]]$. So, $\operatorname{ord}_G K_{Z/Y} = \operatorname{ord}_u \operatorname{Jac}\varphi^\# = 0$ since

$$
\operatorname{Jac}\varphi^\# = \begin{pmatrix} U(u,v) + u\frac{\partial U}{\partial u} & W(u,v) + u\frac{\partial W}{\partial u} \\ u\frac{\partial u}{\partial v} & sv^{s-1}V(v) + v^s\frac{dV}{dv} + u\frac{\partial W}{\partial v} \end{pmatrix}
$$

$$
= U(u,v)\left(sv^{s-1}V(v) + v^s\frac{dV}{dv} \right) \bmod(u).
$$

Once again, a contradiction. $\qquad\square$

Bibliography

[1] J. Fernández de Bobadillas and M. Pereira, The Nash problem for surfaces. *Ann. of Math.*, 2003–2029, 2012.

[2] T. de Fernex, Three-dimensional counter-examples to the Nash problem. *Compositio Math.*, 149:1519–1534, 2013.

[3] T. de Fernex, The space of arcs of an algebraic variety. Algebraic geometry: Salt Lake City 2015, 169–197, *Proc. Sympos. Pure Math.*, 97.1, *Amer. Math. Soc.*, Providence, RI, 2018.

[4] T. de Fernex and R. Docampo, Terminal valuations and the Nash problem. *Invent. Math.*, 203:303–331, 2016.

[5] S. Ishii and J. Kollár, The Nash problem on the arc families of singularities. *Duke Math. J.*, 120:601–620, 2003.

[6] J. M. Johnson and J. Kollár, Arcs spaces of ca-type singularities. *J. Singularities*, 7:238–252, 2013.

[7] M. Lejeune-Jalabert and A. Reguera, Exceptional divisors that are not uniruled belong to the image of the Nash map. *J. Inst. Math. Jussieu*, 11:273–287, 2012.

[8] J. Lipman, Rational singularities, with applications to algebraic surfaces and unique factorization. *Publ. IHES*, 36:195–279, 1969.

[9] D. Mumford, The topology of normal singularities of an algebraic surface and a criterion for simplicity. *Publ. IHES*, 1961, 5–22.

[10] J. Nash, Arc structure of singularities. *Duke Math. J.* 81:31–38, 1995.

[11] A. Reguera, A curve selection lemma in spaces of arcs and the image of the Nash map. *Compositio Math.* 142:119–130, 2006.

11

The Nash Problem from Geometric and Topological Perspective

J. Fernández de Bobadilla[*,‡] **and M. Pe Pereira**[†,§]

*BCAM-Ikerbasque, Academic Collaborator of UPV/EHU
Avenida de Mazarredo 14, 48009 Bilbao, Spain*

†*Facultad de Ciencias Matematicas, UCM, Plaza de Ciencias
3, Ciudad Universitaria, 28040 Madrid, Spain*
‡*jbobadilla@bcamath.org*
§*maria.pe@mat.ucm.es*

We survey the proof of the Nash conjecture for surfaces and show how geometric and topological ideas developed in previous articles by the authors influenced it. Later, we summarize the main ideas in the higher dimensional statement and proof by de Fernex and Docampo. We end the paper by explaining later developments on generalized Nash problem and on Kollár and Nemethi holomorphic arcs.

1. Introduction

The Nash problem [**19**] was formulated in the 1960s (but published later) in an attempt to understand the relation between the structure of resolution of singularities of an algebraic variety X over a field of characteristic 0 and the space of arcs (germs of parameterized curves) in the variety. He proved that the space of arcs centered at the singular locus (endowed with an infinite-dimensional algebraic variety structure) has finitely many irreducible components and proposed to study the relation of these components with the essential irreducible components of the exceptional set of a resolution of singularities.

An irreducible component E_i of the exceptional divisor of a resolution of singularities is called *essential*, if given any other resolution, the birational transform of E_i to the second resolution is an irreducible component of the exceptional divisor. Nash defined a mapping from the set of irreducible components of the space of arcs centered at the singular locus to the set of essential components of a resolution as follows: he assigns to each component W of the space of arcs centered at the singular locus the unique component of the exceptional set which meets the lifting of a generic arc of W to the resolution. Nash established the injectivity of this mapping. For the case of surfaces, it seemed plausible for him that the mapping is also surjective and posed the problem as an open question. He also proposed to study the mapping in the higher dimensional case. Nash resolved the question positively for the surface A_k singularities and in analyzing the higher dimensional A_k singularities, he could not prove the bijectivity for A_4.

As a general reference for the Nash problem, the reader may look at [19, 9].

Bijectivity of the Nash mapping was shown for many classes of surfaces (see [6, 9–11, 14, 15, 18–20, 22–24, 26, 28, 29]). The techniques leading to the proof of each of these cases are different in nature, and the proofs are often complicated. It is worthwhile to note that even for the case of the rational double points not solved by Nash a complete proof had to be awaited until 2011: see [20], where the problem is solved for any quotient surface singularity and also [23, 26] for the cases of D_n and E_6. In [3], it is shown that the Nash problem for surfaces only depends on the topological type of the singularity. In 2012, the authors established in the affirmative the Nash question for the general surface case [4]. The proof we found was of a topological nature, and it is essential to work with convergent arcs and their convergent deformations. This motivated Kollár and Nemethi to pursue the study of convergent arcs and deformations in [13]. The topological ideas of [3, 20] also had an impact on the generalized Nash problem; in [5], Popescu-Pampu and the authors show that the generalized Nash problem is of topological nature and explore the relation and applications of this problem to Arnol'd classical adjacency problem.

It is well known that birational geometry of surfaces is much simpler than in higher dimension. This fact reflects on the Nash problem: Ishii and Kollár showed in [9] a four-dimensional example with a non-bijective Nash mapping. In the same paper, they showed the bijectivity of the Nash mapping for toric singularities of arbitrary dimension. Other advances in the higher dimensional case include [25, 6, 16]. In 2013, de Fernex [1] found the first counterexamples to the Nash question; further counterexamples and a deeper understanding of how they appear were provided by Johnson and Kollár in [7]. There it was proved that the threefold A_4

$$x^2 + y^2 + z^2 + w^5 = 0,$$

the example that Nash left unfinished, was indeed a counterexample! In 2016, de Fernex and Docampo [2] proved that terminal divisors are at the image of the Nash map. Since at the surface case, terminal and essential divisors are precisely the same, this seems to be the correct higher dimensional generalization. It would, however, remain to be characterized which essential non-terminal divisors are at the image of the Nash map.

For other modern review articles concerning the Nash problem, the reader may consult [27, 12, 8].

In this chapter, we explain how geometric and topological techniques contributed to the development of the proof of the Nash conjecture and how they relate with other viewpoints and further developments.

Sections 2–5 explain our proof of the two-dimensional case in a nontechnical way, pointing to the main new ideas appearing in it. We present a proof for the case in which the minimal resolution has a strict normal crossings exceptional divisor. In this case, all new essential ideas already appear, but the amount of technicalities can be reduced drastically. We include enough pictures so that the reader can grasp what is going on in an easy and intuitive way.

In Section 6, we emphasize the notion of returns, which was discovered in [20] and was crucial for the development of the general proof. We also take the opportunity to comment on deformation techniques that were useful to establish the hard cases of E_6, E_7 and E_8.

In Section 7, we explain the relation of our proof with the higher dimensional one of [2]. We do it by giving a short exposition of their proof that we believe condense all main ideas.

In Section 8, we summarize our contribution with Popescu-Pampu on the generalized Nash problem [5] and its impact on Arnol'd classical adjacency problem. Here, we use the techniques of [3] to show that the generalized Nash problem is of topological nature.

Finally, in Section 9, we explain the relation of our ideas with further developments of more geometric–topological nature by Kollár and Nemethi [13].

2. The Idea of the Proof for Surfaces

Let (X, O) be a surface singularity defined over an algebraically closed field of 0 characteristic. Let

$$\pi : (\tilde{X}, E) \to (X, O)$$

be the minimal resolution of singularities, which is an isomorphism outside the exceptional divisor $E := \pi^{-1}(O)$. Consider the decomposition $E = \bigcup_{i=0}^{r} E_i$ of E into irreducible components. These irreducible components are the essential components of (X, O).

Given any irreducible component E_i, we denote by N_{E_i} the Zariski closure in the arc space of X of the set of non-constant arcs whose lifting to the resolution is centered at E_i. These Zariski closed subsets are irreducible and each irreducible component of the space of arcs is equal to some N_{E_i} for a certain component E_i. The Nash mapping is the map assigning to each irreducible component N_{E_i} the exceptional divisor E_i. Injectivity is immediate. The Nash problem is about determining whether the Nash mapping is bijective.

The Nash mapping is not bijective if and only if there exist two different irreducible components E_i and E_j of the exceptional divisor of the minimal resolution, such that we have the inclusion $N_{E_i} \subset N_{E_j}$ (see [19]). Such inclusions were called *adjacencies* in [3].

An application of the Lefschetz principle allows one to reduce to the case in which the base field is $\mathbb{C}$. Details are provided in [4]. We make this assumption for the rest of the paper. Moreover, the case of a non-normal surface follows from the normal surface case easily (see [4, Section 6]). Then, we assume (X, O) to be a complex normal surface singularity.

The idea of the proof is as follows. We reason by contradiction. Let (X, O) be a normal surface singularity and

$$\pi : \tilde{X} \to (X, O)$$

be the minimal resolution of singularities. Assume that the Nash mapping is not bijective. Then, by a theorem of [3], there exists a convergent wedge

$$\alpha : (\mathbb{C}^2, O) \to (X, O)$$

with certain precise properties (see Definition 3.1). As in [20], taking a suitable representative, we may view α as a uniparametric family of mappings

$$\alpha_s : \mathcal{U}_s \to (X, O)$$

from a family of domains $\mathcal{U}_s$ to X with the property that each $\mathcal{U}_s$ is diffeomorphic to a disk. For any s, we consider the lifting

$$\tilde{\alpha}_s : \mathcal{U}_s \to \tilde{X}$$

to the resolution. Note that $\tilde{\alpha}_s$ is the normalization mapping of the image curve.

On the other hand, if we denote by Y_s the image of $\tilde{\alpha}_s$ for $s \neq 0$, then we may consider the limit divisor Y_0 in $\tilde{X}$ when s approaches 0. This limit divisor consists of the union of the image of $\tilde{\alpha}_0$ and certain components of the exceptional divisor of the resolution whose multiplicities are easy to compute. We prove an upper bound for the Euler characteristic of the normalization of any reduced deformation of Y_0 in terms of the following data: the topology of Y_0, the multiplicities of its components and the set of intersection points of Y_0 with the generic member Y_s of the deformation. Using this bound, we show that the Euler characteristic of the normalization of Y_s is strictly smaller than one. This contradicts the fact that the normalization is a disk.

In the following three sections, we fill the details of the above sketch.

3. Turning the Problem into a Problem of Convergent Wedges

The germ (X, O) is embedded in an ambient space $\mathbb{C}^N$. Denote by B_ϵ the closed ball of radius ϵ centered at the origin and by $\mathbb{S}_\epsilon$ its boundary sphere. Take a *Milnor radius* ϵ_0 for (X, O) in $\mathbb{C}^N$, i.e., we choose $\epsilon_0 > 0$, such that for a certain representative X and any radius $0 < \epsilon \leq \epsilon_0$, we have that all the spheres $\mathbb{S}_\epsilon$ are transverse to X and $X \cap \mathbb{S}_\epsilon$ is a closed subset of $\mathbb{S}_\epsilon$ (see [17] for a proof of its existence). In particular, $X \cap B_{\epsilon_0}$ has conical structure. From now on, we will denote by X_{ϵ_0} the *Milnor representative* $X \cap B_{\epsilon_0}$ and by $\tilde{X}_{\epsilon_0}$ the resolution of singularities $\pi^{-1}(X_{\epsilon_0})$ (see Figure 1).

We recall some terminology and results from [3]. Consider coordinates (t, s) in the germ $(\mathbb{C}^2, O)$. A *convergent wedge* is a complex analytic germ

$$\alpha : (\mathbb{C}^2, O) \to (X, O),$$

which sends the line $V(t)$ to the origin O. Given a wedge α and a parameter value s, the arc

$$\alpha_s : (\mathbb{C}, 0) \to (X, O)$$

is defined by $\alpha_s(t) = \alpha(t, s)$. The arc α_0 is called *the special arc* of the wedge. For small enough $s \neq 0$, the arcs α_s are called *generic arcs*.

Any non-constant arc

$$\gamma : (\mathbb{C}, 0) \to (X, O)$$

admits a unique lifting to $(\tilde{X}, O)$ that we denote by $\tilde{\gamma}$.

DEFINITION 3.1 ([3]). A convergent wedge α *realizes an adjacency* $N_{E_i} \subset N_{E_j}$ (with $j \neq i$) if and only if the lifting $\tilde{\alpha}_0$ of the special arc

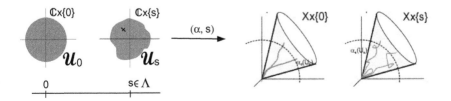

Figure 1. A wedge representative $\alpha : \mathcal{U} \to X \times \Lambda$ and the representatives $\alpha_0|_{\mathcal{U}_0}$ and $\alpha_s|_{\mathcal{U}_s}$.

meets E_i transversely at a non-singular point of E and the lifting $\tilde{\alpha}_s$ of a generic arc satisfies $\tilde{\alpha}_s(0) \in E_j$.

Our proof is based on the following theorem, which is the implication "(1)$\Rightarrow$(a)" of Corollary B of [3].

THEOREM 3.2 ([3]). *An essential divisor E_i is in the image of the Nash mapping if there is no other essential divisor $E_j \neq E_i$, such that there exists a convergent wedge realizing the adjacency $N_{E_i} \subset N_{E_j}$.*

The proof in [3] of this theorem has two parts. The first consists of proving that if there is an adjacency, then there exists a *formal wedge*:

$$\alpha : \mathrm{Spec}(\mathbb{C}[[t,s]]) \to (X, O),$$

realizing the adjacency. For this, first, we use a theorem of Reguera [30], which produces wedges defined over large fields. Then, a specialization argument is performed to produce a wedge defined over the base field $\mathbb{C}$. This was done independently in [16]. The second part is an argument based on Popescu's Approximation Theorem, which produces the convergent wedge from the formal one.

Then, to prove the Nash Conjecture, we reason by contradiction and by Theorem 3.2, we assume that there exists a convergent wedge $\alpha : (\mathbb{C}^2, O) \to (X, O)$ that realizes some adjacency $N_{E_0} \subset N_{E_j}$.

4. Reduction to an Euler Characteristic Estimate

Following [20], we shall work with representatives rather than germs in order to get richer information about the geometry of the possible wedges.

Shrinking ϵ is necessary, we can choose a Milnor representative of α_0, say $\alpha_0|_U$, with U diffeomorphic to a disk, such that $\alpha_0|_U^{-1}(\partial X_\epsilon) = \partial U$ and the mapping $\alpha_0|_U$ is transverse to any sphere $\mathbb{S}_{\epsilon'}$ for any $0 < \epsilon' \leq \epsilon$.

Moreover, we can consider U, such that for a positive and small enough δ, the mapping α is defined in $U \times D_\delta$. Note also that we can assume $\alpha_0|_U$ injective and consequently $\alpha_s|_U$ generically one to one for s small enough (see [4] for details).

We consider the mapping

$$\beta : (\mathbb{C}^2, (0,0)) \to (\mathbb{C}^N \times \mathbb{C}, (O, 0))$$

given by $\beta(t,s) := (\alpha(t,s), s)$ and its restriction

$$\beta|_{U \times D_\delta} : U \times D_\delta \to X \times D_\delta.$$

We denote by pr the projection of $U \times D_\delta$ onto the second factor.

The following lemma is proved using transversality arguments, together with Ehresmann Fibration Theorem, and the method is nowadays classical in Singularity Theory. Details are provided in [4].

LEMMA 4.1. *After possibly shrinking δ, we have that there exists $\epsilon > 0$, such that, defining*

$$\mathcal{U} := \beta|_{U \times D_\delta}^{-1}(X_\epsilon \times D_\delta),$$

we have the following:

(a) *the restriction $\beta|_\mathcal{U} : \mathcal{U} \to X_\epsilon \times D_\delta$ is a proper and finite morphism of analytic spaces;*

(b) *the set $\beta(\mathcal{U})$ is a two-dimensional closed analytic subset of $X_\epsilon \times D_\delta$;*

(c) *for any $s \in D_\delta$, the restriction $\beta|_{U \times \{s\}}$ is transverse to $\mathbb{S}_\epsilon \times \dot{D}_\delta$;*

(d) *the set $\mathcal{U}$ is a smooth manifold with boundary $\beta|_\mathcal{U}^{-1}(\partial X_\epsilon \times D_\delta)$;*

(e) *for any $s \in D_\delta$, the intersection $\mathcal{U} \cap (\mathbb{C} \times \{s\})$ is diffeomorphic to a disk.*

We will denote by $\mathcal{U}_s$ the fiber $pr|_\mathcal{U}^{-1}(s)$. The fact that every $\mathcal{U}_s$ is a disk is a key in the proof as it was in the final step of the proof of the main result of [20].

Now, we consider the image $H := \beta(\mathcal{U})$. For every $s \in D_\delta$, the fiber H_s, by the natural projection onto D_δ, is the image of the representative

$$\alpha_s|_{\mathcal{U}_s} : \mathcal{U}_s \to X_\epsilon.$$

Given the minimal resolution of singularities

$$\pi : \tilde{X}_\epsilon \to X_\epsilon,$$

we consider the mapping

$$\sigma : \tilde{X}_\epsilon \times D_\delta \to X_\epsilon \times D_\delta$$

defined by $\sigma(x, s) = (\pi(x), s)$. Note that the mapping σ is an isomorphism outside $E \times D_\delta$. We denote by Y the strict transform of H by σ in $\tilde{X}_\epsilon \times D_\delta$ that is the analytic Zariski closure in $\tilde{X}_\epsilon \times D_\delta$ of

(4.2) $$\sigma^{-1}(H \setminus (\{O\} \times D_\delta)).$$

The space (4.2) is an irreducible surface, thus so is its closure Y. Since $\tilde{X}_\epsilon \times D_\delta$ is a smooth threefold, the surface Y considered with its reduced structure is a Cartier divisor (that is, a codimension 1 analytic subset whose sheaf of ideals is locally principal). We denote by Y_s the intersection $Y \cap (\tilde{X} \times \{s\})$.

The indeterminacy locus of the mapping $\sigma^{-1} \circ \beta|_\mathcal{U}$ has codimension 2, hence reducing ϵ and δ if necessary, we can assume that the origin $(0, 0) \in \mathcal{U}$ is the only indeterminacy point. Denote by

$$\tilde{\beta} : \mathcal{U} \setminus \{(0, 0)\} \to \tilde{X}_\epsilon \times D_\delta$$

the restriction of $\sigma^{-1} \circ \beta|_\mathcal{U}$ to its domain of definition $\mathcal{U} \setminus \{(0, 0)\}$. Observe that we have the equality

$$\tilde{\beta}(\mathcal{U} \setminus \beta^{-1}(\{O\} \times D_\delta)) = \sigma^{-1}(H \setminus (\{O\} \times D_\delta)).$$

Consequently, Y is the analytic Zariski closure of $\tilde{\beta}(\mathcal{U} \setminus \{(0,0)\})$ and moreover, we have the equality

$$(4.3) \qquad Y \cap (\tilde{X}_\epsilon \times (D_\delta \setminus \{0\})) = \tilde{\beta}(\mathcal{U} \setminus \mathcal{U}_0).$$

For any $s \in D_\delta$, there exists a unique lifting

$$\tilde{\alpha}_s : \mathcal{U}_s \to \tilde{X}_\epsilon,$$

such that $\alpha_s = \pi \circ \tilde{\alpha}_s$. Obviously, for $s \neq 0$, we have the equality $\tilde{\beta}(t) = (\tilde{\alpha}_s(t), s)$ for any $t \in \mathcal{U}_s$. This, together with equality (4.3), implies the equality

$$(4.4) \qquad Y_s = \tilde{\alpha}_s(\mathcal{U}_s).$$

Since Y is reduced, perhaps shrinking δ, we can assume that Y_s is reduced. Since α_s is proper and generically one to one, and $\mathcal{U}_s$ is smooth, we have that the mapping

$$\tilde{\alpha}_s : \mathcal{U}_s \to Y_s$$

is the normalization of Y_s.

Now, we describe the divisor Y_0. It is clear that all the components except $\tilde{\alpha}_0(\mathcal{U}_0)$ live above the origin that is the only indeterminacy point of $\tilde{\beta}$, i.e., the divisor Y_0 decomposes as a sum

$$(4.5) \qquad Y_0 = Z_0 + \sum_{i=0}^{r} a_i E_i,$$

where we have denoted $Z_0 := \tilde{\alpha}_0(\mathcal{U}_0)$. This divisor Y_0 has the following properties (Figure 2):

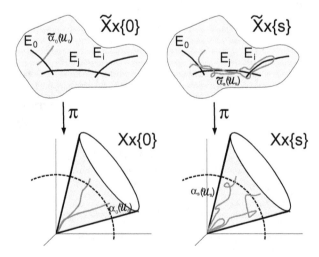

Figure 2. The lifting to the resolution of the special and generic arc of a wedge.

(i) it is reduced at $Z_0 \setminus E$ since σ is an isomorphism outside $E \times D_\delta$ and H_0 is reduced out of the origin;

(ii) Z_0 intersects transversely E_0 in a smooth point;

(iii) all the a_i's are non-negative since the divisor Y_0 is effective;

(iv) some a_i are strictly non-zero, in particular a_0, since α realizes an adjacency and then $\tilde{\beta}|_{\mathcal{U}}$ has indeterminacy.

Assuming the existence of a wedge realizing an adjacency, we have found a deformation Y_s of some Y_0 (as in (4.5) and satisfying (i)–(iv)) that has the following properties:

(a) Y_s is reduced for $s \neq 0$ small enough;

(b) its normalization, i.e., $\mathcal{U}_s$, is diffeomorphic to a disk;

(c) its boundary, i.e., $\tilde{\alpha}_s(\partial \mathcal{U}_s)$, is an $\mathbb{S}^1$ that degenerates to the boundary of Y_0, i.e., $\tilde{\alpha}_0(U) \cap \partial \tilde{X}_\epsilon$;

(d) Y_s meets $E_j \neq 0$.

The remaining part of the proof consists in proving that the Euler characteristic of the normalization of such a deformation Y_s of Y_0 is less than or equal to 0, which contradicts (b).

5. The Euler Characteristic Estimates

To simplify the computation of the Euler characteristic estimates, we assume the minimal resolution of (X, O) has as exceptional divisor a simple normal crossings divisor. This is the first case that we discovered. The general case is technically more elaborate, but follows essentially the same ideas. It may be checked in [**4**].

Let Y_0 be a Cartier divisor as in (4.5) that satisfies (i)–(iv). Consider a deformation Y_s of Y_0 satisfying (a)–(d). Let $n : \mathcal{U}_s \to Y_s$ be its normalization.

We consider a tubular neighborhood of Y_0 inside $\tilde{X}$ as a union of the following sets (Figures 3 and 4):

[h!]

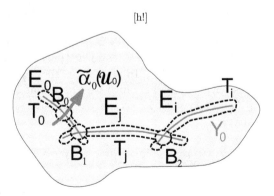

Figure 3. Adapted tubular neighborhood of Y_0.

Figure 4. The normalization $\mathcal{U}_s$ of Y_s inside the tubular neighborhood of Y_0 is a disk. In the picture, we see the result of cutting Y_s along the boundary of the Milnor balls B_i around the normal crossings of Y_0. Each piece is either $n^{-1}(B_i)$ or $n^{-1}(T_j)$. The exterior piece that we call A satisfies that $n(A)$ is contained in B_0.

- a Milnor ball $B_0 := B(Z_0 \cap E, \epsilon_0)$ for Y_0 around the meeting point of Z_0 and E;
- Milnor balls $B_1, \ldots, B_k$ centered at each of the singular points of E^{red};
- tubular neighborhoods $T_1, \ldots, T_r$ contained in $\tilde{X} \backslash \bigcup_{j=0}^{k} B_j$ around each $E_i \setminus \bigcup_{j=0,\ldots,k} B_j$ such that there exist strong deformation retracts

$$\zeta_i : T_i \to E_i \Bigg\backslash \bigcup_{j=0,\ldots,k} B_j$$

(see [**4**] for technical details).

For s small enough, we have

$$Y_s \subset B_0 \cup \bigcup_{j=1}^{k} B_j \cup \bigcup_{i=0}^{r} T_i$$

By the choice of the Milnor balls, we have that for s small enough, we have transversality of Y_s and the boundaries of the B_j's and T_j's (see [**4**] for technical details).

We are going to give an estimate for $\chi(\mathcal{U}_s)$ splitting $\mathcal{U}_s$ as the union of $n^{-1}(B_j)$ and $n^{-1}(T_i)$. Note that $n^{-1}(B_j)$ and $n^{-1}(T_j)$ are respectively the normalization of $Y_s \cap B_j$ and $Y_s \cap T_j$. In particular, they are disjoint unions of Riemann surfaces with boundary. Since we know that $\mathcal{U}_s$ is a disk and B_i has transversal boundary with Y_s, we have a decomposition of $\mathcal{U}_s$ as in Figure 4. Furthermore,

$$(5.1a) \qquad \chi(\mathcal{U}_s) = \chi(n^{-1}(B_0)) + \sum_{j=1}^{k} \chi(n^{-1}(B_j)) + \sum_{i=0}^{r} \chi(n^{-1}(T_j)).$$

We will separately give estimates for each of the summands on the right-hand side of the equality (Figure 6).

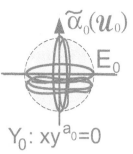

Figure 5. Counting the maximal number of disk images in $Y_s \cap B_0$ as a reduced deformation of $Y_0 \cap B_0$ of equation $xy^{a_0} = 0$.

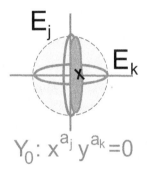

Figure 6. Counting the maximal number of disk images in $Y_s \cap B_i$ as a reduced deformation of $Y_0 \cap B_i$ of equation $x^{a_j} y^{a_k} = 0$.

5.1. Bound in B_0

The set $Y_0 \cap B_0$ is defined by $f_0(x, y) = xy^{a_0} = 0$, where x and y are the coordinates of B_0. The divisor $Y_s \cap B_0$ is defined by some deformation $f_s(x, y) = 0$, where f_s is a 1-parameter holomorphic deformation of f_0, such that f_s is reduced for $s \neq 0$.

We observe that $n^{-1}(B_0)$ is a disjoint union of Riemann surfaces with boundary. The only connected orientable surface with boundary which has positive Euler characteristic is the disk. Hence, $\chi(\mathcal{U}_s)$ is bounded above by the number of connected components of $n^{-1}(B_0)$ that are disks.

There are at the most as many disks in $n^{-1}(B_0)$ as boundary components in $n^{-1}(B_0)$ which are at the most $a_0 + 1$ since they degenerate to the boundary components of $\{xy^{a_0} = 0\} \cap B_0$. But by (c) the component of $n^{-1}(B_0)$ whose boundary degenerates to the boundary of $x = 0$ in B_0, say A, is in fact the exterior component of $\mathcal{U}_s$ (see Figure 4) which cannot be a disk, unless it is the whole disk (and this is not possible because Y_s goes outside B_0 and meets E_j by (d)). Then, A has more than one boundary component. Then, $n^{-1}(B_0)$ has at the most $a_0 - 1$ disks and we have

(5.1b) $$\chi(n^{-1}(B_0)) \leq \#disks \leq a_0 - 1.$$

5.2. Bound in the balls B_i, $i \neq 0$

We have $Y_0 \cap B_i$ defined by $f_0(x, y) = x^{a_j} y^{a_k} = 0$, where x and y are the coordinates in B_i.

Again, the Euler characteristic of $n^{-1}(B_j)$ which is a disjoint union of Riemann surfaces with boundary is bounded from above by the number of disks. Whenever there is a disk D in $n^{-1}(B_i)$, since its boundary degenerates either to the boundary of $x^{a_j} = 0$ or $y^{a_k} = 0$ in B_i, we have that $n(D)$ will meet at least once either $y = 0$ or $x = 0$ in B_0. Then,

$$\chi(n^{-1}(B_i)) \leq \#Y_s \cap Y_0 \cap B_i \leq \sum_{p \in B_i} I_p(Y_s, E).$$

Summing up the estimates of all the balls B_i with $i \neq 0$, we have

$$\sum_{i=1}^{k} \chi(n^{-1}(B_i)) \leq Y_s \cdot E = \sum_{k} Y_s \cdot E_k.$$

Note that $Y_s \cdot E$ counts the returns (see Section 6) with multiplicity.

Now, we can use that the intersection multiplicity is stable by deformation, i.e., $Y_s \cdot E = Y_0 \cdot E$ to get

$$\sum_{i=1}^{k} \chi(n^{-1}(B_i)) \leq Y_s \cdot E = Y_0 \cdot E = \left(Z_0 + \sum_{i=0}^{r} a_i E_i \right) \cdot E$$

(5.1c) $$= 1 + \sum_{i,k=0,\dots,r} a_i E_i \cdot E_k.$$

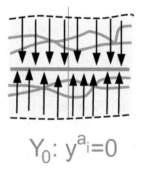

$$Y_0 \colon y^{a_i} = 0$$

Figure 7. Bounding the Euler characteristic of the normalization of $Y_s \cap T_i$ as a reduced deformation of $Y_0 \cap T_i$ of equation $y^{a_i} = 0$.

5.3. Bound in every T_i

To estimate $\chi(n^{-1}(Y_s \cap T_i))$, we consider the composition

$$\zeta_i \circ n : n^{-1}(Y_s \cap T_i) \to E_i \setminus \bigcup_j B_j.$$

Although there are some technicalities to be taken into account, for Euler characteristic computations, one may think that it is a holomorphic branched cover of Riemann surfaces of degree a_i (the reader may find in [4] a completely detailed proof of this). Then, by the Riemann–Hurwitz formula, we get

$$(5.1d) \qquad \chi(n^{-1}(T_i)) \leq a_i \chi \left(E_i \setminus \bigcup_j B_j \right).$$

We denote by g_i the genus of E_i. For $i \neq 0$, it is clear that $\sum_{k \neq j} E_i \cdot E_k$ counts the number of boundary components of $E_i \setminus \bigcup_j B_j$. The number of boundary components of $E_0 \setminus \bigcup B_i$ is $1 + \sum_{k \neq 0} E_0 \cdot E_k$ since E_0 meets also B_0.

Then, summing up the estimates (5.2) for all $i = 0, \ldots, r$, we have

$$(5.2) \qquad \begin{aligned} \sum_{i=0}^{r} \chi(n^{-1}(T_i)) \;\leq\;& a_0 \left(2 - 2g_0 - 1 - \sum_{k \neq 0} E_0 \cdot E_k \right) \\ &+ \sum_{j \neq 0} a_i \left(2 - 2g_i - \sum_{k \neq i} E_i \cdot E_k \right). \end{aligned}$$

5.4. Final estimate

Putting in (5.1a) the estimates (5.1b), (5.1c and (5.2), we get that

$$(5.1e) \qquad \chi(\mathcal{U}_s) \leq \sum_i a_i (2 - 2g_i + E_i \cdot E_i).$$

By negative definiteness, for any $0 \leq i \leq r$, the self-intersection $E_i \cdot E_i$ is a negative integer. Observe that, since $\pi : \tilde{X} \to X$ is the minimal resolution, for any $0 \leq i \leq r$, if $E_i \cdot E_i$ is equal to -1, then either the divisor E_i is singular or it has positive genus (otherwise, it is a smooth rational divisor with self-intersection equal to -1 and the resolution is non-minimal by the Castelnuovo contractibility Criterion). Since we are assuming that every E_i is smooth, we get that $2 - 2g_i + E_i \cdot E_i \leq 0$ for all $i = 0, \ldots, r$.

REMARK 5.2. Note that the right-hand side is the adjunction formula in the surface case, which computes the degree of the relative canonical sheaf at each irreducible component of the exceptional divisor. This serves as an inspiration for the higher dimensional proof of de Fernex and Docampo [2].

Suppose that α is a wedge that does not lift to the minimal resolution. This is equivalent to the existence of indeterminacy of the mapping $\pi^{-1}\circ\alpha$. This implies the inequality $a_0 > 0$ (which is fact is equivalent) and bound (5.1b).

On the other hand, the reader may observe that if the wedge α lifts to the minimal resolution, then the arguments leading to the estimate (5.1b) break down.

The rest of the estimates appearing in our proof are valid in complete generality. So, we conclude the following.

REMARK 5.3. Our proof shows that if α is a wedge such that the lifting to the minimal resolution of the special arc meets the exceptional divisor in a transverse way, then there is a lifting of α to the minimal resolution.

6. The Returns of a Wedge and Deformation Theoretic Ideas

Before the proof of the Nash conjecture in [4], the second author proved in her PhD the conjecture for the quotient surface singularities [21, 20]. In that proof, it is shown that, despite the local nature of arcs, at least semi-local techniques were needed in order to study the arc space and the existence of certain families of arcs or wedges.

In particular, it was observed that for a representative $\alpha|_U$ of a wedge, returns might be non-avoidable. We recall that a *return* is a point in $\alpha_s^{-1}(0)$ different from the origin for $s \neq 0$. A return $p \in \alpha_s^{-1}(0)$ is identified with the associated arc that consists in viewing $\alpha_s|_{U_s}$ as a germ at p. If one thinks of an arc as the image of a parameterization "starting at the singular point", the returns are the points where the parameterization passes again through the singular point. The contribution of the returns is crucial in the Euler characteristic estimates needed in our proof of the Nash conjecture (see Section 5.2).

The study of returns had a direct impact in the application of the valuative criterion to rule out adjacencies $N_E \subset N_F$ in [20]. The valuative criterion was first studied in [28, 22]. Given an exceptional prime divisor D over $(X,0)$, we denote by ord_D the associated divisorial valuation. The valuative criterion says that if there exists a germ $g \in \mathcal{O}_X$, such that $\mathrm{ord}_E(g) < \mathrm{ord}_F(g)$ where E and F are exceptional prime divisors, then the adjacency $N_E \subset N_F$ is not possible. Now, taking into account the returns, we can say that if we have an inequality $\mathrm{ord}_E(g) < \mathrm{ord}_F(g) + \mathrm{ord}_{F'}(g)$, then there is no wedge realizing the adjacency $N_E \subset N_F$ with a return with lifting by F' (nor a wedge realizing the adjacency $N_E \subset N_{F'}$ with a return by F).

This idea is applied in [20] more conveniently for the pullback of the wedges by the quotient map $q : (\mathbb{C}^2, 0) \to (X, 0)$ for a quotient surface singularity.

In [4], this criterion is completely understood in Section 3.2 as follows. Consider a wedge realizing an adjacency $N_{E_0} \subset N_{E_j}$ as in Section 4. Since the divisor Y_s defined in (4.3)–(4.4) is a deformation of the divisor Y_0, we have the equality

$$(6.1) \qquad Y_0 \cdot E_i = Y_s \cdot E_i$$

for any i. Recall notation $Y_0 = Z_0 + \sum_i a_i E_i$. Denote by b_i the intersection product of Y_s . E_i and by M the matrix of the intersection form in $H_2(\tilde{X}_\epsilon, \mathbb{Z})$ with respect to the basis $\{[E_0], \ldots, [E_r]\}$. Then, (6.1) can be expressed as follows:

$$(6.2) \qquad M(a_0, \ldots, a_r)^t = (1 - b_0, b_1, \ldots, b_r)^t.$$

The number b_i is the number of *returns* of the wedge through the divisor E_i counted with appropriate multiplicity.

An important observation is that all the entries of the inverse matrix M^{-1} are non-positive (see [4, Lemma 10]).

The equality (6.2) can be used to prove that wedges realizing certain adjacencies with certain prescribed returns b_i do not exist: the existence of such a wedge is impossible if the solution $a_0, \ldots, a_n$ of (6.2) has either a negative or a non-integral entry.

Moreover, to finish the proof in [20] for the E_8 singularities, further arguments using deformation theory were needed. There, wedges realizing an adjacency $N_{E_0} \subset N_{E_j}$ with a given special arc are seen as δ-constant deformations of the curve parameterized by the special arc. Then, the versal deformation of the curve parameterized by the special arc is computed. The codimensions of the δ-constant stratum and the codimension of the stratum of curves with the topological type of the generic curve of a family parameterized by a wedge representative with prescribed returns are computed. The inequality that these codimensions satisfy is not compatible with the existence of such a wedge (see [20, Proposition 4.5]).

7. The Proof by de Fernex and Docampo for the Higher Dimensional Case

De Fernex and Docampo figured out an algebro-geometric proof of the Nash conjecture based on bounds of coefficients of suitable relative canonical sheaves [2]. This enabled them to formulate and prove a correct statement of the Nash correspondence in higher dimension.

A *terminal valuation* is a divisorial valuation on X, such that there exists a terminal minimal model $\pi : Y \to X$ of X, such that the center of the valuation is a divisor in Y. The centers of terminal valuations are essential divisors. The main theorem of de Fernex and Docampo is as follows.

THEOREM 7.1 (de Fernex, Docampo). *Terminal valuations are at the image of the Nash map.*

The beginning of the proof is similar to the surface case: assuming that the result is false, they derive the existence of a wedge, such that its special arc lifts to Y in a transverse way to the center of a terminal valuation, but that cannot be lifted to Y. Afterwards, assuming the existence of such a wedge, they bound a coefficient for a relative canonical sheaf in two different ways and produce a contradiction. So, in fact, they prove the following.

THEOREM 7.2 (de Fernex, Docampo). *Let $\pi : Y \to X$ be a terminal model of X. Any wedge α, such that its special arc lifts to Y in a transverse way to the center of a terminal valuation, admits a lifting to Y.*

We refer to the original paper for a complete explanation of their proof. Here, instead, we explain the main ideas of the proof in the context of surfaces. In this case, $Y = \tilde{X}$ where $\tilde{X}$ is the unique terminal model which is the minimal resolution $\pi : \tilde{X} \to X$.

On the one hand, it is easier to digest, and all main ideas appear in this case. On the other hand, by doing it, we derive a precise set of numerical equalities (see (7.5)) that are satisfied for any non-constant wedge $\alpha : (\mathbb{C}^2, O) \to X$ not lifting to the minimal resolution and such that $\pi^{-1} \circ \alpha$ is a meromorphic map with an only indeterminacy point, regardless of whether this wedge has special arc lifting transversely or not (this means any wedge that is used in practice). This set of equalities have not been observed before. If one assumes that the special arc of the wedge lifts transversely, one may derive a chain of inequalities giving the contradiction in a straightforward way from this set of equalities.

Let $\alpha : (\mathbb{C}^2, O) \to X$ be any wedge so that $\pi^{-1} \circ \alpha$ is a rational map from $(\mathbb{C}^2, O)$ to $\tilde{X}$ and such that the special arc of the wedge. Let $\sigma : Z \to (\mathbb{C}^2, O)$ be the minimal sequence of blow-ups at points resolving the indetermination of $\pi^{-1} \circ \alpha$. Let $\beta : Z \to \tilde{X}$ be the map, such that $\pi \circ \beta = \alpha \circ \sigma$. De Fernex and Docampo shift the computation from $\tilde{X}$ to Z.

Let $F = \sum_{i=1}^{m} F_i$ and $E = \sum_{i=1}^{n} E_i$ be the decomposition in irreducible components of σ and π, respectively. In $(\mathbb{C}^2, O)$, we consider coordinates (t, s) so that t is the arc variable and s is the deformation parameter. The special arc of the wedge is then $\alpha(t, 0)$. We order the components so that F_1 is the unique component where the strict transform of $V(s)$ meets.

Denote by $K_Z = \sum_i a_i F_i$ the canonical divisor of Z. It is the only representative of the canonical class K_Z supported at the exceptional divisor. Each a_i is positive, and a simple observation on the behavior of the canonical divisor under blow-up shows the following important remark.

REMARK 7.3. The number a_1 is the number of blowing-up centers touching the strict transform of $V(s)$.

Since β is a morphism between smooth spaces, the relative canonical class $K_{Z/\tilde{X}} := K_Z - \beta^* K_{\tilde{X}}$ may be represented by the divisor associated with the Jacobian of β. This is an effective divisor. When we write $K_{Z/\tilde{X}}$, we mean such a divisor. We decompose it as

$$K_{Z/\tilde{X}} = K_{Z/\tilde{X}}^{\text{exc}} + K_{Z/\tilde{X}}^{\text{hor}},$$

where $K_{Z/\tilde{X}}^{\text{exc}}$ is the part with support on the exceptional divisor of π and $K_{Z/\tilde{X}}^{\text{hor}}$ the complement. We have the equality

(7.4) $$K_Z - K_{Z/\tilde{X}}^{\text{exc}} = K_{Z/\tilde{X}}^{\text{hor}} + \beta^* K_{\tilde{X}}.$$

The left-hand side is a divisor concentrated in the exceptional set of σ.

In order to express the right-hand side as a divisor concentrated in the exceptional set, we let M be the intersection matrix of the collection of divisors $\{F_i\}$ in Z. Since σ is a sequence of blow-ups, we have that M is unimodular and that its inverse M^{-1} is the matrix whose ith column $(m_{1,i}, \ldots, m_{n,i})^t$ is obtained as follows: let the C_i be the curve in $(\mathbb{C}^2, O)$ given by the image of σ of a cuvette transverse to F_i. Then, its total transform to Z is

$$\sigma^* C_i = \sum_j -m_{j,i} F_j.$$

As a consequence, we obtain that all the entries of M^{-1} are strictly negative (this is a general phenomenon which is well known, see, for example, [4, Lemma 10] for the proof for general normal surface singularities).

If we express $K_{Z/\tilde{X}}^{\text{exc}} = \sum_i b_i F_i$, and define the intersection numbers,

$$c_i := K_{Z/\tilde{X}}^{\text{hor}} \cdot F_i,$$

$$d_i := \beta^* K_{\tilde{X}} \cdot F_i,$$

Equality (7.4) becomes the following system of numerical equalities:

(7.5) $$(a_1 - b_1, \ldots, a_n - b_n)^t = M^{-1}(c_1 + d_1, \ldots, c_n + d_n).$$

This numerical equality works for any resolution $\pi : \tilde{X} \to X$ (non-necessarily minimal) and any wedge α, so it may have applications in other problems. One instance could be the generalized Nash problem explained in Section 8.

Observe that, since $K_{Z/\tilde{X}}^{\text{hor}}$ has no component included in the exceptional divisor, each c_i is non-negative.

If we assume now that π is the minimal resolution, we have

$$d_i = \beta^* K_{\tilde{X}} \cdot F_i = K_{\tilde{X}} \cdot \beta_*(F_i),$$

which is non-negative by adjunction formula, using the fact that $\pi : \tilde{X} \to X$ is the minimal resolution.

This means that the right-hand side in equation (7.5) is *non-positive*.

In order to prove Theorem 7.2 for the surface case, we assume that the wedge has the special arc lifting transversely to the exceptional divisor and estimate the coefficient $a_1 - b_1$ on the left-hand side of equation (7.5).

By Remark 7.3, if the wedge α does not lift to $\tilde{X}$, then a_1 is strictly positive and integral. Since the right-hand side of equation (7.5) is non-positive, in order to finish the proof, it is enough to prove the strict inequality

$$b_1 < 1.$$

Let $\rho : Z \to Z'$ and $\beta' : Z' \to \tilde{X}$ be such that the factorization $\beta = \beta' \circ \rho$ consists in collapsing all non-dicritical components of the exceptional divisor in Z (a component F_i is non-dicritical if $\beta(F_i)$ is a point). The surface Z' has sandwiched singularities, which are rational and $\mathbb{Q}$-Gorenstein. Then, the canonical divisor $K_{Z'}$ is $\mathbb{Q}$-Cartier. Therefore, we may define the relative canonical class $K_{Z/Z'}$. This class has a unique representative as a $\mathbb{Q}$-divisor supported in the exceptional set of ρ, such that all its coefficients are non-positive.

We have the equality $K_{Z/\tilde{X}} = K_{Z/Z'} + \rho^* K_{Z'/\tilde{X}}$. By the non-positivity of the coefficients of $K_{Z/Z'}$ we get

$$b_1 = \mathrm{ord}_{F_1}(K_{Z/\tilde{X}}) = \mathrm{ord}_{F_1}(K_{Z/Z'} + \rho^* K_{Z'/\tilde{X}}) \leq \mathrm{ord}_{F_1}(\rho^* K_{Z'/\tilde{X}}).$$

We make an abuse of language and denote the components of the exceptional divisor of Z' by the same name that they have in Z. In order to estimate b_1, we enumerate $\{F_{i_1}, \ldots, F_{i_l}\}$ the components of the exceptional set of Z', which contain the image by ρ of F_1. This is a subset of the components of F not collapsed by ρ. Observe that if F_1 is dicritical, then this set of components has F_1 as a unique element, and the estimate that we will prove right away becomes much easier.

The special arc of the wedge α lifts transversely through an irreducible component of E. We enumerate the components so that this component is E_1. Then, for each of the components F_{i_j}, we have that $\beta(F_{i_j}) = E_1$.

Before proving our final estimate, we need the following observation:

The following equality holds : $\mathrm{ord}_{F_{i_j}}(K_{Z'/\tilde{X}}) = \mathrm{ord}_{F_{i_j}}((\beta')^* E_1) - 1$.

This holds because $K_{Z'/\tilde{X}}$ is given at smooth points by the divisor associated with the jacobian of β', and at a generic point of F_{i_j}, the mapping β can be expressed in local coordinates as $\beta(u, v) = (u^a, v)$, where $a = \mathrm{ord}_{F_{i_j}}((\beta')^* E_1)$.

The last estimate we need is

$$\mathrm{ord}_{F_1}(\rho^* K_{Z'/Y}) = \mathrm{ord}_{F_1}\left(\sum_{j=1}^{l} \mathrm{ord}_{F_{i_j}}(K_{Z'/\tilde{X}})\rho^* F_{i_j}\right)$$

$$= \sum_{j=1}^{l} \mathrm{ord}_{F_1}(\mathrm{ord}_{F_{i_j}}((\beta')^* E_1) - 1)\rho^* F_{i_j}$$

$$< \sum_{j=1}^{l} \mathrm{ord}_{F_1}(\mathrm{ord}_{F_{i_j}}(((\beta')^*E_1))\rho^*F_{i_j})$$

$$= \mathrm{ord}_{F_1}\rho^*(\beta')^*E_1 = \mathrm{ord}_{F_1}\beta^*E_1 = 1.$$

The last equality holds because the special arc lifts transversely by E_1. This concludes the proof.

8. The Generalized Nash Problem and the Classical Adjacency Problem

Let X be a normal surface singularity. *The Generalized Nash Problem* consists in characterizing the pair of divisors E, F appearing in resolutions of X, such that the adjacency $N_F \subset N_E$ holds.

For our proof of the Nash conjecture, it is essential to construct a holomorphic wedge α, as we have explained before. This was achieved in [3]. The technique developed to achieve this gave, as a byproduct, a proof of the fact that the validity of the Nash conjecture only depends on the topology of the link of the surface singularity, or equivalently, in the combinatorics of the minimal good resolution. The same technique could be adapted to prove that the generalized Nash problem is a topological problem in the following sense.

Since the generalized Nash problem is wide open even in the case in which X is smooth, we concentrate on this case. To any two exceptional divisors E and F of a sequence of blow-ups at the origin of X, we may associate a decorated graph as follows: consider the minimal sequence of blow-ups of $\pi : Y \to X$, where both E and F appear. Decorate the dual graph of the exceptional divisor of π attaching to each vertex the weight given by the self-intersection of the corresponding divisor. Finally, add labels E and F to the vertices corresponding to the divisors E and F, respectively. In [5], we proved the following.

THEOREM 8.1. *Let (E_1, F_1) and (E_2, F_2) be two pairs of divisors having the same associated graph. Then, the adjacency $N_{F_1} \subset N_{E_1}$ is satisfied if and only if the adjacency, $N_{F_2} \subset N_{E_2}$ is satisfied.*

As a consequence, we could improve the discrepancy obstruction for adjacencies, see [5, Corollary 4.17 and 4.19]. Furthermore, we get a nice structure of nested Nash sets in the arc space of $\mathbb{C}^2$ and we made some conjectures about it, see [5, Conjectures 1 and 2].

For the sake of completeness, we summarize very briefly the other main result of [5].

Given a prime divisor E over the origin of X, we consider its associated valuation ν_E. It is easy to see that if we have the adjacency $N_F \subset N_E$, then the inequality $\nu_E \leq \nu_F$ holds. However, this criterion is not enough to characterize the Nash adjacencies (see Section 6). Our second main result is a characterization of the previous inequality in terms of deformations of plane curves.

We say that a plane curve germ C is *associated with* E in a model $\pi :$ $Y \to \mathbb{C}^2$, where E appears if its strict transform by π meets E transversely at a point which does not meet the singular set of the exceptional divisor of π. A deformation g_s of function germs is a holomorphic function depending holomorphically on a parameter s. It is linear if it is of the form $g_0 + sh$ for g_0 and h holomorphic.

THEOREM 8.2. *Let E and F be prime divisors over the origin of $\mathbb{C}^2$ and let S be the minimal model containing both divisors. The following conditions are equivalent:*

(1) $\nu_E \leq \nu_F$.
(2) *There exists a deformation g_s with g_0 associated with F in S and g_s associated with E in S, for $s \neq 0$ small enough.*
(3) *There exists a linear deformation g_s with g_0 associated with F in S and g_s associated with E in S, for $s \neq 0$ small enough.*

In fact, in [5], we prove a more general version which allows non-prime divisors E and F.

This theorem provides a very easy way of producing adjacencies of plane curve singularities. Using this, we were able to recover most Arnol'd adjacencies, see [5, Section 3.4] for detailed explanations.

9. Holomorphic Arcs

In the proof of the Nash conjecture for surfaces, we study arcs and wedges from a convergent viewpoint and take representatives. In this sense, a wedge is for us a deformation of holomorphic maps from a disc to a representative of the singularity. At a generic parameter, the preimage of the singular point in general contains more points in the disc than just the origin. These points are unavoidable and we call them *returns* following [20] (see Section 6).

Kollár and Nemethi started in [13] the systematic study of convergent arc spaces as opposed to the classical formal arc spaces. We briefly summarize their main results and questions.

Let D denote the closed unit disk. A holomorphic map defined on D is the restriction to D of a holomorphic map in an open neighborhood of D. Let X be a singularity.

DEFINITION 9.1. A *complex analytic* arc is a holomorphic map $\gamma :$ $D \to X$, such that the preimage of the singular set does not intersect the boundary ∂D. A *short complex analytic arc* is a complex analytic arc, such that the preimage of the singular set is just 1 point. A *deformation of a complex analytic arc* parameterized by an analytic space Λ is a holomorphic map

$$\alpha : D \times \Lambda \to X,$$

such that for any $s \in \Lambda$, the restriction $\alpha_s := \alpha|_{D \times \{s\}}$ is a complex analytic arc. If all the arcs appearing are short complex analytic arcs, we say that α is a *deformation of short complex analytic arcs*.

REMARK 9.2. What we do in Section 4 is to derive from a convergent wedge a deformation of a complex analytic arc parameterized by a disk Λ. The special arc at this deformation is a short arc, but the generic arcs α_s are not short arcs, in general, due to the existence of returns. The topological analysis of that deformation of complex analytic arcs yields the proof of the Nash conjecture.

Denote by $\mathrm{Arc}(X)$ and $\mathrm{ShArc}(X)$ the sets of convergent analytic arcs in X. Kollár and Nemethi give natural metrics on these spaces, which endow them with a topology. Given an arc $\gamma : D \to X$, since the preimage of the singular set $\mathrm{Sing}(X)$ is disjoint from the circle ∂D, the restriction $\gamma|_{\partial D}$ defines an element of the fundamental group modulo conjugation $\pi_1(X \setminus \mathrm{Sing}(X))/(conjugation)$. Since this element does not change by continuous deformation of the arc γ, we have defined "winding number maps":

$$\pi_0(\mathrm{Arc}(X)) \to \pi_1(X \setminus \mathrm{Sing}(X))/(conjugation),$$

$$\pi_0(\mathrm{ShArc}(X)) \to \pi_1(X \setminus \mathrm{Sing}(X))/(conjugation).$$

The main result in [**13**] concerns short arcs:

THEOREM 9.3 (Kollár, Nemethi). *The winding number map*

$$\pi_0(\mathrm{ShArc}(X)) \to \pi_1(X \setminus \mathrm{Sing}(X))/(conjugation)$$

is injective for any normal surface singularity X. It is bijective for quotient surface singularities.

For general normal surface singularities, the winding number map is far from being surjective, but its image is described in [**13**] in terms of the combinatorics of the resolution, or what is the same, the topology of the link.

On the other hand, the winding number map for long arcs, which is the one that is more related with the original Nash question, is not well understood.

PROBLEM 9.4 (Kollár, Nemethi). Is the winding number map

$$\pi_0(\mathrm{Arc}(X)) \to \pi_1(X \setminus \mathrm{Sing}(X))/(conjugation)$$

injective?

In [**13**], many other open problems are proposed. Some interesting ones are concerned with the definition of a "finite type" holomorphic atlas in $\mathrm{Arc}(X)$ (see [**13**, Conjecture 72]) and with the existence of a curve selection lemma in $\mathrm{Arc}(X)$.

Acknowledgments

The first author is partially supported by IAS and by ERCEA 615655 NMST Consolidator Grant, MINECO by the project reference MTM2016-76868-C2-1-P (UCM), by the Basque Government through the BERC 2018–2021 program and Gobierno Vasco Grant IT1094-16, by the Spanish Ministry of Science, Innovation and Universities: BCAM Severo Ochoa accreditation SEV-2017-0718 and by Bolsa Pesquisador Visitante Especial (PVE) — Ciencias sem Fronteiras/CNPq Project number: 401947/2013-0. The second author is partially supported by MINECO by the project MTM 2017–89420 and MTM2016-76868-C2-1-P, by CNPq Project Ciencias sem Fronteiras number 401947/2013-0, and by ERCEA 615655 NMST Consolidator Grant.

Bibliography

[1] T. de Fernex, Three-dimensional counter-examples to the Nash problem. *Compositio Math.*, 149:1519–1534, 2013.

[2] T. de Fernex and R. Docampo, Terminal valuations and the Nash problem. *Invent. Math.*, 203:303–331, 2016.

[3] J. Fernández de Bobadilla, The Nash problem for surface singularities is a topological problem. *Adv. Math.*, 230(1):131–176, 2012.

[4] J. Fernández de Bobadilla and M. Pe Pereira, The Nash problem for surfaces. *Ann. of Math.*, 176:2003–2029, 2012.

[5] J. Fernández de Bobadilla, M. Pe Pereira and P. Popescu-Pampu, On the generalized Nash problem for smooth germs and adjacencies of curve singularities. *Adv. Math.*, 320: 1269–1317, 2017.

[6] P. González Pérez, Bijectiveness of the Nash Map for Quasi-Ordinary Hypersurface Singularities. *Int. Math. Res. Not.*, 2007(19): Art. ID rnm076, 2007.

[7] J. Johnson and J. Kollár, Arc spaces of cA-type singularities. *J. Singul.*, 7:238–252, 2013.

[8] J. Johnson and J. Kollár, Arcology. *Amer. Math. Monthly*, 123(6):519–541, 2016.

[9] S. Ishii and J. Kollár, The Nash problem on arc families of singularities. *Duke Math. J.*, 120(3):601–620, 2003.

[10] S. Ishii, Arcs, valuations and the Nash map. *J. Reine Angew. Math.*, 588:71–92, 2005.

[11] S. Ishii, The local Nash problem on arc families of singularities. *Ann. Inst. Fourier, Grenoble*, 56:1207–1224, 2006.

[12] J. Kollár, Nash work in algebraic geometry, *Bull. Amer. Math. Soc. (N.S.)*, 54(2):307–324, 2017.

[13] J. Kollár and A. Nemethi, Holomorphic arcs on singularities. *Invent. Math.*, 200(1):97–147, 2015.

[14] M. Lejeune-Jalabert, *Arcs Analytiques et Résolution Minimale des Singularities des Surfaces Quasi-homogenes*. Lecture Notes in Mathematics, Vol. 777 Springer, 1980, pp. 303–336.

[15] M. Lejeune-Jalabert and A. Reguera-López, Arcs and wedges on sandwiched surface singularities. *Amer. J. Math.*, 121:1191–1213, 1999.

[16] M. Lejeune-Jalabert and A. Reguera-López, Exceptional divisors which are not uniruled belong to the image of the Nash map, Preprint, 2008, arxiv:08011.2421.

[17] J. Milnor, *Singular Points of Complex Hypersurfaces, Annals of Mathematics Studies*, Vol. 61. Princeton University Press, Princeton, NJ; University of Tokyo Press, Tokyo, 1968.

[18] M. Morales, The Nash problem on arcs for surface singularities. Preprint, 2006, arXiv:math.AG/0609629.

[19] J. Nash, Arc structure of singularities. A celebration of John F. Nash, Jr., *Duke Math. J.*, 81(1):31–38, 1995.

[20] M. Pe Pereira, Nash problem for quotient surface singularities. *J. London Math. Soc.*, (2), 87(1):177–203, 2013.

[21] M. Pe Pereira, On Nash problem for quotient surface singularities. Phd Thesis, Universidad Complutense de Madrid, 2011.

[22] C. Plénat, À propos du problème des arcs de Nash. *Ann. Inst. Fourier (Grenoble)*, 55(3):805–823, 2005.

[23] C. Plénat, The Nash problem of arcs and the rational double points D_n. *Ann. Inst. Fourier (Grenoble)*, 58(7):2249–2278, 2008.

[24] C. Plénat and P. Popescu-Pampu, A class of non-rational surface singularities for which the Nash map is bijective. *Bull. Soc. Math. France*, 134(3):383–394, 2006.

[25] C. Plénat and P. Popescu-Pampu, Families of higher dimensional germs with bijective Nash map. *Kodai Math. J.*, 31(2):199–218, 2008.

[26] C. Plénat and M. Spivakovsky, The Nash problem of arcs and the rational double point E_6. *Kodai Math. J.*, 35:173–213, 2012.

[27] C. Plénat and M. Spivakovsky, Nash problem and its solution. *Newsletter EMS*, 88:17–24, 2013.

[28] A. Reguera-López, Families of arcs on rational surface singularities. *Manuscripta Math.*, 88(3):321–333, 1995.

[29] A. Reguera-López, Image of the Nash map in terms of wedges. *C. R. Math. Acad. Sci. Paris*, 338(5):385–390, 2004.

[30] A. Reguera-López, A curve selection lemma in spaces of arcs and the image of the Nash map. *Compositio Math.*, 142:119–130, 2006.

12

Motivic and Analytic Nearby Fibers at Infinity and Bifurcation Sets

Lorenzo Fantini[*,‡] **and Michel Raibaut**[†,§]

[*]*CNRS, Sorbonne Université, Université Paris Diderot, Institut de Mathématiques de Jussieu-Paris Rive Gauche, IMJ-PRG, F-75005 Paris, France*
[†]*Laboratoire de Mathématiques Univ. Grenoble Alpes, Université Savoie Mont Blanc, Bâtiment Chablais, Campus Scientifique, Le Bourget du Lac, 73376 Cedex, France*
[‡]*lorenzo.fantini@imj-prg.fr*
[§]*michel.raibaut@univ-smb.fr*

In this chapter, we use motivic integration and non-archimedean analytic geometry to study the singularities at infinity of the fibers of a polynomial map $f\colon \mathbb{A}_{\mathbb{C}}^d \to \mathbb{A}_{\mathbb{C}}^1$. We show that the motive $S_{f,a}^\infty$ of the motivic nearby cycles at infinity of f for a value a is a motivic generalization of the classical invariant $\lambda_f(a)$, an integer that measures a lack of equisingularity at infinity in the fiber $f^{-1}(a)$. We then introduce a non-archimedean analytic nearby fiber at infinity $\mathcal{F}_{f,a}^\infty$ whose motivic volume recovers the motive $S_{f,a}^\infty$. With each of $S_{f,a}^\infty$ and $\mathcal{F}_{f,a}^\infty$, one can naturally associated a bifurcation set; we show that the first one always contains the second one, and that both contain the classical topological bifurcation set of f if f has isolated singularities at infinity.

1. Introduction

Let U be a smooth and irreducible complex algebraic variety and let $f\colon U \to \mathbb{A}_{\mathbb{C}}^1$ be a dominant map. It is well known that there exists a finite subset B of $\mathbb{A}_{\mathbb{C}}^1(\mathbb{C})$ such that the map

$$f|_{U \setminus f^{-1}(B)}\colon U \setminus f^{-1}(B) \longrightarrow \mathbb{A}_{\mathbb{C}}^1 \setminus B$$

is a locally trivial C^∞-fibration (see, for example, [**33**]). The smallest such subset B is called the *topological bifurcation set* of f and is denoted by B_f^{top}. The topological bifurcation set of f contains the discriminant disc(f) of f, which is the set $f(\mathrm{crit}(f)) \subset \mathbb{A}^1_\mathbb{C}(\mathbb{C})$ image of its set of critical points, but the inclusion might be strict since f need not be proper. For example, the map $f \colon \mathbb{A}^2_\mathbb{C} \to \mathbb{A}^1_\mathbb{C}$ defined by the polynomial $f(x,y) = x(xy-1)$ is smooth, but has bifurcation set $\{0\}$; its fiber over 0 is the only one to be disconnected. This is related to the fact that, loosely speaking, $f(x,y)$ may have some "critical points at infinity"; one way of seeing this is to observe that the compactification of the fiber $f^{-1}(a)$ in $\mathbb{P}^2_\mathbb{C}$ will have a multibranch singularity for $a = 0$ and a cuspidal singularity otherwise.

Assume that $f \colon \mathbb{A}^d_\mathbb{C} \to \mathbb{A}^1_\mathbb{C}$ is a polynomial map. In some special cases, knowing the (compactly supported) Euler characteristic of a fiber $f^{-1}(a)$ is sufficient to determine whether a belongs to B_f^{top}. Indeed, if we denote by $f^{-1}(a_{\mathrm{gen}})$ a general fiber of f, we have

$$(1.1) \qquad B_f^{\mathrm{top}} = \left\{ a \in \mathbb{A}^1_\mathbb{C}(\mathbb{C}) \mid \chi_c\big(f^{-1}(a)\big) \neq \chi_c\big(f^{-1}(a_{\mathrm{gen}})\big) \right\}$$

in the case of curves, that is, whenever $d = 2$, see [**41**], and, more generally, whenever f has *isolated singularities at infinity* (that is, the closure of a general fiber $f^{-1}(a_{\mathrm{gen}})$ of f in $\mathbb{P}^d_\mathbb{C}$ has isolated singularities), see [**31**].

If the polynomial $f \in \mathbb{C}[x_1, \dots, x_d]$ has isolated singularities in $\mathbb{A}^d_\mathbb{C}$ (which does not necessarily mean that it has also isolated singularities at infinity), then Artal Bartolo–Luengo–Melle-Hernández proved in [**2**] that the Euler characteristic of each fiber $f^{-1}(a)$ can be computed in terms of the Euler characteristic of $f^{-1}(a_{\mathrm{gen}})$ and of some local numerical invariants, as we will now explain. To account for the singularities of the fibers, let $\mu_a(f)$ be the sum of the Milnor numbers of the singular points of $f^{-1}(a)$, so that a belongs to the discriminant of f if and only if $\mu_a(f) > 0$, and set $\mu(f) = \sum_{a \in \mathbb{A}^1_\mathbb{C}} \mu_a(f)$. On the other hand, to keep track of the behavior of f at infinity, consider $\lambda_a(f) = \mu\big(\mathbb{P}^n, \overline{f^{-1}(a)}, D\big) - \mu\big(\mathbb{P}^d, \overline{f^{-1}(a_{\mathrm{gen}})}, D\big)$, where $\overline{f^{-1}(a)}$ is the closure $f^{-1}(a)$ in $\mathbb{P}^d$, D is the divisor cut out by the homogeneous part of f of degree $\deg(f)$ in the hyperplane at infinity of $\mathbb{P}^d$, and $\mu\big(\mathbb{P}^n, \overline{f^{-1}(b)}, D\big)$ is the generalized Milnor number of Parusiński from [**30**]. The invariants $\lambda_a(f)$ vanish for all but finitely many values of a, therefore we can set $\lambda(f) = \sum_{a \in \mathbb{A}^1_\mathbb{C}} \lambda_a(f)$. By [**2**, Theorem 1.7], we then have the following equalities:

$$(1.2) \qquad \chi_c\big(f^{-1}(a)\big) = \chi_c\big(f^{-1}(a_{\mathrm{gen}})\big) + (-1)^d (\mu_a(f) + \lambda_a(f)),$$

$$(1.3) \qquad \chi_c\big(f^{-1}(a_{\mathrm{gen}})\big) = 1 + (-1)^{d-1} (\mu(f) + \lambda(f)).$$

Similar results were obtained in [**39**, **32**], and in the case of curves in [**37**, **41**, **6**]. In particular, whenever condition (1.1) also holds, one can deduce that

$$B_f^{\mathrm{top}} = \left\{ a \in \mathbb{A}_{\mathbb{C}}^1(\mathbb{C}) \mid \lambda_a(f) \neq 0 \right\} \cup \mathrm{disc}(f).$$

The invariant $\lambda_a(f)$ gives some rudimentary measure of the lack of equi-singularity of f at infinity at the value a. Since it is defined using Euler characteristics, it is natural to expect to be able to generalize $\lambda_a(f)$ to a motive in a Grothendieck ring of $\mathbb{C}$-varieties by using Denef and Loeser's motivic integration. In that direction, and with the goal of studying singularities at infinity, the second author defined in [34] a motive $S_{f,a}^\infty$, the so-called *motivic nearby cycles at infinity of f for the value a*, living in a localization $\mathcal{M}_{\mathbb{C}}^{\hat{\mu}}$ of the Grothendieck ring of $\mathbb{C}$-varieties with a good $\hat{\mu}$-action, and related to some other constructions due to Matsui–Takeuchi–Tibar in [23, 24, 38]. In order to do so, he used motivic integration techniques, following the work of Denef–Loeser [8], Bittner [3], and Guibert–Loeser–Merle [15]. This allowed him to define a *motivic bifurcation set B_f^{mot}* of f as the union of the discriminant of f with the set of values whose motivic nearby cycles at infinity do not vanish. This is a finite set and it coincides with B_f^{top} if f is a convenient and non-degenerate polynomial with respect to its Newton polygon at infinity (in the sense of [19]), as is shown in [34, Theorem 4.8]. However, no explicit link between $S_{f,a}^\infty$ and $\lambda_a(f)$ was established in [34].

The first contribution of this chapter is to prove that $S_{f,a}^\infty$ is indeed a motivic generalization of the numerical invariant $\lambda_a(f)$, whenever the latter is defined. In particular, we obtain the following result.

THEOREM. *Let f be a polynomial in $\mathbb{C}[x_1, \ldots, x_d]$ with isolated singularities in $\mathbb{A}_{\mathbb{C}}^d$. For any value a in $\mathbb{A}_{\mathbb{C}}^1(\mathbb{C})$, we have*

$$\chi_c\left(S_{f,a}^\infty\right) = (-1)^{d-1} \lambda_a(f).$$

We obtain this result as a consequence of a more general equality between Euler characteristic of motives, proven in Theorem 3.3, which in particular does not require f to have isolated singularities. As an application, we deduce the following fact.

THEOREM. *Let f be a polynomial in $\mathbb{C}[x_1, \ldots, x_d]$. Then we have*

$$\left\{ a \in \mathbb{A}_{\mathbb{C}}^1(\mathbb{C}) \mid \chi_c\left(f^{-1}(a)\right) \neq \chi_c\left(f^{-1}(a_{\mathrm{gen}})\right) \right\} \subset B_f^{\mathrm{mot}}.$$

In particular, B_f^{mot} contains B_f^{top} whenever condition (1.1) is satisfied, for example, if f has isolated singularities at infinity.

Let us briefly explain the definition of $S_{f,a}^\infty$ and give a short outline of the proof of the theorems. We can work in a slightly more general setting and assume that k is a field of characteristic zero containing all roots of unity, U is a smooth k-variety, and $f \colon U \to \mathbb{A}_k^1$ is a dominant morphism. We consider a compactification $(X, \hat{f})$ of (U, f), i.e., a k-variety X containing U as a dense open subset, together with a proper extension $\hat{f}$ of f to X. As is usual in motivic integration, the motive $S_{f,a}^{\infty}$ is defined as a limit of a generating series $\sum \operatorname{mes}(X_i) T^i$, where $\operatorname{mes}(X_i)$ is the motivic measure of a subset X_i of the scheme of arcs $\mathcal{L}(X)$ of X. Namely, X_i consists of arcs that have origin at infinity, that is in $X \setminus U$, order of contact i along the fiber $\hat{f}^{-1}(a)$, and are only mildly tangent to $X \setminus U$. As such, $S_{f,a}^{\infty}$ can be expressed as the difference of the images in $\mathcal{M}_k^{\hat{\mu}}$ of $S_{\hat{f}-a,U}$, the motivic nearby cycles of $\hat{f} - a$ supported by U, a motive defined by Guibert–Loeser–Merle [15] using a motivic integration procedure as above but imposing no condition on the origin of arcs, by the motivic nearby fiber S_{f-a} of Denef–Loeser [8], which is constructed from arcs with origin in U. We show in Theorem 3.3 that the resulting equality generalizes formula (1.2) by taking the Euler characteristics and using realization results for $S_{\hat{f}-a,U}$ and S_{f-a} from [15, 8]. The two theorems we stated above follow then from simple computations.

The second contribution of this chapter is the construction of a non-archimedean analytic version of $S_{f,a}^{\infty}$. Namely, the *analytic nearby fiber* $\mathcal{F}_{f,a}^{\infty}$ *of f for the value a* is the analytic space over the field $K = k((t))$ (endowed with a given t-absolute value) defined as

$$\mathcal{F}_{f,a}^{\infty} = (U_K)^{\mathrm{an}} \setminus \left(\widehat{U_R}\right)^{\beth},$$

where $\left(\widehat{U_R}\right)^{\beth}$ is the space of analytic nearby cycles of [28], that is, the analytic space associated to the formal completion of U along the fiber $f^{-1}(a)$, and $(U_K)^{\mathrm{an}}$ is the analytification of the base change of U to K via the map $f - a$. This definition is analogous to the one of the motive $S_{f,a}^{\infty}$, since if $(X, \hat{f})$ is a compactification of (U, f), then $\mathcal{F}_{f,a}^{\infty}$ consists of those points in the analytification $(X_K)^{\mathrm{an}}$ of X_K that have origin on $X_K \setminus U_K$ but are not completely contained in $X_K \setminus U_K$. One advantage of $\mathcal{F}_{f,a}^{\infty}$ over $S_{f,a}^{\infty}$ is that, while the latter is defined in terms of an extension of f to a compactification of U, and only later proven to be independent of such a choice, the former is defined exclusively in terms of f.

We then construct a motivic specialization of $\mathcal{F}_{f,a}^{\infty}$, its *Serre invariant* $\overline{S}(\mathcal{F}_{f,a}^{\infty})$, and use it to define a third bifurcation set, the *Serre bifurcation set* B_f^{ser} of f, as the union of the discriminant of f with the set of values a such that $\overline{S}(\mathcal{F}_{f,a}^{\infty})$ does not vanish.

Using the decomposition of $\mathcal{F}_{f,a}^{\infty}$ as a difference, and realization results for the volume and Serre invariant of analytic spaces from [28, 5, 16], in Sections 4.4 and 4.5, we prove the following results.

THEOREM. *Let U be a smooth connected k-variety and let $f: U \to \mathbb{A}_k^1$ be a dominant morphism. For any value a in $\mathbb{A}_k^1(k)$, we have:*

(1) $\mathrm{Vol}\big(\mathcal{F}_{f,a}^{\infty}\big) = \mathbb{L}^{-(\dim U - 1)} S_{f,a}^{\infty}$ *in* $\mathcal{M}_k^{\hat{\mu}}$;

(2) $\overline{S}\big(\mathcal{F}_{f,a}^{\infty}\big) = S_{f,a}^{\infty} \bmod(\mathbb{L} - 1)$ *in* $\mathcal{M}_k^{\hat{\mu}}/(\mathbb{L} - 1)$;

(3) *The inclusions*

$$\big\{a \in \mathbb{A}_{\mathbb{C}}^1(\mathbb{C}) \,\big|\, \chi_c\big(f^{-1}(a)\big) \neq \chi_c\big(f^{-1}(a_{\mathrm{gen}})\big)\big\} \subset B_f^{\mathrm{ser}} \subset B_f^{\mathrm{mot}}.$$

We expect that a deep study of the geometry (and étale cohomology) of the analytic nearby fibers at infinity will lead to a better understanding of various phenomena of equisingularity at infinity, and we plan to study this topic further in an upcoming project.

Notation

In this chapter, we will freely use the following notations:

- k is a field of characteristic zero that contains all the roots of unity.
- A k-variety is a separated k-scheme of finite type.
- $\hat{\mu}$ is the projective limit $\varprojlim \mu_n$, where for any positive integer n we denote by μ_n the group scheme of nth roots of unity.
- R is the discrete valuation ring $k[[t]]$, endowed with the t-adic valuation.
- $K = k((t))$ is the fraction field of R, endowed with a fixed t-adic absolute value.
- For any integer n, we denote by $K(n) = K((t^{1/n}))$ the unique extension of degree n of K obtained by joining a nth root of t to K, and by $R(n)$ the normalization of R in $K(n)$.
- Given a k-variety X and a morphism $f \colon X \to \mathbb{A}_k^1 = \mathrm{Spec}\, k[t]$, we set $X_R = X \times_{\mathbb{A}_k^1} \mathrm{Spec}\, R$ and $X_K = X \times_{\mathbb{A}_k^1} \mathrm{Spec}\, K$.

2. Motivic Integration and Nearby Cycles

In this section, we review some basic constructions in motivic integration that will be used throughout the paper. We refer to [**9, 10, 20, 22, 15, 14**] thorough discussion of these notions.

2.1. Grothendieck rings

We say that an action of μ_n on a k-variety X is *good* if every μ_n-orbit is contained in some open affine subscheme of X, and that an action of $\hat{\mu}$ on X is *good* if it factors through a good μ_n-action for some n. If S is a k-variety, we denote by $\mathrm{Var}_S^{\hat{\mu}}$ the category of S-varieties with a good $\hat{\mu}$-action, that is, the category whose objects are the k-varieties X endowed with a good action of $\hat{\mu}$ and with a morphism $X \to S$ that is equivariant with respect to the trivial $\hat{\mu}$-action on S, and whose morphisms are $\hat{\mu}$-equivariant morphisms over S.

We denote by $K_0(\mathrm{Var}_S^{\hat{\mu}})$ the Grothendieck ring of $\mathrm{Var}_S^{\hat{\mu}}$. It is defined as the abelian group generated by the isomorphism classes $[X, \sigma]$ of the elements (X, σ) of $\mathrm{Var}_S^{\hat{\mu}}$, with the relations $[X, \sigma] = [Y, \sigma|_Y] + [X \setminus Y, \sigma|_{X \setminus Y}]$ if Y is a closed subvariety of X stable under the action of $\hat{\mu}$, and moreover $[X \times \mathbb{A}_k^n, \sigma] = [X \times \mathbb{A}_k^n, \sigma']$ if σ and σ' are two liftings of the same $\hat{\mu}$-action on X to an affine action on $X \times \mathbb{A}_n^k$. There is a natural ring structure on $K_0(\mathrm{Var}_S^{\hat{\mu}})$, the product being induced by the fiber product over S. In the rest of this chapter, we will simply write $[X]$ for $[X, \sigma_X]$, as no risk of confusion will arise.

We denote by $\mathcal{M}_S^{\hat{\mu}}$ the localization of $K_0(\mathrm{Var}_S^{\hat{\mu}})$ at the element $\mathbb{L}_S = [\mathbb{A}_S^1]$, i.e., the class of the affine line over S endowed with the trivial action. If $S = \mathrm{Spec}\, k$, we simply write $\mathcal{M}^{\hat{\mu}}$ and $\mathbb{L}$ for $\mathcal{M}_S^{\hat{\mu}}$ and $\mathbb{L}_S$, respectively, and if we only consider varieties with trivial $\hat{\mu}$-action, we obtain analogous Grothendieck rings $\mathcal{M}_k$ and $\mathcal{M}_S$.

If $f : S' \to S$ is a morphism of k-varieties, then the composition with f on the left induces a *direct image* group morphism

$$f_! \colon \mathcal{M}_{S'}^{\hat{\mu}} \to \mathcal{M}_S^{\hat{\mu}},$$

while taking the fiber product with S' over S induces an *inverse image* ring morphism

$$f^* \colon \mathcal{M}_S^{\hat{\mu}} \to \mathcal{M}_{S'}^{\hat{\mu}}.$$

We denote by $\mathcal{M}_S^{\hat{\mu}}[[T]]_{\mathrm{rat}}$ the $\mathcal{M}_S^{\hat{\mu}}$-submodule of $\mathcal{M}_S^{\hat{\mu}}[[T]]$ generated by 1 and by the finite products of terms $p_{e,i}(T) = \mathbb{L}^e T^i / (1 - \mathbb{L}^e T^i)$, with e in $\mathbb{Z}$ and i in $\mathbb{Z}_{>0}$. The formal series in $\mathcal{M}_S^{\hat{\mu}}[[T]]_{\mathrm{rat}}$ are said to be *rational*. There exists a unique $\mathcal{M}_S^{\hat{\mu}}$-linear morphism $\lim_{T \to \infty} \colon \mathcal{M}_S^{\hat{\mu}}[[T]]_{\mathrm{rat}} \to \mathcal{M}_S^{\hat{\mu}}$ such that for any finite subset $(e_i, j_i)_{I \in i}$ of $\mathbb{Z} \times \mathbb{Z}_{>0}$, we have $\lim_{T \to \infty} \left(\prod_I p_{e_i, j_i}(T) \right) = (-1)^{|I|}$.

2.2. Arcs on varieties

Let X be a k-variety of dimension d. We denote by $\mathcal{L}_n(X)$ the *space of n-jets* of X, i.e., the k-scheme of finite type whose functor of points is the following: for every k-algebra L, the L-points of $\mathcal{L}_n(X)$ are the morphisms $\mathrm{Spec}(L[s]/(s^{n+1})) \to X$. In particular, $\mathcal{L}_0(X)$ is canonically isomorphic to X itself. Right compositions with the canonical projections $L[s]/(s^{n+1}) \to L[s]/(s^n)$ yield morphisms $\mathcal{L}_n(X) \to \mathcal{L}_{n-1}(X)$, making $\{\mathcal{L}_n(X)\}_n$ into a projective system. These morphisms are $\mathbb{A}_k^d$-bundles when X is smooth of pure dimension d. The *arc space* of X, denoted by $\mathcal{L}(X)$, is the projective limit of the system $\{\mathcal{L}_n(X)\}_n$; we denote by $\pi_n \colon \mathcal{L}(X) \to \mathcal{L}_n(X)$ the canonical projections. The arc space of X is a k-scheme that is generally not of finite type. For every finite extension k' of k, the k'-rational points of $\mathcal{L}(X)$ parameterize the morphisms $\mathrm{Spec}\, k'[[s]] \to X$. If $\varphi \in \mathcal{L}(X)$ is an arc on X, the point $\pi_0(\varphi) \in X$ is called the *origin* of φ on X. Observe that μ_n acts canonically on $\mathcal{L}_n(X)$ and on $\mathcal{L}(X)$ by setting $\lambda.\varphi(s) = \varphi(\lambda s)$.

Assume that we have a morphism $f\colon X \to \mathbb{A}_k^1 = \operatorname{Spec} k[t]$. With a L-arc $\varphi\colon \operatorname{Spec} L[[s]] \to X$ on X, we can then associate its *order* $\operatorname{ord} f(\varphi) := \operatorname{ord}_s(\varphi^\#(f)) \in \mathbb{Z}_{\geq 0} \cup \{+\infty\}$, and its *angular component* $\overline{\mathrm{ac}}\, f(\varphi) \in L$, defined as the leading coefficient of the power series $\varphi^\#(f) \in L[[s]]$; by convention, $\overline{\mathrm{ac}}\, f(\varphi) = 0$ if $\varphi^\#(f) = 0$.

More generally, if F is a closed subvariety of X of coherent ideal sheaf $\mathcal{I}_F$, then the *contact order* of an arc $\varphi \in \mathcal{L}(X)$ along F is $\operatorname{ord}_F(\varphi) := \inf_g \operatorname{ord} \varphi^\#(g)$, where g runs among the local sections of $\mathcal{I}_F$ at the origin of φ. It is greater than zero if and only if the origin of φ lies in F, while it takes the value infinity if φ is an arc on F. For example, if we have a morphism $f\colon X \to \mathbb{A}_k^1$ as above, then the contact order of φ along the fiber $f^{-1}(0)$ is precisely $\operatorname{ord} f(\varphi)$.

2.3. Motivic nearby cycles

Let X be a purely dimensional k-variety, let $f\colon X \to \mathbb{A}_k^1$ be a morphism, and denote by $X_0(f)$ the zero locus of f in X. By work of Denef–Loeser [**8, 11**]Bittner [**3**], and Guibert–Loeser–Merle [**15**], there exists a group morphism

$$\mathcal{S}_f\colon \mathcal{M}_X \longrightarrow \mathcal{M}_{X_0(f)}^{\hat{\mu}}$$

called the *motivic nearby cycles morphism* of f. In particular, for every open immersion $i\colon U \to X$, we obtain a motive $\mathcal{S}_{f,U} := \mathcal{S}_f(i)$ in $\mathcal{M}_{X_0(f)}^{\hat{\mu}}$, the *motivic nearby cycles of f supported on U*.

We will recall the construction of the motive $\mathcal{S}_{f,U}$ as done in [**15**], restricting to the case where U is a smooth dense open subvariety of X, since this case will play an important role in the rest of this chapter. We denote by $F = X \setminus U$ the complement of U in X, and by passing to a resolution of the singularities of the pair (X, F) we can assume without loss of generality that X is itself smooth. For any two positive integers n and δ, we consider the subscheme

$$X_n^\delta(f, U) := \{\varphi \in \mathcal{L}(X) \mid \operatorname{ord} f(\varphi) = n,\ \overline{\mathrm{ac}}\, f(\varphi) = 1,\ \operatorname{ord}_F(\varphi) \leq n\delta\}$$

of $\mathcal{L}(X)$. Since $\operatorname{ord} f(\varphi) > 0$, we have that $\pi_0(X_n^\delta(f, U)) \subset X_0(f)$, and so for any $m \geq n$, the image $\pi_m(X_n^\delta(f, U)) \subset \mathcal{L}_m(X)$ is endowed with an action of μ_n given by $\lambda.\varphi(t) = \varphi(\lambda t)$, giving rise to a class in $\mathcal{M}_{X_0(f)}^{\hat{\mu}}$; we denote this motive by $[X_{n,m}^\delta(f, U)]$. Since X is smooth, we have an equality

$$[X_{n,m}^\delta(f, U)]\mathbb{L}^{-md} = [X_{n,n}^\delta(f, U)]\mathbb{L}^{-nd} \in \mathcal{M}_{X_0(f)}^{\hat{\mu}}.$$

This element of $\mathcal{M}_{X_0(f)}^{\hat{\mu}}$ is called the *motivic measure* of $X_n^\delta(f, U)$ and denoted by $\operatorname{mes}(X_n^\delta(f, U))$.

For any δ, consider the generating series

$$Z_{f,U}^\delta(T) := \sum_{n\geq 1} \mathrm{mes}\big(X_n^\delta(f,U)\big)T^n \in \mathcal{M}_{X_0(f)}^{\hat\mu}[[T]].$$

By giving an explicit formula for this power series in terms of an embedded resolution of f and of F, Guibert–Loeser–Merle show that it is rational (see [**15**, Proposition 3.8]). One can then consider the limit $-\lim_{T\to\infty} Z_{f,U}^\delta(T)$ as an element of $\mathcal{M}_{X_0(f)}^{\hat\mu}$. Moreover, they prove that the limit does not depend on the choice of δ, provided that it is big enough. This limit is the motive $\mathcal{S}_{f,U}$ that we wanted to define.

REMARKS 2.1.

(1) Assume that X is smooth and take $U = X$. Then, in the construction above, the condition on the tangency of arcs to F disappears, and there is therefore no need to show that the generating series does not depend on δ. The resulting motive $S_f := S_{f,X}$ is the *motivic nearby cycles* defined by Denef–Loeser, and the procedure we sketched is the original construction from [**8**, **11**] It follows from the formula expressing S_f in terms of a log-resolution of $(X, f^{-1}(0))$ that if 0 is not in the discriminant of f, then $S_f = [f^{-1}(0)]$.

(2) It follows from the construction of Guibert–Loeser–Merle that if X and F are smooth, then

$$S_f\big([i_F\colon F \to X]\big) = i_{F!}S_{f\circ i_F,F} = i_{F!}S_{f_{|F}}.$$

In particular, by additivity of the morphism S_f, we obtain an equality

$$S_{f,U} = S_f - i_{F!}S_{f_{|F}}.$$

(3) Assume that k is the field of complex numbers. Bittner [**3**, §8] and Guibert–Loeser–Merle [**15**, Proposition 3.17], extending a result of Denef–Loeser [**8**, Theorem 4.2.1], show that the motivic nearby cycles morphism is indeed a motivic version of the nearby cycles functor, which justifies the terminology. More precisely, they show that the following diagram is commutative:

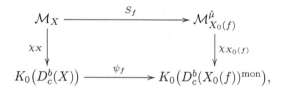

where $K_0\big(D_c^b(X)\big)$ and $K_0\big(D_c^b(X_0(f))^{\mathrm{mon}}\big)$ are the Grothendieck rings of the derived categories of bounded constructible complex sheaves on X and $X_0(f)$, respectively (the latter being endowed with a quasi-unipotent action), ψ_f is the nearby cycle functor at the level of sheaves (see [1]), and χ_X and $\chi_{X_0(f)}$ are the realization morphisms which assign to a variety $p\colon Y \to X$ or $(p\colon Y \to X_0(f), \sigma)$ the corresponding classes $[Rp_! \underline{\mathbb{Q}}_Y]$ and $[Rp_! \underline{\mathbb{Q}}_Y, \sigma]$, respectively. Moreover, the direct image functors at the level of motives and at the level of sheaves are compatible. For instance, if $i\colon U \to X$ is an open immersion and $f\colon X \to \mathbb{A}^1_{\mathbb{C}}$ is a morphism, then we have

$$\chi_{X_0(f)}\left(S_{f,U}\right) = \left[\psi_f\left(Ri_! \underline{\mathbb{Q}}_U\right)\right]$$

and

$$\chi_{\mathbb{C}}\left(f_! S_{f,U}\right) = \left[R_{f!}\psi_f\left(Ri_! \underline{\mathbb{Q}}_U\right)\right].$$

3. Motivic Nearby Cycles at Infinity and the Motivic Bifurcation Set

In this section, we study the motivic nearby cycles at infinity $S^\infty_{f,a}$ for a fiber $f^{-1}(a)$ of a morphism $f\colon U \to \mathbb{A}^1_k$, a motive introduced by the second author in [34]. We prove that this object is a motivic generalization of the invariants $\lambda_a(f)$ mentioned in the introduction. We then deduce that under a natural assumption, that is when f has isolated singularities at infinity, the motivic bifurcation set of f, which is defined in [34] as the set of those values a such that $S^\infty_{f,a}$ does not vanish, contains the topological bifurcation set of f.

3.1. Motivic nearby cycles at infinity

We begin by recalling some constructions of the second author from [34, §4]. Let U be a smooth connected algebraic k-variety and let $f\colon U \to \mathbb{A}^1_k$ be a dominant morphism.

By Nagata compactification theorem, there exists a *compactification* $(X, i, \hat{f})$ of f, by which we mean the data of a k-variety X, an open dominant immersion $i\colon U \to X$, and a proper map $\hat{f}\colon X \to \mathbb{A}^1_k$ such that the following diagram is commutative:

In the following, we will identify U with its image $i(U)$ in X, drop the immersion i from the notation $(X, i, \hat{f})$, and denote by F the closed subvariety $X \setminus U$ of X.

DEFINITION 3.1 (Motivic nearby cycles at infinity). For any a in $\mathbb{A}_k^1(k)$, we call *motivic nearby cycles at infinity of f for the value a* the motive

$$S_{f,a}^\infty = \hat{f}_!\big(S_{\hat{f}-a,U} - i_! S_{f-a}\big) \in \mathcal{M}_k^{\hat{\mu}}.$$

It is shown in [**34**, Theorem 4.2] that the motive $S_{f,a}^\infty$ does not depend on the chosen compactification of f. While the proof in [**34**] relies on a computation on a suitable resolution of the singularities of the pair $\big(X, \hat{f}^{-1}(a) \cup F\big)$, this fact will also naturally follow from the results discussed in Section 4.

REMARK 3.2. The motive $S_{\hat{f}-a,U} - i_! S_{f-a} \in \mathcal{M}_{\hat{f}^{-1}(a)}^{\hat{\mu}}$ can be obtained using motivic integration in the following way. Given integers $\delta > 0$ and $n \geq 1$, consider the set of arcs

$$X_n^{\delta,\infty}(\hat{f} - a) = \left\{ \varphi \in \mathcal{L}(X) \ \middle| \ \begin{array}{l} \varphi(0) \in F, \ \mathrm{ord}_F(\varphi) \leq n\delta \\ \mathrm{ord}(\hat{f} - a)(\varphi) = n, \ \overline{\mathrm{ac}}(\hat{f} - a)(\varphi) = 1 \end{array} \right\},$$

and the zeta function

$$Z_{\hat{f}-a,U}^{\delta,\infty}(T) = \sum_{n \geq 1} \mathrm{mes}\big(X_n^{\delta,\infty}(\hat{f} - a)\big) T^n \in \mathcal{M}_{\hat{f}^{-1}(a)}^{\hat{\mu}}[[T]].$$

Then, for any δ large enough, we have

$$S_{\hat{f}-a,U} - i_! S_{f-a} = -\lim_{T \to \infty} Z_{\hat{f}-a,U}^{\delta,\infty}(T) \in \mathcal{M}_{\hat{f}^{-1}(a)}^{\hat{\mu}}.$$

The motive $S_{f,a}^\infty$ can be expected to be a motivic analogue of the invariants $\lambda_a(f)$ discussed in the introduction. However, no link between the two was established in [**34**]. That such a connection exists is a consequence of the next Theorem.

THEOREM 3.3. *Let f be a polynomial in $\mathbb{C}[x_1, \ldots, x_d]$ and let a be an element of $\mathbb{A}_\mathbb{C}^1(\mathbb{C})$. Then, we have the following conditions:*

(1) $\chi_c\big(S_{f,a}^\infty\big) = \chi_c\big(f^{-1}(a_{\mathrm{gen}})\big) - \chi_c\big(f_! S_{f-a}\big).$

(2) *If a is not a critical value of f, then* $\chi_c\big(S_{f,a}^\infty\big) = \chi_c\big(f^{-1}(a_{\mathrm{gen}})\big) - \chi_c\big(f^{-1}(a)\big).$

(3) *If f has isolated singularities in $\mathbb{A}_\mathbb{C}^1$, then $\chi_c\big(S_{f,a}^\infty\big) = (-1)^{d-1} \lambda_a(f).$*

PROOF. By applying the Euler characteristic to the definition of $S_{f,a}^\infty$, we obtain

$$\chi_c\big(\hat{f}_! S_{\hat{f}-a,U}\big) = \chi_c\big(f_! S_{f-a}\big) + \chi_c\big(S_{f,a}^\infty\big),$$

therefore we need to establish the following equality:

$$\chi_c\big(\hat{f}_! S_{\hat{f}-a,U}\big) = \chi_c\big(f^{-1}(a_{\mathrm{gen}})\big).$$

By [**15**, Proposition 3.17] (or [**3**, §8]), as noted in Remark 2.1, for any value a, the motive $\hat{f}_! S_{\hat{f}-a,U}$ realizes on the class $[R_{\hat{f}_!}\psi_{\hat{f}-a}(Ri_!\underline{\mathbb{Q}}_U)]$. Using the proper base change for the nearby cycles sheaves via the application $t - a$ in $\mathbb{A}^1_k$ (see [**12**, Proposition 4.2.11]), we deduce that

$$[R_{\hat{f}_!}\psi_{\hat{f}-a}(Ri_!\underline{\mathbb{Q}}_U)] = [\psi_{t-a}((R\hat{f} \circ Ri)_!\underline{\mathbb{Q}}_U)] = [\psi_{t-a}(Rf_!\underline{\mathbb{Q}}_U)].$$

Taking the Euler characteristics χ_c for the motives and χ for the sheaves, we obtain

$$\chi_c(\hat{f}_! S_{\hat{f}-a,U}) = \chi([R_{\hat{f}_!}\psi_{\hat{f}-a}(Ri_!\underline{\mathbb{Q}}_U)]) = \chi([\psi_{t-a}(Rf_!\underline{\mathbb{Q}}_U)])$$

(3.4) $$[3pt] = \chi_c(f^{-1}(a_{\text{gen}})),$$

where the last equality follows from [**18**, (2.5.6); **12**, (2.3.26)]. Hence, we obtain

$$\chi_c(S^{\infty}_{f,a}) = \chi_c(f^{-1}(a_{\text{gen}})) - \chi_c(f_! S_{f-a}),$$

proving part (1).

In particular, if $a \in \mathbb{A}^1_{\mathbb{C}}(\mathbb{C})$ is not a critical value of f, we have $f_! S_{f-a} = [f^{-1}(a)]$, and therefore

$$\chi_c(f^{-1}(a_{\text{gen}})) - \chi_c(f^{-1}(a)) = \chi_c(S^{\infty}_{f,a}),$$

which proves part (2).

Now, assume that f has isolated singularities. Then, the following equalities of motives over $f^{-1}(a)$ hold:

$$S_{f-a} = [f^{-1}(a) \setminus \text{crit}(f) \to f^{-1}(a)] + \sum_{x \in f^{-1}(a) \cap \text{crit}(f)} S_{f,x}$$

$$= [f^{-1}(a) \to f^{-1}(a)] + \sum_{x \in f^{-1}(a) \cap \text{crit}(f)} (S_{f,x} - [x \to f^{-1}(a)])$$

where, for any critical point x, the motive $S_{f,x}$ is the motivic Milnor fiber of f at the point x, which is the motive constructed analogously as S_f but only using arcs with origin x. This can be shown by subdividing the arcs defining S_{f-a} according to whether their origin falls in the smooth part of the fiber or not, see [**34**, §4.4] for more details.

Applying the direct image $f_!$ and taking the Euler characteristic of both sides of the last equation, we obtain

$$\chi_c(f_! S_{f-a}) = \chi_c(f^{-1}(a)) + \sum_{x \in f^{-1}(a) \cap \text{crit}(f)} (-1)^{d-1}\mu(f,x),$$

where $\mu(f,x)$ is the Milnor number of f at x, since it follows from results of Denef–Loeser (for instance from [**8**, Theorem 4.2.1]) that $\chi_c(S_{f,x} - 1) = (-1)^{d-1}\mu(f,x)$.

Finally, by applying to this formula (1.2) and the previous part of the theorem, we obtain the equality

$$\chi_c(S_{f,a}^\infty) = (-1)^{d-1}\lambda_a(f),$$

showing part (3) and thus concluding the proof of the theorem. $\square$

3.2. Motivic bifurcation set

In the light of the previous result, it is natural to define a motivic version of the topological bifurcation set of f as follows.

DEFINITION 3.1 ([**34**, Définition 4.6]). The *motivic bifurcation set* of f is the set defined as

$$B_f^{\mathrm{mot}} = \{a \in \mathbb{A}_k^1(k) \mid S_{f,a}^\infty \neq 0\} \cup \mathrm{disc}(f),$$

where $\mathrm{disc}(f)$ denotes the discriminant of f.

The set B_f^{mot} is finite, as is proven in [**34**, Théorème 4.13] via a computation on a suitable resolution of a compactification $(X, \hat{f})$ of f.

If we assume that k is the field of complex numbers, the first part of Theorem 3.3 implies the following result about the motivic bifurcation set of f.

COROLLARY 3.2. *Let f be a polynomial in $\mathbb{C}[x_1, \ldots, x_d]$. Then we have*

$$\{a \in \mathbb{A}_{\mathbb{C}}^1(\mathbb{C}) \mid \chi_c(f^{-1}(a)) \neq \chi_c(f^{-1}(a_{\mathrm{gen}}))\} \subset B_f^{\mathrm{mot}}.$$

PROOF. If a belongs to the discriminant of f then a belongs to B_f^{mot} by definition, therefore we can assume that a is not in $\mathrm{disk}(f)$. Now, since $\chi_c(f^{-1}(a)) \neq \chi_c(f^{-1}(a_{\mathrm{gen}}))$, it follows from the second part of Theorem 3.3 that $\chi_c(S_{f,a}^\infty)$ is non-zero, therefore $S_{f,a}^\infty$ is non-zero as well, that is, a belongs to B_f^{mot}. $\square$

REMARK 3.3. In particular, whenever $B_f^{\mathrm{top}} = \{a \in \mathbb{A}_{\mathbb{C}}^1(\mathbb{C}) \mid \chi_c(f^{-1}(a)) \neq \chi_c(f^{-1}(a_{\mathrm{gen}}))\}$, then B_f^{top} is included in B_f^{mot}. For example, as observed in the introduction, this holds in the case of plane curves, or more generally whenever f has isolated singularities at infinity, by [**31**]. Observe also that Theorem 3.3 does not require f to have isolated singularities in $\mathbb{A}_{\mathbb{C}}^d$, therefore it applies also to situations where the invariants $\lambda_a(f)$ are not defined.

REMARK 3.4. In this remark, we explain how to lift the equality (1.3) to an equality in $\mathcal{M}_k^{\hat{\mu}}$. In order to do this, we need to recall the notion of global motivic zeta function introduced in [**34**]. For any $n \geq 1$ and $\delta \geq 1$, we consider

$$X_n^{\text{global},\delta}(\hat{f}) = \left\{ \varphi(t) \in \mathcal{L}(X) \;\middle|\; \begin{array}{c} \text{ord}_F\big(\varphi(t)\big) \le n\delta, \text{ ord}\left(\hat{f}\big(\varphi(t)\big) \right. \\ \left. -\hat{f}\big(\varphi(0)\big)\right) = n \\ \overline{\text{ac}}\left(\hat{f}\big(\varphi(t)\big) - \hat{f}\big(\varphi(0)\big)\right) = 1 \end{array} \right\}.$$

The *global motivic zeta function*, defined as

$$Z_{\hat{f},U}^{\text{global},\delta}(T) = \sum_{n \ge 1} \text{mes}\big(X_{n,a}^{\text{global},\delta}(\hat{f})\big) T^n,$$

is a rational series for δ large enough (see [**34**, Theorem 4.10]), so that we can set

$$S_{\hat{f},U}^{\text{global}} = -\lim_{T \to \infty} Z_{\hat{f},U}^{\text{global},\delta}(T) \in \mathcal{M}_X^{\hat{\mu}}.$$

Moreover, again as in [**34**, Theorem 4.10], if f has finitely many critical points, then by decomposing the sets $X_n^{\text{global},\delta}$ according to the origin of the arcs, we obtain the following decomposition:

$$S_{\hat{f},U}^{\text{global}} = [U \setminus \text{crit}(f) \to X] + \sum_{x \in \text{crit}(f)} i_! S_{f,x} + \sum_{a \in B_f^{\text{mot}}} \big(S_{\hat{f}-a,U} - i_! S_{f-a}\big),$$

and applying the direct image $\hat{f}_!$ we obtain the following equality in $\mathcal{M}_{\mathbb{A}_k^1}^{\hat{\mu}}$:

$$\hat{f}_! S_{\hat{f},U}^{\text{global}} = \big[U \setminus \text{crit}(f) \to \mathbb{A}_k^1\big] + \sum_{x \in \text{crit}(f)} f_! S_{f,x} + \sum_{a \in B_f^{\text{mot}}} \hat{f}_!\big(S_{\hat{f}-a,U} - i_! S_{f-a}\big).$$

We claim that whenever k is the field of complex numbers this equality generalizes equation (1.3). Indeed, by pushing forward via the structure morphism $p \colon \mathbb{A}_{\mathbb{C}}^1 \to \text{Spec } \mathbb{C}$ of $\mathbb{A}_{\mathbb{C}}^1$ and taking the Euler characteristic, we obtain

$$\chi_c(p_! \hat{f}_! S_{\hat{f},U}^{\text{global}}) = 1 + (-1)^{d-1} \sum_{x \in \text{crit}(f)} \mu(f,x) + \sum_{a \in B_f^{\text{mot}}} \chi_c\big(S_{f,a}^{\infty}\big).$$

Now, observe that $\hat{f}_! S_{\hat{f},U}^{\text{global}} \in \mathcal{M}_{\mathbb{A}_{\mathbb{C}}^1}^{\hat{\mu}}$ is a motive over the affine line $\mathbb{A}_{\mathbb{C}}^1$. Its fiber over a value a is the motive $\hat{f}_! S_{\hat{f}-a,U}$, hence all fibers have Euler characteristics equal to the Euler characteristic of $f^{-1}(a_{\text{gen}})$ by formula (3.4). Since the Euler characteristic of the base $\mathbb{A}_{\mathbb{C}}^1$ is equal to 1, we obtain the equality

$$\chi_c(p_! \hat{f}_! S_{\hat{f},U}^{\text{global}}) = \chi_c\big(f^{-1}(a_{\text{gen}})\big).$$

Indeed, as in classical topology, if $g \colon V \to \mathbb{A}^1$ is a surjective morphism whose fiber has all the same Euler characteristic c, then the Euler characteristic of V is c as well. This can be seen using constructible functions and integration

against Euler characteristic (see, for instance, [40] or [12, Remark 4.1.32])
or more generally Cluckers–Loeser motivic integration [7]. Thus, we obtain

$$\chi_c\big(f^{-1}(a_{\text{gen}})\big) = 1 + (-1)^{d-1} \sum_{x \in \text{crit}(f)} \mu(f,x) + \sum_{a \in B_f^{\text{mot}}} \chi_c\big(S_{f,a}^{\infty}\big),$$

which, since $\chi_c\big(S_{f,a}^{\infty}\big) = (-1)^{d-1}\lambda_a(f)$ by part (3) of Theorem 3.3, yields

$$\chi\big(f^{-1}(a_{\text{gen}})\big) = 1 + (-1)^{d-1}\big(\mu(f) + \lambda(f)\big),$$

which is precisely identity (1.3) cited in the introduction.

4. Analytic Nearby Fiber at Infinity and the Serre Bifurcation Set

In this section, after recalling some constructions in non-archimedean geometry and motivic integration, we define a non-archimedean analytic version
of the nearby fiber at infinity and study its properties.

4.1. Non-archimedean analytic spaces

We will briefly recall some basic notions of non-archimedean analytic geometry. While we chose to adopt the point of view of Berkovich, for the
purpose of this chapter, one could also work with rigid analytic spaces. We
refer the reader to [25] and to the references therein for a more thorough
discussion of these theories.

With any K-variety X is associated an analytic space over K, its *analytification* X^{an}. It is a locally ringed space, whose points, in analogy with
the theory of schemes, are the morphisms $\text{Spec}\, K' \to X$, where K' is a valued field extension of K, modulo the relation that identifies two morphisms
$\text{Spec}\, K' \to X$ and $\text{Spec}\, K'' \to X$ if they both factor through a third morphism $\text{Spec}\, K''' \to X$, where K''' is an intermediate valued field extension
of both $K|K'$ and $K|K''$.

On the other hand, if X is a separated scheme of finite type over the
valuation ring R of K, then we can also attach to it a formal scheme over
R, its *formal completion* $\widehat{X}$. As a locally ringed space, $\widehat{X}$ is isomorphic
to the inverse limit of the schemes $X \otimes_R R/(t^n)$; its underlying topologi-
cal space is X_k and its sheaf of functions is $\varprojlim \mathcal{O}_{X \otimes_R R/(t^n)}$. The *analytic
space associated with* $\widehat{X}$ (sometimes also called the generic fiber of $\widehat{X}$), denoted by $\widehat{X}^{\beth}$, is the compact subspace of $(X_K)^{\text{an}}$ consisting of those points
$x \colon \text{Spec}\, K' \to X_K$ that extend to an R-morphism $\tilde{x} \colon \text{Spec}\, R' \to X$, where
R' is the valuation ring of K'. If such an extension exists, then it is unique
by the valuative criterion of separatedness, therefore we obtain a morphism
$\text{sp}_{\widehat{X}} \colon \widehat{X}^{\beth} \to X_k$, called *specialization*, that is defined by sending a point x
as above to the image through $\tilde{x}$ of the closed point of $\text{Spec}\, R'$.

REMARK 4.1. Let X be a separated scheme of finite type over R and let U be an open subscheme of X. Then, the following results follow directly from the definitions:

(1) $(U_K)^{\mathrm{an}}$ is an open subspace of the K-analytic space $(X_K)^{\mathrm{an}}$, and we have
$$(X_K)^{\mathrm{an}} \setminus (U_K)^{\mathrm{an}} = (X_K \setminus U_K)^{\mathrm{an}}.$$

(2) If X is proper over R, then the inclusion of $\widehat{X}^{\beth}$ in $(X_K)^{\mathrm{an}}$ is an isomorphism by the valuative criterion of properness.

(3) The open immersion $U \to X$ induces an isomorphism $\widehat{U}^{\beth} \cong \mathrm{sp}_{\widehat{X}}^{-1}(U)$.

REMARK 4.2. Let X be a k-variety, let $f \colon X \to \mathbb{A}_k^1 = \operatorname{Spec} k[t]$ be a morphism, and set $X_R = X \times_{\mathbb{A}_k^1} \operatorname{Spec} R$ and $X_K = X \times_{\mathbb{A}_k^1} \operatorname{Spec} K$. Then, the k-arcs on X with origin in $f^{-1}(0)$ correspond to the totally ramified points of $(\widehat{X_R})^{\beth}$. Indeed, if $\varphi \colon \operatorname{Spec} k[[s]] \to X$ is an arc on X with $\mathrm{ord}_f(\varphi) = n > 0$, then φ factors through X_K and the composition
$$\operatorname{Spec} k[[s]] \xrightarrow{\varphi} X_K \longrightarrow \operatorname{Spec} K$$
identifies $K' = \mathrm{Frac}(k[[s]]) \cong K(n)$ with a totally ramified degree n extension of K, inducing a K'-point x of X_K^{an} whose specialization $\mathrm{sp}_{\widehat{X_R}}(x)$ coincides with $\pi_0(\varphi)$, hence a point of $(\widehat{X_R})^{\beth}$. Conversely, with a $K(n)$-point of $(\widehat{X_R})^{\beth}$, a morphism $\operatorname{Spec} R(d) \to X_K$ is associated, whose composition with the projection $X_K \to X$ is a k-arc on X such that $\mathrm{ord}_f(\varphi) = n$, as $R(n) \cong k[[t^{1/n}]]$. For a more precise statement, we refer the reader to [**25**, 6.1.2] or [**28**, 9.1.2].

More generally, if $\mathcal{X}$ is a (separated and quasi-compact) formal R-scheme of finite type, that is, a quasi-compact locally ringed space that is locally of the form $\widehat{X}$ for some separated R-scheme of finite type X, then by gluing the associated K-analytic spaces, we obtain a compact K-analytic space $\mathcal{X}^{\beth}$. We say that $\mathcal{X}$ is a *formal model* of $\mathcal{X}^{\beth}$, and we say that $\mathcal{X}$ is generically smooth if $\mathcal{X}^{\beth}$ is a smooth K-analytic space. It follows from a celebrated theorem of Raynaud (see [**25**, §4.10]) that if X is a K-variety, then every compact subspace of X^{an} admits a formal model; in particular, this is true for a distinguished class of compact subspaces of X^{an}, its affinoid domains.

For the purpose of motivic integration, it is sufficient to have a formal model for the unramified part of a given analytic space; this is formalized by the theory of weak Néron models. If X is a (separated) K-analytic space, a *weak Néron model* of X is a formal scheme of finite type $\mathcal{X}$ over R together with an open immersion $i \colon \mathcal{X}^{\beth} \to X$ such that for every unramified extension R' of R, the map i induces a bijection $\mathcal{X}(R') \to X(K')$, where K' is the fraction field of R'.

4.2. Motivic integration on formal schemes and analytic spaces

Extending on the work of Sebag [35], Loeser–Sebag [21] and Nicaise–Sebag [27–29], Hartmann [16] defined a theory of equivariant motivic integration on formal schemes of finite type. We will briefly summarize the results we need.

Let n be an integer, let $\mathcal{X}$ be a generically smooth and flat formal R-scheme of finite type endowed with a good μ_n-action on $\mathcal{X}$ (meaning that every μ_n-orbit is contained in an affine formal subscheme of $\mathcal{X}$), and let ω be a μ_n-closed *volume form* on the compact K-analytic space $\mathcal{X}^{\beth}$ (that is, a nowhere vanishing differential form of degree $\dim \mathcal{X}^{\beth}$ on $\mathcal{X}^{\beth}$ satisfying an additional compatibility condition in relation with the μ_n-action, see [16, Definition 6.2, p. 32]). Note that this setting includes the case where there is no action, by taking $n = 1$, in which case the constructions below reduce to those of [35, 21, 28]. Then, there exists a μ_n-equivariant Néron smoothening $h\colon \mathcal{Y} \to \mathcal{X}$, which is an equivariant morphism of formal R-schemes such that $\mathcal{Y}$ is a weak Néron model of $\mathcal{X}^{\beth}$ and h factors via an open immersion through an equivariant morphism $\mathcal{Y}' \to \mathcal{X}$ that induces an isomorphism $\mathcal{Y}'^{\beth} \to \mathcal{X}^{\beth}$. It can be shown that the motive

$$\int_{\mathcal{X}} |\omega| := \sum_{\substack{C \text{ connected} \\ \text{component of } \mathcal{Y}_k}} [C] \mathbb{L}^{-\mathrm{ord}_C\left((h^{\beth})^*(\omega)\right)} \in \mathcal{M}^{\mu_n}_{\mathcal{X}_k}$$

only depends on $\mathcal{X}$ and ω and not on the Néron smoothening h.

Moreover, if X is a smooth K-analytic space that admits a weak Néron model $\mathcal{U}$ (such as, for example, the space $\mathcal{X}^{\beth}$ above, with $\mathcal{U} = \mathcal{Y}$) and X is endowed with a μ_n-action that extends to a good μ_n-action on $\mathcal{U}$, then the image of $\int_{\mathcal{U}} |\omega|$ under the forgetful morphism $\mathcal{M}^{\mu_n}_{\mathcal{U}_k} \to \mathcal{M}^{\mu_n}_k$ only depends on X and ω and not on $\mathcal{U}$, and is denoted by

$$\int_X |\omega| \in \mathcal{M}^{\mu_n}_k.$$

Now, let $\mathcal{X}$ be a generically smooth and flat formal R-scheme of finite type, and assume that $\mathcal{X}^{\beth}$ admits a gauge form ω. Since weak Néron models only see the unramified points of $\mathcal{X}^{\beth}$ (which in the setting of Remark 4.2 correspond only to arcs with contact order 1 along $\mathcal{X}_k$), to see its ramified points, we consider the *volume Poincaré series* of $(\mathcal{X}, \omega)$, which is defined as

$$S(\mathcal{X}, \omega, T) = \sum_{n \geq 1} \int_{\mathcal{X}(n)} |\omega(n)| \, T^n \in \mathcal{M}^{\hat{\mu}}_{\mathcal{X}_k}[[T]],$$

where for any integer n we denote by $\mathcal{X}(n) = \mathcal{X} \otimes_R R(n)$ the base change of $\mathcal{X}$ to $R(n) \cong k[[t^{1/n}]]$, endowed with the natural action of $\mathrm{Gal}(K(n)|K) \simeq \mu_n$. This series is rational, therefore it admits a limit when T goes to infinity. The limit, which does not depend on the choice of ω by [28, Proposition

8.1], is called the *motivic volume* of $\mathcal{X}$ and denoted by

$$\mathrm{Vol}(\mathcal{X}) = -\lim_{T\to\infty} S(\mathcal{X},\omega,T) \in \mathcal{M}_{\mathcal{X}_k}^{\hat{\mu}}.$$

Given a smooth connected k-variety U endowed with a fixed volume form ω and a dominant morphism $f\colon U \to \mathbb{A}^1_k$, by taking a motivic volume of the compact K-analytic space $(\widehat{U_R})^{\beth}$, one can retrieve the motivic nearby cycles S_f, see [**28**, Theorem 9.13; **16**, Proposition 7.7]. More precisely, $(\widehat{U_R})^{\beth}$ comes endowed with a canonical gauge form, its *Gelfand–Leray form* ω/df, and we have

$$\mathrm{Vol}(\widehat{U_R}) = \mathbb{L}^{-(\dim U - 1)} S_f \in \mathcal{M}_{U_k}^{\hat{\mu}}.$$

Its direct image by $f_!$ does not depend on $\widehat{U_R}$, it is therefore called the *motivic volume of* $(\widehat{U_R})^{\beth}$, denoted by $\mathrm{Vol}((\widehat{U_R})^{\beth})$. We have

(4.1) $$\mathrm{Vol}((\widehat{U_R})^{\beth}) = \mathbb{L}^{-(\dim U - 1)} f_! S_f \in \mathcal{M}_k^{\hat{\mu}}.$$

4.3. The motivic volume of the analytification of an algebraic variety

Let U be a smooth connected k-variety endowed with a fixed volume form ω, let $f\colon U \to \mathbb{A}^1_k$ be a dominant morphism, let $(X,\hat{f})$ be a compactification of f, and let F be the closed subset $X \setminus U$. After seeing the results of the previous section, one would expect to be able to retrieve the motive $S_{f,U}$ from the (non-compact) K-analytic space $(X_K)^{\mathrm{an}}$, since the latter contains $(\widehat{U_R})^{\beth}$ and all the points corresponding to the arcs on X that have origin in F but are not arcs in F. This is indeed the case, as shown by Bultot [**5**, §2.4].

We will briefly recall Bultot's construction. Without loss of generality, we can assume that the compactification $(X,\hat{f})$ of f is normal, that F is a Cartier divisor on X, and that we have additional data $(W_\alpha, g_\alpha)_\alpha$, where $\{W_\alpha\}_\alpha$ is a finite cover of $(X_K)^{\mathrm{an}}$ by affinoid domains and, for each α, g_α is an analytic function on W_α such that $W_\alpha \cap (F_K)^{\mathrm{an}}$ is defined by $g_\alpha = 0$ in W_α. For every non-negative integer γ, we set

$$U_\gamma := \bigcup_{\alpha\in A} \{x \in W_\alpha \mid |g_\alpha(x)| \geq |t|^\gamma \}.$$

The U_γ form an increasing sequence of compact analytic domains of $(U_K)^{\mathrm{an}}$ such that

$$(U_K)^{\mathrm{an}} = \bigcup_{\gamma\geq 0} U_\gamma = \bigcup_{\alpha\in A} \{x \in W_\alpha \mid |g_\alpha(x)| > 0\}.$$

In order to see that the motivic volumes of the U_γ stabilize, it is useful to compute them relatively to the formal scheme $\widehat{X_R}$ in order to be able to compare them in the Grothendieck ring $\mathcal{M}_{\mathcal{X}_k}^{\hat{\mu}}$. This is possible since, if $\mathcal{U}$ and $\mathcal{U}'$ are two weak Néron models of U_γ mapping to $\widehat{X_R}$, then there exists a third weak Néron model of U_γ dominating both, hence the images of the integral $\int_{\mathcal{U}} \omega/df$ in $\mathcal{M}_{\mathcal{X}_k}^{\hat{\mu}}$ depends only on U_γ and $\widehat{X_R}$, but not on the choice of $\mathcal{U}$. In particular, the image of the motivic volume $\mathrm{Vol}(\mathcal{U})$ in $\mathcal{M}_{\mathcal{X}_k}^{\hat{\mu}}$ depends on U_γ and not on $\mathcal{U}$; we denote this motive by $\mathrm{Vol}_{\widehat{X_R}}(U_\gamma)$.

Bultot then proves that there exists an integer $\gamma_0 \geq 0$ such that, for every $\gamma \geq \gamma_0$, we have

$$\mathrm{Vol}_{(\widehat{X_R})}(U_\gamma) = \mathrm{Vol}_{(\widehat{X_R})}(U_{\gamma_0}) \in \mathcal{M}_{X_k}^{\hat{\mu}}.$$

This motive does not depend on the sequence (U_γ), we thus denote it by $\mathrm{Vol}_{\widehat{X_R}}((U_K)^{\mathrm{an}})$ and call it *motivic volume of* $(U_K)^{\mathrm{an}}$ *over* $\widehat{X_R}$ (mind that this motive is called motivic volume of U_K over X_k in [5]). Furthermore, Bultot gives a formula computing $\mathrm{Vol}_{\widehat{X_R}}((U_K)^{\mathrm{an}})$ in terms of the combinatorics of a good compactification of $\mathcal{U}$, and by comparing it with the analogous formula of Guibert–Loeser–Merle, he shows that, as expected, we have

$$(4.1) \qquad \mathrm{Vol}_{\widehat{X_R}}((U_K)^{\mathrm{an}}) = \mathbb{L}^{-(\dim U - 1)} S_{\hat{f}, U} \in \mathcal{M}_{X_k}^{\hat{\mu}}.$$

Since the direct image $\hat{f}_! S_{\hat{f}, U}$ in $\mathcal{M}_k^{\hat{\mu}}$ does not depend on the choice of X, we can define the motivic volume of $(U_K)^{\mathrm{an}}$ as

$$\mathrm{Vol}((U_K)^{\mathrm{an}}) = \hat{f}_!\big(\mathrm{Vol}_{\widehat{X_R}}((U_K)^{\mathrm{an}})\big) \in \mathcal{M}_k^{\hat{\mu}}.$$

4.4. Analytic nearby fiber at infinity

As before, let U be a smooth k-variety and let $f\colon U \to \mathbb{A}_k^1$ be a dominant morphism. We can now introduce non-archimedean analytic version of the motivic nearby fibers at infinity of f.

DEFINITION 4.1. Using these notations, we define the *analytic nearby fiber at infinity of* f *for the value* 0 to be the non-archimedean K-analytic space

$$\mathcal{F}_{f,0}^\infty = (U_K)^{\mathrm{an}} \setminus \big(\widehat{U_R}\big)^{\beth}.$$

Similarly, for every a in $\mathbb{A}_k^1(k)$, we define the *analytic nearby fiber at infinity of* f *for the value* a to be the analytic nearby fiber at infinity of $f - a$ for the value 0; we denote it by $\mathcal{F}_{f,a}^\infty$. Observe that these analytic spaces are clearly independent of the choice of a compactification of f.

REMARK 4.2. The definition of $\mathcal{F}_{f,0}^\infty$ is analogous to the one of $S_{f,0}^\infty$, for which one chooses a compactification $(X, \hat{f})$ of (U, f) and considers only arcs with origin in $F = X \setminus U$. Indeed, as we observed in Remark 4.2, the correct analogue for the origin of an arc on X is the specialization $\mathrm{sp}_{\widehat{X_R}}(x)$ of a point x of $(\widehat{X_R})^{\beth}$, and the inclusion of $\mathcal{F}_{f,0}^{\infty}$ in $(\widehat{X_R})^{\beth} = (X_K)^{\mathrm{an}}$ induces an isomorphism

$$\mathcal{F}_{f,0}^{\infty} \cong \mathrm{sp}_{\widehat{X_R}}^{-1}(F) \setminus (F_K)^{\mathrm{an}},$$

since we have $(U_K)^{\mathrm{an}} \cong (X_K)^{\mathrm{an}} \setminus (F_K)^{\mathrm{an}}$ by the first part of Remark 4.1, and $(\widehat{U_R})^{\beth} \cong \mathrm{sp}_{\widehat{X_R}}^{-1}(U)$ by the third part of the same remark. Observe that removing $(F_K)^{\mathrm{an}}$ is necessary to obtain a space that does not depend on the choice of the compactification $(X, \hat{f})$; this independence can also be seen as a consequence of the valuative criterion of properness, as two compactifications can always be dominated by a common third one.

We declare the volume of $\mathcal{F}_{f,0}^{\infty}$ to be

$$\mathrm{Vol}\left(\mathcal{F}_{f,0}^{\infty}\right) = \mathrm{Vol}\left((U_K)^{\mathrm{an}}\right) - \mathrm{Vol}\left((\widehat{U_R})^{\beth}\right) \in \mathcal{M}_k^{\hat{\mu}}.$$

THEOREM 4.3. *Let U be a smooth connected k-variety and let $f \colon U \to \mathbb{A}_k^1$ be a dominant morphism. Then we have an equality*

$$\mathrm{Vol}(\mathcal{F}_{f,0}^{\infty}) = \mathbb{L}^{-(\dim U - 1)} S_{f,0}^{\infty} \in \mathcal{M}_k^{\hat{\mu}}.$$

PROOF. The result follows immediately from the definition of the volume and equalities (4.1) and (4.2). □

REMARK 4.4. Another approach to motivic integration was developed by Hrushovski–Kazhdan [17] with model theoretic methods and gives a group morphism $\mathrm{Vol}^{\mathrm{HK}}$ from the Grothendieck ring of semi-algebraic sets over the valued field K to $\mathcal{M}_k^{\hat{\mu}}$ (see also [26, Theorem 2.5.1]). In particular, the volume of any locally closed subset of $(U_K)^{\mathrm{an}}$ is defined, and the equality $\mathrm{Vol}^{HK}(\mathcal{F}_{f,0}^{\infty}) = \mathrm{Vol}^{\mathrm{HK}}\left((U_K)^{\mathrm{an}}\right) - \mathrm{Vol}^{\mathrm{HK}}\left((\widehat{U_R})^{\beth}\right)$ holds naturally in this context. Moreover, the motivic volumes in [17] can be computed in an analogous way as the volumes in [15, 14] in terms of suitable resolutions of singularities (see [26, Theorem 2.6.1]), and this can be used to show that $\mathrm{Vol}^{\mathrm{HK}}\left((\widehat{U_R})^{\beth}\right) = f_! S_f$, see [26, Corollary 2.6.2] or [13]. One should similarly be able to deduce that $\mathrm{Vol}^{\mathrm{HK}}\left((U_K)^{\mathrm{an}}\right) = \hat{f}_! S_{\hat{f},U}$, so that $\mathrm{Vol}^{\mathrm{HK}}(\mathcal{F}_{f,0}^{\infty})$ is the volume $\mathrm{Vol}(\mathcal{F}_{f,0}^{\infty})$ that we defined above. Observe that, since our definition of the analytic nearby fiber at infinity is independent of the choice of a compactification of f, as is the morphism $\mathrm{Vol}^{\mathrm{HK}}$, this approach would also yield a compactification-independent definition of the motivic nearby cycles at infinity.

4.5. Serre bifurcation set

We will now recall the notion of motivic Serre invariant and use it to define another bifurcation set B_f^{ser} which we call the Serre bifurcation set of f.

Since the Euler characteristic is additive on the category of k-varieties, it gives rise to a morphism $\chi_c \colon \mathcal{M}_k^{\hat{\mu}} \to \mathbb{Z}$. Moreover, since $\chi_c(\mathbb{L} - 1) = \chi_c(\mathbb{G}_{m,k}) = 0$, this morphism factors through a morphism

$$\chi_c \colon \mathcal{M}_k^{\hat{\mu}}/(\mathbb{L} - 1) \longrightarrow \mathbb{Z}.$$

Therefore, when interested in working with the Euler characteristic it is often useful to study the class of a variety modulo $\mathbb{L} - 1$. Now, the constructions made in this section can all be done modulo $\mathbb{L} - 1$, which leads to some simplifications. Indeed, if $\mathcal{X}$ is a generically smooth and flat formal R-scheme of finite type endowed with a μ_n-action, ω is a μ_n-closed gauge form on $\mathcal{X}^{\beth}$, and $h \colon \mathcal{Y} \to \mathcal{X}$ is a μ_n-equivariant Néron smoothening, then in $\mathcal{M}_k^{\hat{\mu}}$, we have

$$\int_{\mathcal{X}} |\omega| \equiv \sum_{\substack{C \text{ connected} \\ \text{component of } \mathcal{Y}_k}} [C]\mathbb{L}^{-\text{ord}_C\left((h^{\beth})^*(\omega)\right)} \equiv \sum_C [C] = [\mathcal{Y}_k] \bmod (\mathbb{L} - 1).$$

This element of $\mathcal{M}_k^{\hat{\mu}}/(\mathbb{L} - 1)$, that only depends on $\mathcal{X}$ and not on ω, is called the *motivic Serre invariant* of $\mathcal{X}$, and denoted by $S(\mathcal{X})$.

REMARK 4.1. The motivic Serre invariant of a generically smooth formal R-scheme of finite type was introduced by Loeser–Sebag [21], and developed by Nicaise–Sebag [27–29], Bultot [5] and Hartmann [16]. It generalizes an invariant introduced by Serre in [36] in order to classify the compact analytic manifolds over a local field L and defined using classical p-adic integration with value in the ring $\mathbb{Z}/(q-1)\mathbb{Z}$, where q is the cardinality of the residue field l of L. Counting l-rational points yields a canonical morphism $\mathcal{M}_l/(\mathbb{L} - 1) \to \mathbb{Z}/(q - 1)\mathbb{Z}$, and Loeser–Sebag showed that the image by this morphism of the motivic Serre invariant $S(X)$ of a smooth and compact L-analytic space X is equal to the classical Serre invariant of the underlying compact manifold.

We then consider the motive

$$(4.2) \qquad \overline{S}(\mathcal{X}) := \text{Vol}(\mathcal{X}) \bmod(\mathbb{L} - 1) \in \mathcal{M}_k^{\hat{\mu}}/(\mathbb{L} - 1),$$

which can also be obtained as the limit of the generating series $-\sum_{n \geq 1} S(\mathcal{X}(n))T^n$. Similarly, if X is a smooth K-analytic space admitting a weak Néron model $\mathcal{U}$ over R, one also sets $S(X) = S(\mathcal{U})$, and we obtain a motive

$$(4.3) \qquad \overline{S}(X) = \text{Vol}(X) \bmod(\mathbb{L} - 1) \in \mathcal{M}_k^{\hat{\mu}}/(\mathbb{L} - 1)$$

that is also the limit of an analogous generating series.

It follows from the results of Bultot that, if $U, X, \hat{f}, \{U_\gamma\}_\gamma$ are as in Section 4.3, then

$$\overline{S}((U_K)^{\mathrm{an}}) := \overline{S}(U_\gamma) \in \mathcal{M}_k^{\hat{\mu}}/(\mathbb{L}-1)$$

only depends on U and f and not on γ, if γ is large enough, nor on X, $\{U_\gamma\}_\gamma$, and $\hat{f}$. We can then define the *Serre invariant* of $\mathcal{F}_{f,0}^\infty$ as

$$\overline{S}(\mathcal{F}_{f,0}^\infty) := \overline{S}((U_K)^{\mathrm{an}}) - \overline{S}(\widehat{U_R}) \in \mathcal{M}_k^{\hat{\mu}}/(\mathbb{L}-1).$$

DEFINITION 4.4. Let U be a smooth k-variety and let $f: U \to \mathbb{A}_k^1$ be a dominant morphism. The *Serre bifurcation set* of f is

$$B_f^{\mathrm{ser}} = \{a \in \mathbb{A}_{\mathbb{C}}^1 \mid \overline{S}\left(\mathcal{F}_{f,a}^\infty\right) \neq 0\} \cup \mathrm{disk}(f).$$

THEOREM 4.5. *Let U be a smooth k-variety and let $f: U \to \mathbb{A}_k^1$ be a dominant morphism. Then we have an equality*

$$\overline{S}(\mathcal{F}_{f,0}^\infty) = S_{f,0}^\infty \bmod(\mathbb{L}-1).$$

in $\mathcal{M}_k^{\hat{\mu}}/(\mathbb{L}-1)$, and B_f^{ser} is contained in B_f^{mot}.

PROOF. By combining the definition of $\overline{S}(\mathcal{F}_{f,0}^\infty)$ and (4.3) and (4.4) with the formulas (4.1) and (4.2), we deduce that the two equalities

$$\overline{S}((U_K)^{\mathrm{an}}) = \mathbb{L}^{-(\dim U - 1)} \hat{f}_!(S_{\hat{f},U}) \bmod(\mathbb{L}-1)$$

and

$$\overline{S}(\widehat{U_R}) = \mathbb{L}^{-(\dim U - 1)} f_!(S_f) \bmod(\mathbb{L}-1)$$

hold in $\mathcal{M}_k^{\hat{\mu}}/(\mathbb{L}-1)$, from which the theorem follows. $\qquad\square$

COROLLARY 4.6. *Let f be a polynomial in $\mathbb{C}[x_1,\ldots,x_d]$. Then we have the inclusions*

$$\{a \in \mathbb{A}_{\mathbb{C}}^1(\mathbb{C}) \mid \chi_c(f^{-1}(a)) \neq \chi_c(f^{-1}(a_{\mathrm{gen}}))\} \subset B_f^{\mathrm{ser}} \subset B_f^{\mathrm{mot}}.$$

PROOF. The result follows from Corollary 3.2 together with the fact that for any value a, the Euler characteristics $\chi_c\left(\overline{S}(\mathcal{F}_{f,a}^\infty)\right)$ and $\chi_c\left(S_{f,a}^\infty\right)$ are equal. $\qquad\square$

REMARK 4.7. In particular, whenever $B_f^{\mathrm{top}} = \{a \in \mathbb{A}_{\mathbb{C}}^1(\mathbb{C}) \mid \chi_c(f^{-1}(a)) \neq \chi_c(f^{-1}(a_{\mathrm{gen}}))\}$, as in the case of plane curves, or more generally whenever

f has isolated singularities at infinity, we have a chain of inclusions $B_f^{\text{top}} \subset B_f^{\text{ser}} \subset B_f^{\text{mot}}$.

Acknowledgments

We are very thankful to David Bourqui, Raf Cluckers, Johannes Nicaise, and Julien Sebag, who organized the conference "*Nash: Schémas des arcs et singularités*" in Rennes in 2016, gave the first of us the opportunity to give a talk there, and proposed us to contribute to the conference proceedings. We are also grateful to Emmanuel Bultot, Pierrette Cassou-Noguès, Alexandru Dimca, Johannes Nicaise, and Claude Sabbah for inspiring discussions, and the anonymous referee for his comments and corrections. This work is partially supported by ANR-15-CE40-0008 (Défigéo).

Bibliography

[1] *Groupes de monodromie en géométrie algébrique. II*, Lecture Notes in Mathematics, Vol. 340 Springer-Verlag, Berlin, 1973. séminaire de Géométrie Algébrique du Bois-Marie 1967–1969 (SGA 7 II), Dirigé par P. Deligne et N. Katz.

[2] E. A. Bartolo, I. L. Velasco and A. Melle-Hernández, Milnor number at infinity, topology and Newton boundary of a polynomial function. *Math. Z.*, 233(4):679–696, 2000.

[3] F. Bittner, On motivic zeta functions and the motivic nearby fiber. *Math. Z.*, 249(1):63–83, 2005.

[4] E. Bultot, Computing zeta functions on log smooth models. *C. R. Math. Acad. Sci. Paris*, 353(3):261–264, 2015.

[5] E. Bultot, Motivic integration and logarithmic geometry. Ph.D. thesis, KU Leuven, 2015. Available on arXiv (arXiv:1505.05688).

[6] P. Cassou-Noguès, Sur la généralisation d'un théorème de Kouchnirenko. *Compositio Math.*, 103(1):95–121, 1996.

[7] R. Cluckers and F. Loeser, Constructible motivic functions and motivic integration. *Invent. Math.*, 173(1):23–121, 2008.

[8] J. Denef and F. Loeser, Motivic Igusa zeta functions. *J. Algebraic Geom.*, 7(3):505–537, 1998.

[9] J. Denef and F. Loeser, Germs of arcs on singular algebraic varieties and motivic integration. *Invent. Math.*, 135(1):201–232, 1999.

[10] J. Denef and F. Loeser, Geometry on arc spaces of algebraic varieties. in *European Congress of Mathematics*. Vol. I (Barcelona, 2000), Progress in Mathematics, Vol. 201, Birkhäuser, Basel, 2001, pp. 327–348.

[11] J. Denef and F. Loeser, Lefschetz numbers of iterates of the monodromy and truncated arcs. *Topology* 41(5):1031–1040, 2002.

[12] A. Dimca, *Sheaves in Topology*, Universitext (Springer, Berlin), 2004.

[13] A. Forey, Virtual rigid motives of semi-algebraic sets. Preprint, 2017, arXiv:1706.07233.

[14] G. Guibert, F. Loeser and M. Merle, Nearby cycles and composition with a nondegenerate polynomial. *Int. Math. Res. Not.*, (31):1873–1888, 2005.

[15] G. Guibert, F. Loeser and M. Merle, Iterated vanishing cycles, convolution, and a motivic analogue of a conjecture of Steenbrink. *Duke Math. J.*, 132(3):409–457, 2006.

[16] A. Hartmann, Equivariant motivic integration on formal schemes and the motivic zeta function. Preprint, 2015, arXiv:1511.08656.

[17] E. Hrushovski, and D. Kazhdan, Integration in valued fields, in *Algebraic Geometry and Number Theory*, Progress in Mathematics, Vol. 253, Birkhäuser Boston, Boston, MA, 2006, pp. 261–405.

[18] M. Kashiwara, and P. Schapira, *Sheaves on manifolds. With a Short History "Les débuts de la théorie des faisceaux" by Christian Houzel*. Springer-Verlag, Berlin, 1990.

[19] A. G. Kouchnirenko, Polyèdres de Newton et nombres de Milnor. *Invent. Math.*, 32(1):1–31, 1976.

[20] F. Loeser, Seattle lectures on motivic integration. In *Algebraic Geometry—Seattle 2005. Part 2*, Proc. Sympos. Pure Math., Vol. 80 Amer. Math. Soc., Providence, RI, 2009, pp. 745–784.

[21] F. Loeser, and J. Sebag, Motivic integration on smooth rigid varieties and invariants of degenerations. *Duke Math. J.*, 119(2):315–344, 2003. .

[22] E. Looijenga, Motivic measures. *Astérisque*, (276):267–297, 2002. Séminaire, Vol. Bourbaki, 1999/2000.

[23] Y. Matsui and K. Takeuchi, Monodromy at infinity of polynomial maps and Newton polyhedra (with an appendix by C. Sabbah). *Int. Math. Res. Not. IMRN*, (8):1691–1746, 2013.

[24] Y. Matsui and K. Takeuchi, Motivic Milnor fibers and Jordan normal forms of Milnor monodromies. *Publ. Res. Inst. Math. Sci.*, 50(2):207–226, 2014.

[25] J. Nicaise, Formal and rigid geometry: an intuitive introduction and some applications, *Enseign. Math.*, (2),54(3–4):213–249, 2008.

[26] J. Nicaise and S. Payne, A tropical motivic Fubini theorem with applications to Donaldson–Thomas theory, Preprint, 2017, arxiv:1703.10228,.

[27] J. Nicaise and J. Sebag, Motivic Serre invariants of curves. *Manuscripta Math.*, 123(2):105–132, 2007.

[28] J. Nicaise and J. Sebag, Motivic Serre invariants, ramification and the analytic Milnor fiber. *Invent. Math.*, 168(1):133–173, 2007.

[29] J. Nicaise and J. Sebag, Motivic invariants of rigid varieties and applications to complex singularities. In *Motivic Integration and Its Interactions with Model Theory and Non-Archimedean Geometry*. Vol. I, London Math. Soc. Lecture Note Ser., Vol. 383, Cambridge University Press, Cambridge, 2011, pp. 244–304.

[30] A. Parusiński, A generalization of the Milnor number. *Math. Ann.*, 281(2):247–254, 1988.

[31] A. Parusiński, On the bifurcation set of complex polynomial with isolated singularities at infinity. *Compositio Math.*, 97(3):369–384, 1995.

[32] A. Parusiński, A note on singularities at infinity of complex polynomials. in *Symplectic Singularities and Geometry of Gauge Fields* (Warsaw, 1995), Banach Center Publ., Vol. 39, Polish Acad. Sci., Warsaw, 1997, pp. 131–141.

[33] F. Pham, Vanishing homologies and the n variable saddlepoint method. In *Singularities, Part 2* (Arcata, Calif., 1981), Proc. Sympos. Pure Math., Vol. 40. Amer. Math. Soc., Providence, RI, 1983, pp. 319–333.

[34] M. Raibaut, Singularités à l'infini et intégration motivique. *Bull. Soc. Math. France* 140(1):51–100, 2012.

[35] J. Sebag, Intégration motivique sur les schémas formels, *Bull. Soc. Math. France* 132(1):1–54, 2004.

[36] J.-P. Serre, Classification des variétés analytiques p-adiques compactes. *Topology* 3:409–412, 1965.

[37] M. Suzuki, Propriétés topologiques des polynômes de deux variables complexes, et automorphismes algébriques de l'espace $\mathbf{C}^2$. *J. Math. Soc. Japan*, 26:241–257, 1974.

[38] K. Takeuchi and M. Tibăr, Monodromies at infinity of non-tame polynomials. *Bull. Soc. Math. France*, 144(3): 477–506, 2016.

[39] M. Tibăr, *Polynomials and vanishing cycles*, Cambridge Tracts in Mathematics, Vol. 170, Cambridge University Press, Cambridge, 2007.

[40] O. Y. Viro, Some integral calculus based on Euler characteristic, in *Topology and geometry—Rohlin Seminar*, Lecture Notes in Mathematics, Vol. 1346, Springer, Berlin, 1988, pp. 127–138.

[41] H. H. Vui, and L. D. Tráng, Sur la topologie des polynômes complexes. *Acta Math. Vietnam.* 9(1):21–32 (1985), 1984.

13

The Néron Multiplicity Sequence
of Singularities

Beatriz Pascual-Escudero[*,‡] **and Julien Sebag**[†,§]

*Centrale Nantes, 1, rue de la Noë
44321 Nantes Cedex 3, France
†Institut de recherche mathématique de Rennes
UMR 6625 du CNRS, Université de Rennes 1
Campus de Beaulieu, 35042 Rennes cedex, France
‡beatriz.pascual.escudero@gmail.com
§julien.sebag@univ-rennes1.fr

Let R be a Henselian local ring. In this chapter, we introduce the *Néron multiplicity sequence* associated with the datum of a pair (V, γ) formed by a flat R-scheme of finite type and a section $\gamma \in \mathscr{V}(R)$. We show that this sequence of integers has a good behavior with respect to étale surgery and provide, in this way, a formal invariant of curve singularities. In the special case of cuspidal plane curve singularity, we also compare it with the Nash multiplicity sequence introduced in [**4**].

1. Introduction

1.1. Let R be a Henselian local ring with fraction field K and residue field k. Let $\mathscr{V}$ be a flat R-scheme of finite type. Let $\gamma \in \mathscr{V}(R)$ be a section of $\mathscr{V}$ that we assume to be *non-degenerate*, i.e., whose generic point $\gamma_\eta \in \mathscr{V}_\eta(K)$ is a smooth point of the generic fiber $\mathscr{V}_\eta$. By using the theory of Néron dilatations, we construct a sequence $(\mathfrak{N}\mathfrak{e}_i)_{i \in \mathbf{N}}$ of non-negative integers, that we call the *Néron multiplicity sequence*, attached to the initial datum of the pair $(\mathscr{V}, \gamma)$. We study its main properties especially with respect to singularity theory (see Section 2). In particular, we show that this sequence is decreasing and stationary, with limit equal to zero. In the specific case of curves, we show that this construction defines a formal invariant of singularities (see Theorem 4.1).

1.2. The analogy with the Nash multiplicity sequence introduced in [4] allows us to interpret our construction as a generalization of Lejeune's one to the broader geometric context of degenerations of singularities. The direct comparison of these two objects in case $R = k[[t]]$ seems to us to be a challenging problem which can improve the general understanding of both situations. We prove indeed that these sequences share in general various common properties; e.g., the lengths of both sequences are the same (i.e., both begin to be constant at the same time) and can be computed from the same spaces (see Section 3). Nevertheless, these two sequences in general differ, we address the following question.

QUESTION 1.1. Does one of these sequences determine the other one?

It seems very plausible to us that the general answer is positive, at least for analytically irreducible curve singularities. We prove that it is indeed the case for quasi-homogeneous plane curve singularities (see Section 4).

2. The Néron Multiplicity Sequence of Singularities

Let $\mathcal{V} =: \mathcal{V}^{(0)}$ be a flat R-scheme of finite type. Let $\gamma^{(0)} := \gamma \in \mathcal{V}(R)$ be a section of $\mathcal{V}$ that we assume to be non-degenerate.

2.1. By [1, §3.6], one can consider the *dilatation* (see [1, §3.2]) $\mathcal{V}^{(1)}$ of $\mathcal{V}$ at $\gamma_s \in \mathcal{V}_s$. By the valuative criterion of properness, one can lift $\gamma^{(0)}$ to $\gamma^{(1)} \in \mathcal{V}^{(1)}(R)$ in a unique way. By iterating this process, and thanks to [1, §3.6/Proposition 4], we deduce that there exists a finite sequence of dilatations:

$$(2.1) \quad (\mathcal{V}^{(0)}, \gamma^{(0)}) \longleftarrow (\mathcal{V}^{(1)}, \gamma^{(1)}) \longleftarrow \cdots \longleftarrow (\mathcal{V}^{(M+1)}, \gamma^{(M+1)}) \longleftarrow \cdots ,$$

where the integer M is the smallest integer, such that the section $\gamma^{(M+1)}$ factorizes through the smooth locus of $\mathcal{V}^{(M+1)}$. For every integer $i \in \mathbf{N}$, we denote by $\mathfrak{Ne}_i$ the so-called *smoothness defect* of $\mathcal{V}^{(i)}$ at $\gamma^{(i)}$ which is a non-negative integer defined to be the following length:

$$\mathfrak{Ne}_i = \lg(\mathrm{Tors}((\gamma^{(i)})^* \Omega^1_{\mathcal{V}^{(i)}/R})).$$

By [1], one knows that $\mathfrak{Ne}_{M+1} = 0$ and that every further step will provide 0 as value for the smoothness defect. The sequence $(\mathfrak{Ne}_i)_{i \in \{0,\dots,M+1\}}$, called the *Néron multiplicity sequence*, is strictly decreasing by [1, §3.6].

The following lemma can be directly deduced from the construction and [1, Lemma 3.3.2].

LEMMA 2.1. *Keep the notation of Section 2. Let $\varphi \in \text{Aut}_k(R)$ be a local automorphism of R. Let us set $\mathscr{V}_\varphi = \mathscr{V} \otimes_R R$ the base change via φ. Then, the Néron multiplicity sequences associated with $(\mathscr{V}, \gamma)$ and $(\mathscr{V}_\varphi, \varphi \circ \gamma)$ coincide.*

REMARK 2.2. In general, the Néron multiplicity sequence depends on the section γ. However, it is easy to see, thanks to [1, Lemma 3.3.2], that two sections γ, γ', whose images in $\mathscr{V}(R/\mathfrak{m}^{1+\mathfrak{N}\mathfrak{e}_0})$ coincide, induce the same Néron multiplicity sequence.

EXAMPLE 2.3. In the direction of Question 1.1, let us stress that we can associate a Néron multiplicity sequence with every k-scheme of finite type $\mathscr{X}$ endowed with a germ of curve $\gamma \in \mathscr{X}(k[[t]])$. Indeed, if $R = k[[t]]$, and if we set $\mathscr{V} := \mathscr{X} \otimes_k R$ (which is a flat R-scheme of finite type), the germ γ can be viewed as a section of $\mathscr{V}$. If γ sends the generic point of $\text{Spec}(R)$ in the smooth locus of $\mathscr{X}$ (as usually this image is denoted γ_η), it follows that the corresponding section of $\mathscr{V}$ is non-degenerate. Then, we apply the former construction to the pair $(\mathscr{V}, \gamma)$.

2.2. This sequence does not vary under étale morphisms. The following statement makes precise this property. We assume from now on, for simplicity, that the field k is separably closed.

PROPOSITION 2.1. *Keep the notation of Section 2. Let $\mathscr{W}$ be a flat R-scheme of finite type. Let $f : \mathscr{W} \to \mathscr{V}$ be an étale morphism. Let $\eta \in \mathscr{W}(R)$ be a section, such that $f \circ \eta = \gamma$. Then, the Néron multiplicity sequence associated with $(\mathscr{W}, \eta)$ and that associated with $(\mathscr{V}, \gamma)$ coincide.*

By Proposition 2.1, we deduce that the Néron multiplicity sequence is invariant by a (strict) henselization of the ring R.

PROOF. Let $\mathscr{B}_\gamma$ (resp. $\mathscr{B}_\eta$) be the blowing-up of $\mathscr{V}$ (resp. $\mathscr{W}$) at γ_s (resp. $f^{-1}(\gamma_s)$). Let $\mathscr{V}^{(1)}$ (resp. $\widetilde{\mathscr{W}^{(1)}}$) be the corresponding dilatations. We have the following commutative diagram of morphisms of R-schemes:

$$(2.2) \qquad
\begin{array}{ccccc}
(\mathscr{W}^{(0)}, \eta^{(0)}) & \longleftarrow & (\mathscr{B}_\eta, \eta^{(1)}) & \longrightarrow & (\widetilde{\mathscr{W}^{(1)}}, \eta^{(1)}) \\
\downarrow{\scriptstyle f^{(0)}:=f} & & \downarrow{\scriptstyle u} & & \\
(\mathscr{V}^{(0)}, \gamma^{(0)}) & \longleftarrow & (\mathscr{B}_\gamma, \gamma^{(1)}) & \longrightarrow & (\mathscr{V}^{(1)}, \gamma^{(1)})
\end{array}$$

By [5, Proposition 8.1.12], we know that $\mathscr{B}_\eta \cong \mathscr{B}_\gamma \times_{\mathscr{V}^{(0)}} \mathscr{W}^{(0)}$ and that u is the first projection. By the universal property of dilatation (see [1, Proposition 3.2.1]) and diagram (2.2), we also deduce that there exists a unique morphism of R-schemes $f^{(1)}\colon \widetilde{\mathscr{W}^{(1)}} \to \mathscr{V}^{(1)}$ which provides a commutative diagram

$$
\begin{array}{ccc}
(\mathscr{W}^{(0)},\eta^{(0)}) & \longleftarrow & (\widetilde{\mathscr{W}^{(1)}},\eta^{(1)}) \\
\Big\downarrow{\scriptstyle f^{(0)}:=f} & & \Big\downarrow{\scriptstyle f^{(1)}} \\
(\mathscr{V}^{(0)},\gamma^{(0)}) & \longleftarrow & (\mathscr{V}^{(1)},\gamma^{(1)})
\end{array}
$$

By [**5**, Lemma 8.1.2], let us stress that $\widetilde{\mathscr{W}^{(1)}} \cong \mathscr{V}^{(1)} \times_{\mathscr{V}^{(0)}} \mathscr{W}^{(0)}$ and that the morphism $f^{(1)}$ is étale by base change (see [**5**, Proposition 4.3.22]). Now, we observe, thanks to the commutativity of the first square of diagram (2.2) and the uniqueness of the lifting, that $u(\eta^{(1)}) = f^{(1)}(\eta^{(1)}) = \gamma^{(1)}$.

By iterating these arguments, we conclude that the dilatations in sequence (2.1) for $\mathscr{W}$ are obtained as the base changes of the corresponding dilatations in sequence (2.1) for $\mathscr{V}$. So, we have a commutative diagram of morphisms of R-schemes:

$$
\begin{array}{ccccccc}
(\mathscr{W}^{(0)},\eta^{(0)}) & \longleftarrow & (\widetilde{\mathscr{W}^{(1)}},\eta^{(1)}) & \longleftarrow & \cdots & \longleftarrow & (\widetilde{\mathscr{W}^{(M+1)}},\eta^{(M+1)}) \\
\Big\downarrow{\scriptstyle f^{(0)}:=f} & & \Big\downarrow{\scriptstyle f^{(1)}} & & & & \Big\downarrow{\scriptstyle f^{(M+1)}} \\
(\mathscr{V}^{(0)},\gamma^{(0)}) & \longleftarrow & (\mathscr{V}^{(1)},\gamma^{(1)}) & \longleftarrow & \cdots & \longleftarrow & (\mathscr{V}^{(M+1)},\gamma^{(M+1)})
\end{array}
$$

where the vertical arrows are étale. By the usual isomorphism, we have

$$
(\eta^{(i)})^*(f^{(i)})^*\Omega^1_{\mathscr{V}^{(i)}/R} \cong (\eta^{(i)})^*\Omega^1_{\widetilde{\mathscr{W}^{(i)}}/R}.
$$

Hence, we conclude that, for every integer $i \in \{0,\dots,M+1\}$, the integer $\mathfrak{N}\mathfrak{e}_i$ equals the smoothness defect at $\eta^{(i)}$ on $\widetilde{\mathscr{W}^{(i)}}$. For every integer $i \in \{0,\dots,M+1\}$, we denote by $\mathscr{W}^{(i)}$ the dilatation constructed for $(\mathscr{W},\eta)$ in diagram (2.1). By the universal property of dilatations (see [**1**, Proposition 3.2.1]), for every integer $i \in \{1,\dots,M+1\}$, there exists a morphism of R-schemes $g^{(i)}\colon \mathscr{W}^{(i)} \to \widetilde{\mathscr{W}^{(i)}}$ which makes the following diagram of morphisms of R-schemes commutative:

$$
\begin{array}{ccccccc}
(\mathscr{W}^{(0)},\eta^{(0)}) & \longleftarrow & (\mathscr{W}^{(1)},\eta^{(1)}) & \longleftarrow & \cdots & \longleftarrow & (\mathscr{W}^{(M+1)},\eta^{(M+1)}) \\
\Big\| & & \Big\downarrow{\scriptstyle g^{(1)}} & & & & \Big\downarrow{\scriptstyle g^{(M+1)}} \\
(\mathscr{W}^{(0)},\eta^{(0)}) & \longleftarrow & (\widetilde{\mathscr{W}^{(1)}},\eta^{(1)}) & \longleftarrow & \cdots & \longleftarrow & (\widetilde{\mathscr{W}^{(M+1)}},\eta^{(M+1)})
\end{array}
$$

Since, for every integer $i \in \{1,\dots,M+1\}$, the morphism $f^{(i)}$ is étale, we observe that $(f_s^{(i)})^{-1}(\gamma_s^{(i)})$ is a disjoint sum of points (which are rational thanks to our assumption on the field k). Thus, for every integer $i \in \{1,\dots,M+1\}$, there exists an open subscheme $\mathscr{U}^{(i)}$ (resp. $\widetilde{\mathscr{U}^{(i)}}$) of $\mathscr{W}^{(i)}$

(resp. $\widetilde{\mathscr{W}^{(i)}}$), containing $\eta^{(i)}$, endowed with an isomorphism $\mathscr{U}^{(i)} \cong \widetilde{\mathscr{U}^{(i)}}$. We conclude the proof once again by applying the usual isomorphisms:

$$(\eta^{(i)})^* \Omega^1_{\mathscr{W}^{(i)}/R} \cong (\eta^{(i)})^* \Omega^1_{\mathscr{U}^{(i)}/R} \cong (\eta^{(i)})^* \Omega^1_{\widetilde{\mathscr{U}^{(i)}}/R} \cong (\eta^{(i)})^* \Omega^1_{\widetilde{\mathscr{W}^{(i)}}/R}. \quad \square$$

COROLLARY 2.3. *Keep the notation and assumptions of Proposition 2.1. Let $\mathscr{U}$ be an open subscheme of $\mathscr{V}$ such that $\gamma \in \mathscr{U}(R)$. Then, the Néron multiplicity sequence associated with $(\mathscr{V}, \gamma)$ coincides with that associated with $(\mathscr{U}, \gamma)$.*

2.3. Let us prove the main theorem of this section. First of all, let us stress that the notion of dilatations for adic noetherian formal schemes has been introduced in [**6**, §2.3]. Since this extended notion of dilatations and the defect of smoothness commute with formal completions (see [**6**, Proposition 2.21]), we deduce the following statement.

THEOREM 2.1. *Let R be a Henselian local ring with residue field k. Let $\mathscr{V}$ be a flat R-scheme of finite type. Let $\gamma \in \mathscr{V}(R)$ which is assumed to be non-degenerate. Then, the associated Néron multiplicity sequence only depends on the formal completion of $\mathscr{V}$ at γ_s (and not on the scheme $\mathscr{V}$ itself).*

2.4. Let us focus on the case of curves (a specific class of examples be studied in Section 4). Let k be an algebraically closed field of characteristic zero. Let $\mathscr{C}$ be a curve analytically irreducible at its singular points. In this situation, for every point $x \in C$, there exists an affine subscheme $\mathscr{U}_x$ of $\mathscr{C}$ such that the normalization $\pi \colon \overline{\mathscr{C}} \to \mathscr{C}$ induces a section $\gamma_x \in \mathscr{U}_x(k[[t]]) \subset \mathscr{C}(k[[t]])$. Such a section, called *primitive*, is unique up to a (continuous) local automorphism of $k[[t]]$. By Lemma 2.1 and Proposition 2.1, we deduce the following statement.

COROLLARY 2.1. *Keep the notation of Section 2.4. The Néron multiplicity sequence associated with $(\mathscr{C}, \gamma_x)$ is a formal invariant of the pointed curve $(\mathscr{C}, x)$.*

3. Nash and Néron Multiplicity Sequences

3.1. Let k be a field. In [4], the notion of *Nash multiplicity sequence* has been introduced. It is a decreasing sequence of positive integers $(\mathfrak{Na}_i)_{i \in \{0,...,N+1\}}$ attached to the initial datum of a pair $(\mathscr{V}, \gamma)$ formed by a k-variety $\mathscr{V}$ and a formal germ of curve $\gamma \in \mathscr{V}(k[[t]])$ (which can be viewed as a section of the $k[[t]]$-variety $\mathscr{V} \otimes_k k[[t]]$). This sequence is built as follows. Let us consider the projection

$$\mathscr{W} \leftarrow \mathscr{W}^{(0)} := \mathscr{W} \otimes_k k[t].$$

The blowing-up of the special point $\gamma_s := \gamma(0)$ in $\mathscr{W}_s^{(0)}$ provides a $\mathscr{W}^{(0)}$-scheme $\mathscr{W}^{(1)} \to \mathscr{W}^{(0)}$ and the valuative criterion of properness a lifting

$\gamma^{(1)}$ of $\gamma^{(0)} := (\gamma, t)$. Then, one can reiterate the process. In this way, one obtains a sequence of spaces:

$$(\mathscr{W}^{(0)}, \gamma^{(0)}) \longleftarrow (\mathscr{W}^{(1)}, \gamma^{(1)}) \longleftarrow \cdots$$

For every integer $i \in \mathbf{N}$, one can consider the multiplicity of $\mathscr{W}^{(i)}$ at $\gamma_s^{(i)}$. If the generic point γ_η is assumed to be smooth, one shows that there exists a minimal integer $N \in \mathbf{N}$, such that, for every integer $j \geq N+1$, we have $\mathfrak{Na}_i := \mathrm{mult}_{\gamma_s^{(j)}}(\mathscr{W}^{(j)}) = 1$ (see [**3**]). In particular, the sequence $(\mathfrak{Na}_i)_{i \in \mathbf{N}}$ is stationary and the germ of curve $\gamma^{(N+1)} \in \mathscr{W}^{(N+1)}(k[[t]])$ is contained in the smooth locus of $\mathscr{W}^{(N+1)}$. One calls $(\mathfrak{Na}_i)_{i \in \{0,\dots,N+1\}}$ the *Nash multiplicity sequence*.

3.2.   Keep the notation of Section 3.1. Let us set $R = k[[t]]$. Let $\mathscr{W}$ be a k-variety, with $\gamma \in \mathscr{W}(R)$ a germ of curve whose generic point is contained in the smooth locus of $\mathscr{W}$. Let $\mathscr{V}^{(0)} := \mathscr{W} \otimes_{k[t]} R =: \mathscr{W}^{(0)} \otimes_{k[t]} R$. Let $\mathscr{B}_{\mathscr{V}^{(0)}}$ be the blowing-up of $\mathscr{V}^{(0)}$ at γ_s. We have the following commutative diagram of morphisms of R-schemes:

$$(3.1) \qquad \begin{array}{ccc} (\mathscr{V}^{(0)}, \gamma^{(0)}) & \longleftarrow & (\mathscr{B}_{\mathscr{V}^{(0)}}, \gamma^{(1)}) \\ \downarrow & & \downarrow{\scriptstyle u} \\ (\mathscr{W}^{(0)}, \gamma^{(0)}) & \longleftarrow & (\mathscr{W}^{(1)}, \gamma^{(1)}) \end{array}$$

Since the completion morphism $k[t] \to R$ is flat (see [**5**, Theorem 1.3.15]), we deduce from [**5**, Proposition 8.1.12] that $\mathscr{B}_{\mathscr{V}^{(0)}} \cong \mathscr{W}^{(1)} \times_{\mathscr{W}^{(0)}} \mathscr{V}^{(0)} \cong \mathscr{W}^{(1)} \otimes_{k[t]} R$. In particular, the morphism u is flat. By iterating this construction, if, for every integer $i \in \{1, \dots, M+1\}$, the R-scheme $\mathscr{V}^{(i)}$ be the dilatation of $\mathscr{V}^{(i-1)}$ at $\gamma_s^{(i-1)}$, we obtain the following commutative diagram of morphisms of schemes:

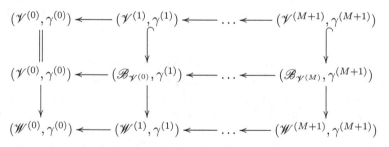

If M is the smallest integer, such that $\gamma^{(M+1)}$ factorizes through the smooth locus of $\mathscr{V}^{(M+1)}$, then, by the very definition of the integer N, we have $M = N$.

REMARK 3.2.   The former remark shows that, in this situation, we have two sequences of non-negative integers of same length computable from the sequence of schemes $(\mathscr{W}^{(i)})_i$, as explained above. However, as proved in Section 4, in general, the Nash and the Néron multiplicity sequences do not coincide, but seem to be comparable (see Theorem 4.1).

4. Cuspidal Plane Curve Singularities

In this section, we compute the Néron multiplicity sequence in the case of cuspidal singularities of curves in terms of the Nash multiplicity sequence. We assume that the characteristic of the field k is zero.

4.1. Let $(\mathscr{C}, x)$ be a pointed k-curve with $x \in \mathscr{C}(k)$. We say that $(\mathscr{C}, x)$ is a *cuspidal (plane curve) singularity* if there exist a pair $(\alpha, \beta) \in \mathbf{N}^2$ of coprime integers, with $\alpha, \beta \geq 2$, and a local isomorphism of k-algebras $\widehat{\mathscr{O}_{\mathscr{C},x}} \cong k[[X, Y]]/\langle X^\beta - Y^\alpha \rangle$. In particular, such a plane curve singularity is analytically irreducible. We denote by $e_0(\mathscr{C})$ the multiplicity of $\mathscr{C}$ at the origin. As in Section 4.1, we denote by γ_0 a primitive parameterization at the origin. We compute the Néron multiplicity sequence of the pair $(\mathscr{C}, \gamma_0)$.

EXAMPLE 4.1. Let us assume that the field k is algebraically closed. By [2], one knows that every irreducible quasi-homogeneous polynomial $f \in k[X, Y] =: \mathbf{A}$ defines at the origin a cuspidal singularity.

4.2. The following statement answers Question 1.1 for this class of plane curve singularities. By Corollary 2.1, we may assume that there exist two coprime integers $(\alpha, \beta) \in \mathbf{N}^2$, with $\alpha \leq \beta$ and $\alpha \geq 2$, such that $\mathscr{C} = \mathrm{Spec}(\mathbf{A}/\langle f \rangle)$, $f = X^\beta - Y^\alpha$ and $\gamma_0 = (t^\alpha, t^\beta)$. Because of our choice, we may assume that the tangent cone of f is of the form Y^α. The defect of smoothness is also given by the partial derivative $\partial_2(f)$ of f that we simply denote by $\partial(f)$.

THEOREM 4.1. *Keep the notation of Section 4.2. Then, for every integer $i \in \{1, \ldots, N\}$, we have*

$$\mathfrak{Ne}_i = -i(\mathfrak{Na}_i - 1) + [f : \partial(f)],$$

where we denote by $[f \cdot \partial(f)]$ the intersection multiplicity of $\mathscr{C}$ and $V(\partial(f))$ at the origin $\mathfrak{o}$.

PROOF. $\circ$ First of all, we observe that $N = \beta - 1$. Indeed, by construction, the curve $\mathscr{C}^{(\beta)}$ is defined, thanks to the quasi-homogeneity of f, by the equation

$$f^{(\alpha)} = f(t^{\beta - \alpha} X + 1, Y).$$

We conclude that the special point $\gamma_s^{(\beta)}$ is smooth in the special fiber of $\mathscr{C}^{(\beta)}$. So, we have $N \leq \beta - 1$. Now, if $f = f_r + \cdots + f_d$ is the homogenous decomposition of f, for every integer $i \in \{0, \ldots, \alpha - 1\}$, we have

$$f(t^i X, t^i Y)$$

$$= \sum_{j=r}^{d} f_j(t^i X, t^i Y)$$

(4.2)
$$= \sum_{j=r}^{d} t^{ji} f_j(X, Y)$$

$$= t^{ri} \left(f_r(X, Y) + t^i f_{r+1}(X, Y) + \cdots + t^{(d-r)i} f_d(X, Y) \right)$$

$$= t^{e_0(\mathscr{C})i} \left(f_r(X, Y) + t^i f_{r+1}(X, Y) + \cdots + t^{(d-r)i} f_d(X, Y) \right).$$

From equation (4.2), we deduce that, for every integer $i \in \{0, \ldots, \alpha - 1\}$, the curve $\mathscr{C}^{(i)}$ is defined by the equation

$$f^{(i)} = f_r(X, Y) + t^i f_{r+1}(X, Y) + \cdots + t^{(d-r)i} f_d(X, Y),$$

and that the section $\gamma^{(i)}$ does not factorize through the smooth locus of $\mathscr{C}^{(i)}$. For every integer $i \in \{0, \ldots, \beta - \alpha\}$, the curve $\mathscr{C}^{(\alpha+i)}$ is defined by the equation

$$f^{(\alpha+i)} = f(t^{(\beta-\alpha)\alpha - \alpha i}(t^i X + 1), Y).$$

So, we have $N \geq \beta - 1$.

∘ In the end, the previous formulas for the polynomials $f^{(i)}$ allow us to obtain the expression of the map L. For every integer $i \in \{1, \ldots, \beta - 1\}$, we deduce by equation (4.2) that

$$\mathrm{ord}_t(\partial(f^{(i)})(\gamma^{(i)})) = -i(e_0(\mathscr{C}) - 1) + \mathrm{ord}_t(\partial(f)(\gamma)). \qquad \square$$

EXAMPLE 4.3. Let us consider $f = X^3 - Y^2$ with $\gamma_0 = (t^2, t^3)$. In this case, we check that the curve $\mathscr{C}^{(1)}$ is defined by the datum of the equation:

$$f^{(1)} = t^2 X^3 - Y^2,$$

with section (t, t^2), the curve $\mathscr{C}^{(2)}$ is defined by the datum of the equation:

$$f^{(2)} = t^2 (X + 1)^3 - Y^2,$$

with section $(0, t)$, and that the curve $\mathscr{C}^{(3)}$ is defined by the datum of the equation:

$$f^{(3)} = (tX + 1)^2 - Y^2,$$

with section $(0, 1)$. With this picture, it is easy to verify our formula by observing that the various multiplicities take the following numerical values:

$$\begin{cases} \mathfrak{Na}_0 = 2 \\ \mathfrak{Na}_1 = 2 \\ \mathfrak{Na}_2 = 2 \\ \mathfrak{Na}_3 = 1 \\ \rule{2cm}{0.4pt} \\ \mathfrak{Ne}_0 = 3 \\ \mathfrak{Ne}_1 = 2 \\ \mathfrak{Ne}_2 = 1 \\ \mathfrak{Ne}_3 = 0. \end{cases}$$

Acknowledgments

We would like to thank Nicaise and the referee for their precise readings and comments which broadly improve the presentation of our work.

Bibliography

[1] S. Bosch, W. Lütkebohmert and M. Raynaud, *Néron Models*, Ergebnisse der Mathematik und ihrer Grenzgebiete (3) [Results in Mathematics and Related Areas (3)], Vol. 21 Springer-Verlag, Berlin, 1990.

[2] L. C. Meireles, On the classification of quasi-homogeneous curves. https://arxiv.org/pdf/1009.1664.pdf.

[3] M. Hickel, Sur quelques aspects de la géométrie de l'espace des arcs tracés sur un espace analytique. *Ann. Fac. Sci. Toulouse Math.* (6) 14(1):1–50, 2005.

[4] M. Lejeune-Jalabert, Courbes tracées sur un germe d'hypersurface. *Amer. J. Math.*, 112(4):525–568, 1990.

[5] Q. Liu, *Algebraic Geometry and Arithmetic Curves*, Oxford Graduate Texts in Mathematics. Vol. 6 Oxford University Press, Oxford, 2002. Translated from the French by Reinie Erné, Oxford Science Publications.

[6] J. Nicaise, A trace formula for rigid varieties, and motivic Weil generating series for formal schemes. *Math. Ann.*, 343(2):285–349, 2009.

14

The Dual Complex of Singularities After de Fernex, Kollár and Xu

Mirko Mauri

*Department of Mathematics, Imperial College London,
180 Queen's Gate, London SW72AZ, UK
m.mauri15@imperial.ac.uk*

This is a brief introduction to the theory of dual complexes of singularities. This survey chapter focuses on the recent results of Ref. [10]. We also offer a brief overview of their applications and related problems outside the theory of singularities.

1. Introduction

Given a simple normal crossing (SNC) divisor, it is possible to construct a CW-complex called *dual complex*, encoding the combinatorial data of how its irreducible components intersect. A recent important work by de Fernex, Kollár and Xu [10] explains how dual complexes are affected by transformations of the minimal model program (MMP). As a result, they have been able to describe the topology of dual complexes in many interesting cases. In this survey chapter, we present some of their ideas. Where possible, we will try to avoid the technicalities of the MMP in order to make this presentation accessible to readers unfamiliar with their subtleties.

1.1. Dual complex of a resolution of singularities

Let $(x \in X)$ be the germ of an isolated singularity over an algebraically closed field of characteristic zero and $f : Y \to X$ be a resolution of singularities, i.e., a projective morphism such that Y is a smooth variety and the exceptional locus $\mathrm{Ex}(f)$ has simple normal crossings. We can consider the dual complex of $\mathrm{Ex}(f)$, denoted by $\mathcal{DR}(x \in X)$ (cf. Definitions 4.1 and 5.1). It has been proved that its homotopy type is independent of the choice of a resolution, see [8, 37, 27, A.4; 41, 38, 2, 33].

The authors of [10] have introduced a refined invariant of singularities. If a canonical divisor K_X is $\mathbb{Q}$-Cartier, they have defined a distinguished homeomorphism class, denoted $\mathcal{DMR}(x \in X)$, which sits inside the homotopy class of $\mathcal{DR}(x \in X)$ and corresponds to the dual complex of minimal partial resolutions, namely dlt modifications. The main properties of these objects are collected in the following theorem.

Main Theorem ([10]). *Let $(x \in X)$ be a germ of an isolated singularity such that K_X is $\mathbb{Q}$-Cartier. Then, the following statements hold:*

(1) *$\mathcal{DMR}(x \in X)$ is independent of the choice of a dlt modification;*
(2) *$\mathcal{DR}(x \in X)$ has the homotopy type of $\mathcal{DMR}(x \in X)$;*
(3) *if X is klt, then $\mathcal{DR}(x \in X)$ is contractible.*

The last assertion means roughly that provided that X is mildly singular, the homotopy type of the dual complexes of its resolutions is the simplest possible. However, if we remove this hypothesis, the dual complex of an arbitrary singularity can be arbitrarily complicated: a construction by Kollár [23] grants that any finite simplicial complex can be realized as the dual complex of a resolution of an isolated singularity of an algebraic space.

Note that for expository reason, the Main Theorem is stated under stronger hypotheses than those in [10, Theorem 28]. There, the singularities of X are not necessarily isolated. Moreover, they are not supposed to be $\mathbb{Q}$-Gorenstein, i.e., K_X is not necessarily $\mathbb{Q}$-Cartier, but it is sufficient that there exists an effective divisor Δ such that $K_X + \Delta$ is $\mathbb{Q}$-Cartier. The interested reader can consult the original source [10].

1.2. Dual complex of a log canonical singularity

In the intermediate case, when weaker hypotheses of regularity (e.g., log canonical singularity) are imposed, the topology of dual complexes is not yet completely understood. In dimension 2, the only graphs which appear as dual complexes of a minimal resolution of a log canonical surface singularity are (contractible) trees, or circles in the case of cusp singularities, see [24, Theorem 4.7]. In higher dimension, a complete classification of the dual complexes of log canonical singularities is unknown and at the moment few results about their topology are available. Among them, it is worthy to mention that Kollár has constructed new examples of log canonical 3-fold singularities whose dual complexes are any prescribed connected real 2-manifold without boundary, see [20, Theorem 1; 21, Theorem 3.47].

1.3. Dual complex of a rational singularity

Klt singularities are rational singularities, see [**24**, Definition 5.8, Theorem 5.22]. However, Main Theorem (3) fails to extend to the larger class of rational singularities. As a counterexample, Payne has constructed a rational 3-fold singularity whose dual complex has the homotopy type of the real projective plane (cf. [**33**, Example 8.3]): it is obtained as a deformation of a cone over a degenerate Enriques surface.

In fact, a cohomological characterization of the dual complexes of isolated rational singularities is known: these dual complexes have the rational homology of a point, i.e., $H_i(\mathcal{DR}(x \in X), \mathbb{Q}) = 0$ for $i > 0$, [**33**, Proposition 8.2], and conversely any connected finite simplicial n-dimensional complex with this property can be realized as the dual complex of a resolution of an isolated rational $(n+1)$-dimensional singularity, see [**21**, Theorem 42; **16**, Theorem 8; **23**, Theorem 3].

1.4. Dual complex of a qdlt singularity

The largest class of singularities for which a well-posed definition of dual complex is known are quotient dlt singularities. They are log canonical singularities which are locally quotients of an snc pair by the action of a finite abelian group preserving each irreducible components of the boundary, see also [**10**, Proposition 34]. For instance, they arise as singularities of the push-forward of an snc or dlt vertical pair via a Fano contraction, see [**10**, Proposition 40; **15**, Lemmas 5.1 and 5.2]. The study of qdlt singularities is reduced to the dlt case in [**10**, Corollary 38].

1.5. History

Preliminary contributions due to [**40**; **19**, Theorem 7.6] paved the ground for the results of [**10**]. They had already shown that if $(x \in X)$ has klt singularities, then the fundamental group of $\mathcal{DR}(x \in X)$ is trivial. Additional comments on the genesis of the Main Theorem can be found in the introduction of [**10**].

1.6. Essential skeleton of a degeneration

Beyond the theory of singularities, the techniques developed in [**10**] have been finding new applications in different contexts. Although a detailed account on them goes beyond the scope of this chapter, for the rest of this introduction, we will mention some of these applications and related open problems.

In mirror symmetry, dual complexes appear in relation to the study of the singular fibers of degenerations of Calabi–Yau varieties, namely proper morphisms of K_X-trivial projective varieties X over a small analytic disk. Kontsevich and Soibelman first defined a dual complex denoted the essential skeleton of the Berkovich analytification X^{an} of X, see [**26**, §3.3]. They conjecture that it should be the basis of two mirror Lagrangian fibrations, whose existence is predicted by the SYZ conjecture [39]. They have also shown in [26, §3.3, Proposition 3] that this dual complex can be identified with the locus where a real-valued function on X^{an}, called weight function, attains its minimum.

This object has been generalized by Mustață and Nicaise [30]. Nicaise and Xu [32] have proved that if X has a semi-ample canonical divisor (e.g., X Calabi–Yau or conjecturally all minimal models), the essential skeleton of X can be identified with the dual complex of the reduced special fiber of a good minimal dlt degeneration. Moreover, Nicaise, Xu and Yu [31] have recently shown that the essential skeleton is indeed the basis of a non-archimedean version of the SYZ fibration. Inspired by [10], Nicaise and Xu [32] have also proved that the essential skeleton is a strong deformation retract of the Berkovich analytification X^{an}. They have also suggested that there should exist a flow in the direction of minimal values of the weight function which defines a retraction onto the essential skeleton. Potentially, the weight function provides an effective method to identify the dual complex of a minimal dlt model without running any MMP, an idea implemented for instance in [5]. In the non-semi-ample case, precisely for X rationally connected (e.g., a Fano variety), it has been proved that the dual complex of a dlt degeneration of X [10, Theorem 41] and its Berkovich analytification [6, Corollary 1.1.4] are contractible.

The actual homeomorphism type of the essential skeleton of a degeneration of a K_X-trivial variety is still unknown and great effort has been made in this direction. By the Beauville–Bogomolov decomposition [3], a smooth projective variety with trivial first Chern class is a product of abelian varieties, Calabi–Yau in the strict sense (i.e., $\pi_1(X) = 1$ and $h^i(X, \mathcal{O}_X) = 0$ for $0 < i < \dim X$) and hyper-Kähler manifolds, up to étale covers. We can consider separately each type in the decomposition; see also [5, Theorem 1.6.1, Proposition 6.1.9] which describes the essential skeleton of finite quotients and products. The case of abelian varieties has been settled by Halle and Nicaise in [13, Proposition 4.3.2]: in this case, the essential skeleton is a finite quotient of a real torus (cf. also [4, Theorem 6.5.1]). The SYZ conjecture and Konstevich–Soibelman predict that the essential skeleton of a maximally unipotent degeneration (i.e., when the essential skeleton has maximal dimension $n := \dim_{\mathbb{C}} X$) is a sphere $\mathbb{S}^n$ or a complex projective space $\mathbb{P}^{n/2}(\mathbb{C})$ respectively for Calabi–Yau or hyper-Kähler manifolds. A cohomological result in this direction has been proved: under the previous hypothesis, the dual complex of a degeneration of Calabi–Yau or hyper-Kähler manifolds has the rational homology respectively of a sphere (cf. [25, §3.4]) or of a complex projective space (cf. [18, Theorem 0.10]). This fact, together with the simply-connectedness of the dual complex, implies that the conjecture holds for Calabi–Yau varieties in dimension ≤ 3 and in dimension ≤ 4 conditional to the existence of a minimal degeneration with snc singular fiber. In [5], Brown and Mazzon have proved the conjecture for an interesting class of hyper-Kähler degenerations, namely degenerations of Hilbert schemes of points of K3 surfaces and generalized Kummer varieties induced by semi-stable degenerations of the associated K3 and abelian surfaces. Using the weight function, they have identified their essential skeletons without computing any explicit resolution of singularities, in the spirit of the weight function approach mentioned above.

1.7. Tropical complex

Constraints on the topology of dual complexes of degenerations come from the intersection theory of the irreducible components of the special fibers. For instance, Carlwright [7] has shown that no topological surface with negative Euler characteristic can be realized as the dual complex of a strictly semi-stable degeneration of surfaces. The obstruction consists in promoting the dual complex to a tropical complex, namely a Δ-complex enriched with numerical labels which account for the facts that the reduced singular fiber is principal and the intersection form on its irreducible components obeys the Hodge index theorem.

As already noted, this restriction does not apply to dual complex of singularities, see [20, 21, 23]. Note also that all topological surfaces of non-negative Euler characteristic can be obtained as dual complex of a strictly semi-stable degeneration: spheres, real projective spaces, tori and Klein bottles can be obtained as degenerations of K3 surfaces, Enriques surfaces, abelian varieties and bielliptic surfaces respectively, see [28, 34, 35, 29].

1.8. Dual complex of a log Calabi–Yau pair

A deeper understanding of the local topology of dual complexes passes through the study of log Calabi–Yau pairs. Indeed, given $(X, \Delta =: \sum_i D_i)$ a dlt pair, the link of the vertex v_{D_j} associated to the divisor D_j in the dual complex $\mathcal{D}(X, \Delta)$ is homeomorphic to the dual complex $\mathcal{D}(D_j, \sum_{i \neq j} D_i|_{D_j})$. If the divisor Δ is the exceptional divisor of a minimal dlt modification of a log canonical singularity or a special fiber of a minimal dlt degeneration of Calabi–Yau varieties, then the pair $(D_j, \sum_{i \neq j} D_i|_{D_j})$ is log Calabi–Yau, i.e., $K_{D_j} + \sum_{i \neq j} D_i|_{D_j} \sim_{\mathbb{Q}} 0$. Kollár and Xu have raised the question in [25, Question 4] whether the dual complex of a log Calabi–Yau pair of maximal intersection (i.e., when the dual complex has maximal dimension $n := \dim_{\mathbb{C}} X - 1$) should be a finite quotient of a sphere $\mathbb{S}^n$. They have proved that it has the rational homology of a sphere and they have confirmed the conjecture in dimension ≤ 4 and in dimension ≤ 5 for snc pairs. Similar methods have been exploited also in [18, §6] to obtain the cohomological results mentioned in Section 1.6. An interesting open problem is to determine the torsion of the homology of the dual complex of log Calabi–Yau pairs in higher dimension, see [25].

1.9. Non-abelian Hodge theory

The study of the topology of the dual complex of log Calabi–Yau pairs has its own interest, independent of the previous conjectures in mirror symmetry. In non-abelian Hodge theory, the asymptotic behavior at infinity of the Corlette–Simpson correspondence can be phrased in terms of properties of dual complexes. More precisely, the dual complex of the boundary of the compactification of a character variety is conjectured to be homotopic equivalent to the sphere at infinity of the corresponding Hitchin moduli space of Higgs bundles (cf. [17, Conjecture 1.1] for a stronger formulation of the conjecture). In particular, this relation should give a geometric realization, at least in degree zero, of the conjectured exchange between the weight filtration of the mixed Hodge structure of a character variety and the perverse Leray filtration induced by the Hitchin map, see [9]. Evidence for the former conjecture has been provided by Simpson [36], who has shown that the dual complex of the boundary of certain SL_2-character varieties on punctured spheres is indeed a sphere. It is clear that for character varieties which admit log Calabi–Yau compactifications, a positive answer to [25, Question 4] would imply the weak formulation of conjecture [17, Conjecture 1.1] which is stated here. For further information, we refer the interested reader to [17, §1.1; 36, §1.2].

1.10. Other surveys

We bring to the attention of the reader the excellent surveys *Links of Complex Analytic Singularities* [21], *The Topology of Log Canonical Singularities* [22, §8.6] and *Interaction Between Singularity Theory and the Minimal Model Program* [42].

1.11. Structure of the chapter

In the earlier sections, we recall some definitions of birational geometry and simplicial topology and fix the essential notations. In Section 4, we define the dual complex of an snc pair. Following [10], we keep track of the changes that the complex undergoes throughout the step of an MMP. A first difficulty arises from the fact that the property of being snc is not stable under the birational transformations of the MMP, thus questioning the very definition of a dual complex. Therefore, in Section 5, the class of divisors for which we can construct a dual complex is enlarged to the so-called dlt pairs. In Section 6, we preliminarily focus on a smaller class of birational transformations, i.e., blow-ups along the strata of an snc divisor, and we quote [37] to describe the resulting dual complex. We enter the details of [10] in the last sections: the simplices of a dual complex are in correspondence with the lc centers of a dlt pair, whose behavior under transformation of the MMP is well studied. Finally, those authors successfully employ subtle MMP techniques to control the homotopy type of dual complexes.

2. Notation: Birational Dictionary

A pair (X, Δ) is the datum of a normal variety X and a $\mathbb{Q}$-divisor Δ. If all coefficients of Δ are in $(0, 1]$ (resp. $(-\infty, 1]$), we say that Δ is a boundary (resp. a sub-boundary). Its support is the union of the prime divisors with non-zero coefficient in Δ and it is denoted by $\mathrm{Supp}\,\Delta$.

Let $f : Y \dashrightarrow X$ be a birational map. The **exceptional locus** of f, denoted $\mathrm{Ex}(f)$, is the locus, contained in Y, where f fails to be an isomorphism. A subset of Y is said to be **contracted**, if it is contained in $\mathrm{Ex}(f)$.

A $\mathbb{Q}$-divisor is $\mathbb{Q}$-**Cartier** if one of its multiples is Cartier. A normal variety X is $\mathbb{Q}$-**factorial** if any Weil divisor of X is $\mathbb{Q}$-Cartier.

Assume now that f is a morphism. Given a pair (X, Δ) such that $K_X + \Delta$ is $\mathbb{Q}$-Cartier, its **log pull-back** via f is the pair (Y, Δ_Y) identified by the relations

$$K_Y + \Delta_Y \sim_{\mathbb{Q}} f^*(K_X + \Delta), \quad f_*\Delta_Y = \Delta.$$

The negative of the coefficient of a prime divisor E in Δ_Y, labeled $a(E, X, \Delta)$, is its **discrepancy**.

A pair (X, Δ) is **log canonical (lc)**, if $a(E, X, \Delta) \geq -1$ for any $f : Y \to X$ birational morphism and for any exceptional divisor E. An irreducible subvariety $Z \subset X$ is an **lc center** if there exists a birational morphism $f : Y \to X$ and an exceptional divisor E whose discrepancy $a(E, X, \Delta)$ equals -1 and whose image coincides with Z.

A pair (X, Δ) is **log smooth** or **simple normal crossing (snc)**, if X is a smooth variety and the support of Δ has simple normal crossings. Given any pair, there is the largest open subset $X^{\mathrm{snc}} \subset X$, called simple normal crossing locus, such that $(X^{\mathrm{snc}}, \Delta|_{X^{\mathrm{snc}}})$ is snc.

A log canonical pair (X, Δ) is dlt, alias **divisorial log terminal**, if none of the lc centers is contained in $X \setminus X^{\mathrm{snc}}$.

A log canonical pair (X, Δ) with no lc center is klt, alias **Kawamata log terminal**.

Two pairs (Y, Δ_Y) and $(Y', \Delta_{Y'})$ are said to be **crepant birational** if Y and Y' are birational and $a(E, Y, \Delta_Y) = a(E, Y', \Delta_{Y'})$ for any divisor E; see [**22**, §2.23.2] for an equivalent definition.

Let (X, Δ) be again a pair such that $K_X + \Delta$ is $\mathbb{Q}$-Cartier and let $f : X \to Z$ be a morphism. A **transformation of a** $(K_X + \Delta)$-**MMP** corresponding to a $(K_X + \Delta)$-negative extremal ray R over Z is a map defined over Z

$$\varphi : X \dashrightarrow Y,$$

such that:

(i) $((K_X + \Delta) \cdot R) < 0$;

(ii) any contracted curve $C \subset X$ is numerically equivalent to a multiple of R, i.e., there exists $m \in \mathbb{N}$ such that $(mR \cdot D) = (C \cdot D)$ for any Cartier divisor $D \subset X$ and any contracted curve $C \subset X$.

In particular, note that

(i) (discrepancies increase) if φ is birational, then $a(E, X, \Delta) \leq a$ $(E, Y, \varphi_* \Delta)$ for any divisor E and the strict inequality holds if the center of E in X is contracted, see [**24**, Lemma 3.38];

(ii) (dlt pairs are stable under MMP) if (X, Δ) is a dlt pair, then $(Y, \varphi_* \Delta)$ is so too, see [**24**, Corollary 3.44].

A $(K_X + \Delta)$-MMP over Z is a sequence of these transformations over Z

$$(X, \Delta) = (X_0, \Delta_0) \xrightarrow{\varphi_1} (X_1, \Delta_1) \xrightarrow{\varphi_2} \cdots \xrightarrow{\varphi_l} (X_l, \Delta_l),$$

which terminates either when the dimension of X_l drops (Mori fiber space) or when there is no more $(K_X + \Delta)$-negative curve over Z, i.e., $K_X + \Delta$ is relatively nef over Z (relative minimal model). Other properties of transformations of MMP will be provided in due course. For a detailed account, we refer the interested reader to [**24**].

3. Notation: Simplicial Complexes

We first recall the definition of a simplicial complex.

Let $\Delta^k := \{(x_0, \ldots, x_k) \in \mathbb{R}^{k+1} \mid \sum_i x_i = 1, \, x_i \geq 0\}$ be the standard k-simplex.

DEFINITION 3.1 ([**14**, §2.1]). A CW-complex D is said to be

• a **Δ-complex** if:
    (i) the characteristic maps $\alpha : \Delta^k \to D$ are embeddings on the interior of the simplex Δ^k;
    (ii) for each $(k-1)$-dimensional face of Δ^k, the restriction of the characteristic maps α to that face are characteristic maps of $(k-1)$-cells;

• a **regular** Δ-complex if in addition the characteristic maps are embeddings;

• a **simplicial** complex if it is a regular Δ-complex and any two k-cells have at most a $(k-1)$-cell in common.

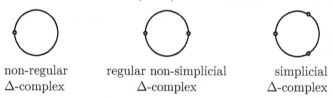

non-regular	regular non-simplicial	simplicial
Δ-complex	Δ-complex	Δ-complex

Note that a CW-complex is prescribed both by the set of its cells and their attaching maps. However, in order to define a regular Δ-complex, the datum of the attaching maps is redundant: it is enough to provide the poset of its cells.

Let D be a regular Δ-complex. If $v \subset D$ is one of its faces, we define

• the **(open) star** of v, denoted $\mathrm{St}(v)$, as the union of the interiors of the cells whose closure intersects v;

- the **closed star** of v, denoted $\overline{\mathrm{St}(v)}$, is the closure of $\mathrm{St}(v)$;
- the **link** of v, denoted $\mathrm{Lk}(v)$, is the difference $\overline{\mathrm{St}(v)} \setminus \mathrm{St}(v)$.

open star of v closed star of v link of v

The figures represent the star subdivision of a standard 2-simplex. In general, a **star subdivision** of D with center $p \in D$ is a regular Δ-complex obtained in the following way (cf. [**10**, §2, Note on terminology]):

(i) the closed cells not containing p are unchanged;

(ii) if v is a closed cell containing p, we replace it with the cells spanned by p and the faces of v not containing p.

Compared to D, the underlying topological space of any of its star subdivisions is unchanged; only the structure of Δ-complex is refined. Up to star subdivisions, any regular Δ-complex can be refined to a simplicial complex.

For future purposes, we recall the definition of the join of topological spaces.

DEFINITION 3.2. The **join** of two topological spaces X and Y is the quotient space of $X \times Y \times [0,1]$ under the identifications $(x, y_1, 0) \sim (x, y_2, 0)$, for all $x \in X$ and $y_1, y_2 \in Y$, and $(x_1, y, 1) \sim (x_2, y, 1)$, for all $x_1, x_2 \in X$ and $y \in Y$.

4. Dual Complex of a Log Smooth Pair

Let (X, Δ) be a log smooth pair.

DEFINITION 4.1. The **dual complex of the simple normal crossing** divisor Δ, denoted $\mathcal{D}(\Delta)$, is a (regular) Δ-complex whose vertices are in correspondence with the irreducible components of Δ and whose k-faces correspond to the irreducible components of the intersection of $k+1$ irreducible components (also called **strata** of Δ).

EXAMPLE 4.2. The dual complex of a two-dimensional log smooth pair is extensively studied and commonly known as dual graph.

EXAMPLE 4.3. Let $X(\Sigma)$ be a smooth toric variety of dimension n with associated fan Σ. Let Δ be its (snc) toric boundary, i.e., the complement of the big torus in $X(\Sigma)$. The dual complex of Δ is the link of the vertex of the fan Σ, i.e., the Δ-complex cut out by Σ onto a sphere centered at the vertex of the fan. In addition, if X is compact, its fan is complete and $\mathcal{D}(\Delta)$ is a triangulation of a sphere of dimension $n-1$.

EXAMPLE 4.4. Given (X_1, Δ_1) and (X_2, Δ_2) log smooth pairs. The dual complex of the product pair $(X_1 \times X_2, X_1 \times \Delta_2 + \Delta_1 \times X_2)$ is naturally isomorphic to the join $\mathcal{D}(\Delta_1) * \mathcal{D}(\Delta_2)$ (cf. [**33**, Lemma 6.2]).

In Section 5, we will extend the definition of dual complex to less regular pairs. Thus, it is important to identify the properties of log smooth pairs necessary for a well-posed definition. This is the content of the following remarks.

REMARK 4.5 (Simple normal crossing variety). The embedding of Δ into X is not relevant to the construction of the dual complex. As a consequence, one can define the dual complex of a simple normal crossing variety. We recall that a simple normal crossing variety Δ is a scheme of pure dimension n with irreducible components D_i having the property that each D_i is smooth and any point $p \in X$ has a Euclidean neighborhood $p \in U_p \subset \Delta$ and an embedding $U_p \hookrightarrow \mathbb{C}^{n+1}$ such that the image U_p is an open subset of the union of coordinates hyperplanes $(z_1 \cdot \ldots \cdot z_r = 0)$ with $r \leq n+1$.

REMARK 4.6 (Attaching map). Note that the poset of the strata of Δ contains all the gluing data to build $\mathcal{D}(\Delta)$. No further specification concerning the attaching maps has to be given. This fact relies on the following property of the simple normal crossing divisor $\Delta = \sum_{i \in I} D_i$:

(i) for every $J \subseteq I$, if $\bigcap_{i \in J} D_i$ is non-empty, then every connected component of $\bigcap_{i \in J} D_i$ is irreducible.

It implies that

(ii) for every $j \in J$, every irreducible component of $\bigcap_{i \in J} D_i$ is contained in a unique irreducible component $\bigcap_{i \in J \setminus \{j\}} D_i$.

Equivalently, in term of simplices, this means that if $k := \#J - 1$, any k-cell is attached to $k+1$ $(k-1)$-cells of the $(k-1)$-skeleton $\mathcal{D}(\Delta)_{k-1}$.

Withdrawing this condition, it is not clear how to glue the cells of the complex together. For instance, consider the divisor $\Delta := D_1 + D_2 + D_3$ where $D_1 := (z = 0)$, $D_2 := (x = y)$, $D_3 := (z = xy)$ in $\mathbb{C}^3$ with coordinates x, y, z, illustrated in Figure 1. The divisor D_2 intersects D_1 and D_3 along irreducible one-dimensional strata, while D_1 and D_3 meet in a couple of intersecting lines $(z = xy = 0)$. These three divisors all intersect at the origin. Since the intersecting lines correspond to two distinct 1-cells, there are at least two ways to glue the 2-cell corresponding to $D_1 \cap D_2 \cap D_3$ to the 1-skeleton. There is no reason to prefer one to the other choice.

REMARK 4.7 (Strata of codimension $k+1$ correspond to k-cells). Note that in the snc case, there exists a bijective correspondence between the strata of Δ of codimension $k+1$ and the k-cells of $\mathcal{D}(\Delta)$. Without the simple normal crossing assumption, different cells corresponding to the same stratum of Δ could occur. For example, consider three lines in a plane intersecting in a single point. The dual complex associated is a triangle, but both the 1-cells and the 2-cell correspond to the unique zero-dimensional stratum.

REMARK 4.8 (Simplicial dual complex). By construction, a dual complex is a regular Δ-complex. It is simplicial if and only if any face is

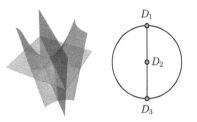

D_1

D_2

D_3

Figure 1. Real trace of the support of D as defined in Remark 4.6 and the 1-skeleton of its putative dual complex.

prescribed by the collection of its vertices, or equivalently if and only if the intersection of any collection of divisors is either empty or irreducible (equivalently connected, due to Remark 4.6). Repeatedly blowing-up the original pair (X, Δ) along strata of Δ, one can always reduce to a simplicial complex. Indeed, a blow-up of a stratum of Δ corresponds to a star subdivision of the corresponding cell in the dual complex, see also Stepanov's lemma (cf. Lemma 6.2). Performing a series of star subdivisions, any regular Δ-complex can be turned into a simplicial one.

5. Dual Complex of a Resolution of Singularities and of a dlt Pair

Let $(x \in X)$ be a germ of an isolated singularity and $f : Y \to X$ a resolution of $(x \in X)$ such that $E := f^{-1}(x)$ has simple normal crossings.

DEFINITION 5.1. The **dual complex of the resolution f of the isolated singularity** $(x \in X)$ is the dual complex of the snc divisor E.

PROPOSITION 5.2 ([**8, 37, 27**, A.4; **41, 38, 2, 33**]). *The homotopy type of the dual complex of a resolution of an isolated singularity is independent of the chosen resolution.*

Hence, we will refer to this homotopy type as the **dual complex of the singularity**, denoted $\mathcal{DR}(x \in X)$ (cf. Section 6 for a proof of Proposition 5.2).

NOTATION 5.3. *The label $\mathcal{DR}(x \in X)$ stands for* Dual complex of a Resolution.

As a result, the homotopy type of the dual complex of the exceptional locus of a resolution is an invariant of the singularity. In contrast, the homeomorphism class of a dual complex is not, since it depends on the choice of a resolution. Consider, for instance, a (minimal) resolution of a surface singularity whose exceptional locus is a cycle of rational curves. The blow-up of a general closed point of one of these rational curves provides another resolution (although no longer minimal). The dual complexes of the two resolutions are respectively a circle and a wedge sum of a circle with a segment: although they have the same homotopy type, the dual complexes are not homeomorphic. In general, blow-ups over general points of exceptional irreducible components are likely to change the homeomorphism class of the dual complex.

Notwithstanding, it is possible to define a distinguished homeomorphism class $\mathcal{DMR}(x \in X)$ inside $\mathcal{DR}(x \in X)$, provided that we limit ourselves to *special* resolutions of singularity. Indeed, the modifications in the previous examples have to be understood as *unnecessary*. The goal is then to contract all the exceptional loci which are *inessential*. The authors of [**10**] have shown that it is sufficient to modify the pair (Y, E) to a *minimal model* (Y', E') relative to X by performing the steps of a so-called $(K_Y + E)$-MMP. It means that the mentioned pairs fit in the following diagram whose arrows are all birational morphisms or maps:

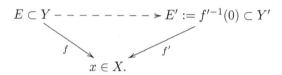

The disadvantage of this procedure is that in high dimension, minimal models are not necessarily smooth or unique. In particular, the property of being log smooth is not stable under the birational transformations of MMP, so that *a priori* the notion of dual complex of the pair (Y', E') is not well posed. However, although not log smooth, the pair (Y', E') is divisorial log terminal (dlt) (cf. Section 2 for the definition of a dlt pair) see also [**24**, Corollary 3.4]. In [**10**], it is observed that the definition of dual complex can be extended to a dlt pair as follows.

Let $(X, \Delta := \sum_{i \in I} a_i D_i)$ be a pair and $\Delta^{=1}$ be the set of irreducible components of Δ whose coefficient a_i is equal to one. If (X, Δ) is a dlt pair, then the dual complex of the pair $(X, \Delta^{=1})$, denoted $\mathcal{D}(\Delta^{=1})$, is constructed as in the simple normal crossing case, see Definition 4.1. In the following, we will check that it is well defined, see (i).

DEFINITION 5.4. The **dual complex of the dlt pair** (X, Δ) is the dual complex of $\Delta^{=1}$:

$$\mathcal{D}(\Delta) := \mathcal{D}(\Delta^{=1}).$$

The minimal models (Y', E') are not necessarily unique but they are characterized by the properties that the pair (Y', E') is dlt and $K_{Y'} + E'$ is f'-nef. A partial resolution $f' : Y' \to X$ satisfying these properties or allowing a slight abuse of language the pair (Y', E') itself, is called a dlt modification of $(X, \emptyset)$.

DEFINITION 5.5. A proper birational morphism $f : Y \to X$ is a **dlt modification** of the pair (X, Δ) if the pair $(Y, f_*^{-1}\Delta + E)$ is dlt, where E

is the sum of all the exceptional divisors with coefficient one, and $K_Y + f_*^{-1}\Delta + E$ is f-nef.

If $K_X + \Delta$ is $\mathbb{Q}$-Cartier, then dlt modifications exist, see, for instance, [**21**, Theorem 1.34].

DEFINITION 5.6. Let $f : Y \to X$ be a dlt modification of the pair (X, Δ). The **dual complex of the dlt modification** f of the pair (X, Δ) is the homeomorphism class of dual complex of the dlt pair $(Y, f_*^{-1}\Delta + E)$:

$$\mathcal{DMR}(x \in (X, \Delta)) := \mathcal{D}(f_*^{-1}\Delta^{=1} + E^{=1}).$$

NOTATION 5.7. The label $\mathcal{DMR}(x \in (X, \Delta))$ stands for *Dual complex of a Minimal dlt partial Resolution*. We define conventionally the dual complex $\mathcal{DMR}(x \in (X, \Delta))$ of a klt pair (X, Δ) to be a point. This choice is compatible with and inspired by the proof of the Main Theorem.

For sake of clarity, we will ignore the boundary Δ until Section 9. Thus, we will simply denote $\mathcal{DMR}(x \in (X, \emptyset))$ by $\mathcal{DMR}(x \in X)$. Here and in the following, we always assume that K_X is $\mathbb{Q}$-Cartier.

Finally, the Main Theorem shows that $\mathcal{DMR}(x \in X)$ is a distinguished homeomorphism class in $\mathcal{DR}(x \in X)$.

Main Theorem [10]. *Let $(x \in X)$ be a germ of an isolated singularity such that K_X is $\mathbb{Q}$-Cartier. The following statements hold:*

(1) $\mathcal{DMR}(x \in X)$ *is independent of the choice of a dlt modification;*
(2) $\mathcal{DR}(x \in X)$ *has the homotopy type of $\mathcal{DMR}(x \in X)$;*
(3) *if X is klt, then $\mathcal{DR}(x \in X)$ is contractible.*

Before proceeding further, we show why the definition of the dual complex of a dlt pair is well posed and why it involves only the irreducible components of Δ which appear with coefficient one.

(i) The dual complex $\mathcal{D}(\Delta^{=1})$ coincides with the dual complex $\mathcal{D}(\Delta^{=1}|_{X^{\mathrm{snc}}})$ of the restriction of $\Delta^{=1}$ to the snc locus X^{snc}. Indeed, the strata of $\Delta^{=1}$ are lc centers, hence they are contained in X^{snc} by definition of dlt. This means that $\mathcal{D}(\Delta^{=1})$ is actually the dual complex of a log smooth pair of an open subset of X. Since we have ignored the irreducible components of Δ with coefficient <1, the definition of the dual complex of a dlt pair is well posed, because that of a log smooth pair is, see Remarks 4.6 and 4.7.

(ii) One could first naively guess that the dual complex of a dlt pair should be the dual complex of its support. However, this complex is usually not well defined. Indeed, the support of a dlt pair can be arbitrary singular so that the issue of Remark 4.6 can affect the construction.

Even assuming that $\mathrm{Supp}(\Delta)$ remains snc under some MMP transformations, some undesired phenomena can occur if we allow coefficients <1. For instance, Lemma 7.1 can fail, as we show in the following example. Let $f : X \to \mathbb{P}^2$ be the blow-up of a point with exceptional divisor E, and C_1 and C_2 be the strict transforms of two smooth conics, intersecting transversely in the blown-up point. The pair $(X, \Delta := \epsilon C_1 + \epsilon C_2)$ is snc, hence dlt, for $0 < \epsilon < 1$. The blow-up f is the divisorial contraction of X corresponding to the $(K_X + \epsilon C_1 + \epsilon C_2)$-negative extremal ray E for $\epsilon < 1/2$. Indeed, we have

$$((K_X + \epsilon C_1 + \epsilon C_2) \cdot E) = ((f^* K_{\mathbb{P}^2} + E + \epsilon C_1 + \epsilon C_2) \cdot E) = -1 + 2\epsilon < 0.$$

Note that although $f_* \Delta$ is snc, $\mathcal{D}(\mathrm{Supp}(\Delta))$ and $\mathcal{D}(\mathrm{Supp}(f_* \Delta))$ do not have the same homotopy type: the former is contractible, but the latter is a circle.

(iii) Case 3 in Stepanov's lemma (cf. Lemma 6.2) is the only source of PL-surgeries, induced by blow-ups along admissible loci, which do not induce homeomorphisms between the corresponding dual complexes. One of the main advantages of ignoring irreducible components of Δ with coefficient < 1 is that the transformations described in Case 3 never affect $\mathcal{D}(\Delta^{=1})$. Indeed, the discrepancy of the exceptional divisor of the blow-up along an admissible locus Z is

$$a(E, X, \Delta) = -1 + \mathrm{codim}_X Z - \#\{D_i : Z \subset D_i\}.$$

Hence, it is strictly bigger than -1 in the hypothesis of Case 3, namely if Z is contained in $\mathrm{Supp}(\Delta)$, but it is not a stratum of Δ. This implies that the divisor E does not appear in the log pull-back Δ' of Δ with coefficient $= 1$, so that the dual complexes $\mathcal{D}(\Delta'^{=1})$ and $\mathcal{D}(\Delta^{=1})$ coincide. This is one of the main reasons why $\mathcal{DR}(x \in X)$ defines just a homotopy type, whereas $\mathcal{DMR}(x \in X)$ identifies a homeomorphism class.

(iv) There exists a bijective correspondence between the lc centers of the pair (X, Δ) and the cells of the dual complex $\mathcal{D}(\Delta^{=1})$. Hence, one can describe the topology of $\mathcal{D}(\Delta^{=1})$ through the well-studied birational geometry of lc centers of dlt pairs, see, for instance, [**22**, §4.2].

6. Proof of Proposition 5.2 and Main Theorem (1)

In Section 7, we prove the independence of the dual complex $\mathcal{DR}(x \in X)$ of the choice of a resolution (Proposition 5.2). We invoke the Weak Factorization Theorem (Theorem 6.5) to assert that two such resolutions differ by a chain of blow-ups and blow-downs along admissible loci. We are left to describe the transformations that a dual complex undergoes via a blow-up. This is the content of Stepanov's lemma (cf. Lemma 6.2).

In a similar fashion, following [**10**, Proposition 11], we adapt the arguments to prove Main Theorem (1).

DEFINITION 6.1. Let (X, Δ) be a log smooth pair. An **admissible locus** with respect to (X, Δ) is a smooth irreducible subvariety $Z \subset X$ that has only simple normal crossings with Δ, i.e., for every point $z \in Z \cap \Delta$ there exists a regular system of local parameters $z_1, \ldots, z_n \in \mathfrak{m}_z$ in the maximal ideal of the local ring $\mathcal{O}_{X,z}$ such that Δ_{red} and Z are locally given by the local equations $z_{i_1} \ldots z_{i_r} = 0$ and $z_{j_1} = \cdots = z_{j_s} = 0$ respectively for some r, $s \leq n$.

LEMMA 6.2 (Stepanov's lemma [**37**]). *Let $f : \mathrm{Bl}_Z(X) \to X$ be the blow-up of an admissible locus Z with respect to the log pair $(X, \Delta := \sum_{i \in I} D_i)$. Let $E := f^{-1}(Z)$ be the exceptional divisor and $D_i' := f_*^{-1} D_i$ be the birational transform of the irreducible component D_i and analogously $\Delta' := f_*^{-1} \Delta$. Then $\Delta' \cup E$ is an snc divisor. Moreover,*

Case 1. if Z is not contained in $\mathrm{Supp}(\Delta)$, then $\mathcal{D}(\Delta') = \mathcal{D}(\Delta)$;

Case 2. if Z is a stratum, then $\mathcal{D}(\Delta' \cup E)$ is obtained from $\mathcal{D}(\Delta)$ by star subdivision of the cell corresponding to Z;

Case 3. if Z is not a stratum of Δ, but $Z \subset \mathrm{Supp}(\Delta)$, then $\mathcal{D}(\Delta' \cup E)$ is obtained from $\mathcal{D}(\Delta)$ by gluing the cone over $\mathcal{D}(\Delta'|_E)$ to $\mathcal{D}(\Delta)$ as follows.

Let Δ_Z be the smallest stratum of Δ containing Z, which is an irreducible component of $\bigcap_{i \in J} D_i$ for some subset J of I, and v_Z be the corresponding cell. The dual complex $\mathcal{D}(\Delta'|_E)$ equals the join $v_Z * \mathcal{D}(\sum_{i \in I \setminus J} D_i|_Z)$. The natural inclusion $\tau_L : \mathcal{D}(\sum_{i \in I \setminus J} D_i|_Z) \to \mathrm{Lk}(v_Z)$ induces an inclusion $\tau_S : v_Z * \mathcal{D}(\sum_{i \in I \setminus J} D_i|_Z) \to \overline{\mathrm{St}}(v_Z)$. Then,

$$\mathcal{D}(\Delta' \cup E) = \mathcal{D}(\Delta) \sqcup \mathrm{Cone}\left(v_Z * \mathcal{D}\left(\sum_{i \in I \setminus J} D_i|_Z \right) \right) \Big/ \sim_{\tau_S},$$

where the relation $\sim_{\tau_S}$ identifies $v_Z * \mathcal{D}(\sum_{i \in I \setminus J} D_i|_Z)$, intended as a subcomplex of its cone, with its image in $\mathcal{D}(\Delta)$ via the map τ_S.

Since the cone over the join retracts onto the join, $\mathcal{D}(\Delta' \cup E)$ retracts onto $\mathcal{D}(\Delta)$.

REMARK 6.3. The result follows from a direct computation, for which we refer the interested reader to [**37**]. The hypothesis that Z is an admissible locus cannot be omitted. Without this assumption, the pair $(\mathrm{Bl}_Z(X), \Delta')$ is not necessarily dlt and the definition of its dual complex may not be well posed and Stepanov's lemma (cf. Lemma 6.2) may fail.

For instance, let Δ be the arrangement of three hyperplanes $(xyz = 0) \subset \mathbb{C}^3$ with coordinates (x, y, z) and Z be the line spanned by the vector $(1, 1, 1)$. Then, the strict transforms of the three hyperplanes meet along a line. This latter line is a stratum for the pair $(\mathrm{Bl}_Z(X), \Delta')$, but it does not belong to its snc locus. Hence, the pair $(\mathrm{Bl}_Z(X), \Delta')$ is not dlt and the definition of dual complex is not well posed.

COROLLARY 6.4 ([**10**, 9.(4)]). *Let* (X, Δ) *be a log smooth pair, where* Δ *is a sub-boundary. Assume* $(\mathrm{Bl}_Z X, \Delta_Z)$ *is the log pull-back of the pair* (X, Δ) *via a blow-up of* X *along an admissible locus* Z. *Then,* $\mathcal{D}(\Delta_Z^{=1})$ *and* $\mathcal{D}(\Delta^{=1})$ *are homeomorphic.*

PROOF. If Z is not a stratum (Cases 1 and 3 in Stepanov's lemma), $\Delta_Z^{=1}$ equals $\Delta^{=1}$, and $\mathcal{D}(\Delta_Z^{=1})$ and $\mathcal{D}(\Delta^{=1})$ are isomorphic.

On the other hand, if Z is a stratum (Case 2 in Stepanov's lemma), $\Delta_Z^{=1}$ equals $\Delta^{=1} \cup E$. The corresponding dual complexes differ by a star subdivision: the triangulation is refined but the homeomorphism class is unchanged. □

Combined with the following celebrated theorem due to [**1**], Stepanov's lemma implies Proposition 5.2.

THEOREM 6.5 (Weak Factorization Theorem [**1**]). *Let* (X_1, Δ_1) *and* (X_2, Δ_2) *be log smooth pairs, and* $f : X_1 \dashrightarrow X_2$ *be a birational map which restricts to an isomorphism on the open set* $U := X_1 \setminus \Delta_1 \simeq X_2 \setminus \Delta_2$. *Then, the map* f *can be factored as a series of blow-ups and blow-downs along admissible loci disjoint from* U, *namely there exists a sequence of blow-ups and blow-downs between complete non-singular algebraic varieties*

$$X_1 = V_0 \xrightarrow{\varphi_0} V_1 \xrightarrow{\varphi_1} \cdots \xrightarrow{\varphi_{l-1}} V_{l-1} \xrightarrow{\varphi_l} V_l = X_2,$$

where

(1) $f = \varphi_l \circ \varphi_{l-1} \circ \cdots \circ \varphi_2 \circ \varphi_1$;
(2) *either* φ_i *or* φ_i^{-1} *is a morphism obtained by blowing up an admissible locus disjoint from* U;
(3) *there exists an index* m *such that for any* $i \leq m$ *the map* $V_i \dashrightarrow X_1$ *is a morphism and for any* $i \geq m$ *the map* $V_i \dashrightarrow X_2$ *is a morphism*;
(4) *if* X_1 *and* X_2 *are projective, then all the* V_i *are projective too.*

PROOF OF PROPOSITION 5.2. Two different resolutions are isomorphic away from the snc exceptional locus. Combining the Weak Factorization Theorem and Stepanov's lemma, we conclude. □

The proof of Main Theorem (1) is slightly more complicated. It results from the following lemmas.

LEMMA 6.6. *Two dlt modifications of an lc pair* (X, Δ) *are crepant birational.*

PROOF. The following proof is a variation of the proof of [**24**, Theorem 3.52]. Let $f : Y \to X$ and $f' : Y' \to X$ be dlt modifications of the lc pair

(X, Δ). Let W be any common resolution of both Y and Y'. Consider the diagram

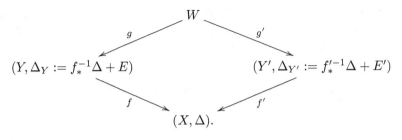

Write

$$K_W + g_*^{-1} f_*^{-1}\Delta + g_*^{-1}E = g^*(K_Y + \Delta_Y) + \sum_{F_i \subseteq \mathrm{Ex}\, g} a(F_i, Y, \Delta_Y) F_i,$$

$$K_W + g_*'^{-1} f_*'^{-1}\Delta + g_*'^{-1}E' = g'^*(K_{Y'} + \Delta_{Y'}) + \sum_{F_i \subseteq \mathrm{Ex}\, g'} a(F_i, Y', \Delta_{Y'}) F_i.$$

Subtracting term by term, we conclude

$$g'^*(K_{Y'} + \Delta_{Y'}) - g^*(K_Y + \Delta_Y)$$

$$= g_*'^{-1}E' - g_*^{-1}E + \sum_{F_i \subseteq \mathrm{Ex}\, g} a(F_i, Y, \Delta_Y) F_i - \sum_{F_i \subseteq \mathrm{Ex}\, g'} a(F_i, Y', \Delta_{Y'}) F_i$$

$$= Z - Z' + \sum_{F_i \subseteq \mathrm{Ex}\, g \cap \mathrm{Ex}\, g'} (a(F_i, Y, \Delta_Y) - a(F_i, Y', \Delta_{Y'})) F_i,$$

where Z and Z' are effective divisors whose supports do not contain the exceptional locus of g' and g, respectively. They are defined as follows:

$$Z := \sum_{F_i \subseteq \mathrm{Ex}\, g \backslash \mathrm{Ex}\, g'} (a(F_i, Y, \Delta_Y) + 1) F_i \geq 0,$$

$$Z' := \sum_{F_i \subseteq \mathrm{Ex}\, g' \backslash \mathrm{Ex}\, g} (a(F_i, Y', \Delta_{Y'}) + 1) F_i \geq 0.$$

Since $K_{Y'} + \Delta_{Y'}$ is f'-nef, $g'^*(K_{Y'} + \Delta_{Y'})$ is g-nef. Since $\mathrm{Supp}(Z')$ is not g-exceptional, then

$$g'^*(K_{Y'} + \Delta_{Y'}) + Z' \sim_{g,\mathbb{Q}} Z + \sum_{F_i \subseteq \mathrm{Ex}\, g \cap \mathrm{Ex}\, g'} (a(F_i, Y, \Delta_Y) - a(F_i, Y', \Delta_{Y'})) F_i$$

is g-nef and g-exceptional. By the negativity lemma (cf. [**24**, Lemma 3.39]), we conclude that $a(F_i, Y, \Delta_Y) \leq a(F_i, Y', \Delta_{Y'})$. By symmetry, $a(F_i, Y, \Delta_Y) = a(F_i, Y', \Delta_{Y'})$. $\square$

REMARK 6.7. Note that if $f : Y \to X$ is a thrifty dlt modification of the pair (X, Δ) (cf. [**21**, Corollary 1.36] for a definition), the discrepancies of all the f-exceptional divisors are -1. Hence, the boundary $f_*^{-1}\Delta + E$

is the log pull-back of Δ. In that case, Lemma 6.6 follows easily by the commutativity of the diagram above.

LEMMA 6.8 (**10**, Proposition 11). *Suppose that the pairs* (X_i, Δ_i) *are dlt and crepant birational, for* $i = 1, 2$. *Then, the dual complexes* $\mathcal{D}(\Delta_i^{=1})$ *are naturally homeomorphic to each other.*

PROOF. Perform log resolutions $f : (Y_i, \Delta_{Y_i}) \to (X_i, \Delta_i)$, which are isomorphisms on the snc locus of (X_i, Δ_i). Then, $\mathcal{D}(\Delta_{Y_i}^{=1})$ and $\mathcal{D}(\Delta_{X_i}^{=1})$ are homeomorphic. Indeed,

$$\mathcal{D}(\Delta_{X_i}^{=1}) = \mathcal{D}(\Delta_{X_i}^{=1}|_{X_i^{snc}}) = \mathcal{D}(\Delta_{Y_i}^{=1}|_{f^{-1}(X_i^{snc})}) = \mathcal{D}(\Delta_{Y_i}^{=1}),$$

where the last equality is a consequence of the fact that no lc center of Δ_{Y_i} can lay on the exceptional locus by the definition of dlt pair. Thus, we may assume that the pairs (X_i, Δ_i) are snc and crepant. Note that Δ_i are in general sub-boundaries.

By the Weak Factorization Theorem [**1**], there exists a sequence of blow-ups and blow-downs

$$X_1 = V_0 \xrightarrow{\varphi_0} V_1 \xrightarrow{\varphi_1} \cdots \xrightarrow{\varphi_{l-1}} V_{l-1} \xrightarrow{\varphi_l} V_l = X_2,$$

enjoying the properties described in Theorem 6.5. In particular, there exists an index m such that for any $i \leq m$, the map $V_i \dashrightarrow X_1$ is a morphism and for any $i \geq m$ the map $V_i \dashrightarrow X_2$ is a morphism.

Let Θ_i be the log pull-backs of Δ_1 for $i \leq m$ and the log pull-backs of Δ_2 for $i \geq m$. Since (X_i, Δ_i) are crepant, the two log pull-backs on V_m coincide.

By Corollary 6.4, $\mathcal{D}(\Theta_i)$ differ by compositions of star subdivisions or their inverse; so $\mathcal{D}(\Delta_i^{=1})$ do as well.   □

PROOF OF MAIN THEOREM 1. Combine Lemmas 6.6 and 6.8.   □

7. Running MMP

In Section 6, we have studied how dual complexes change via blow-ups along admissible loci. We are now interested in their behavior under more general MMP transformations, see Section 2 for the basic notions.

LEMMA 7.1 ([**10**, Lemma 16]). *Let* (X, Δ) *be a dlt pair and* $f : X \dashrightarrow Y$ *be a birational transformation of a* $(K_X + \Delta)$-*MMP. Set* $\Delta_Y := f_* \Delta$. *Then, the dual complex* $\mathcal{D}(\Delta_Y^{=1})$ *is naturally a subcomplex of* $\mathcal{D}(\Delta^{=1})$.

PROOF. We first remark that $\mathcal{D}(\Delta_Y^{=1})$ is well defined, since the property of being dlt is stable under MMP transformations (cf. [**24**, Corollary 3.44]).

Since discrepancies strictly increase for each divisor whose center is contracted (cf. [**24**, Lemma 3.38]), the lc centers of (Y, Δ_Y) are the birational push-forwards of non-contracted lc centers of (X, Δ). In terms of simplices, the cells of $\mathcal{D}(\Delta_Y^{=1})$ are in bijection with those cells of $\mathcal{D}(\Delta^{=1})$ which correspond to non-contracted strata of $\Delta^{=1}$.   □

In general, the inclusion $\mathcal{D}(\Delta^{=1}) \subseteq \mathcal{D}(\Delta_Y^{=1})$ is not a homotopy equivalence (cf. [10, Example 17]). However, de Fernex, Kollár and Xu have given conditions for this to happen. In these cases, the dual complexes retract one onto the other, preserving the piecewise linear structure, since the retractions are realized as a composition of elementary collapses.

DEFINITION 7.2. Let D be a regular cell complex. Let v be a cell in D and w one of its faces.

- The couple (v, w) is said to be a **free pair** if v is the only cell of which w is a strict face.
- Assume that (v, w) is a free pair. The **elementary collapse** (D, v, w) is a regular complex obtained from D by removing the interior of v and w.
- A composition of elementary collapses is called a **collapse**.
- A regular complex D is **collapsible** if it can be collapsed to a point.

THEOREM 7.3 (**10**, Theorem 19). *Assume the following:*

(1) (X, Δ) *is a dlt pair;*

(2) $f : X \dashrightarrow Y$ *is a birational transformation of a* $(K_X + \Delta)$*-MMP corresponding to a* $(K_X + \Delta)$*-negative extremal ray R;*

(3) *there is a prime divisor $D_0 \subset \Delta^{=1}$ such that $(D_0 \cdot R) > 0$.*

Then, $\mathcal{D}(\Delta^{=1})$ *collapses to* $\mathcal{D}(\Delta_Y^{=1})$.

PROOF (SKETCH). By Lemma 7.1, the faces of $\mathcal{D}(\Delta_Y^{=1})$ are identified with the non-contracted lc centers of Δ. Fixing D_0, one can show that the subcomplex corresponding to contracted strata can be obtained via a sequence of elementary collapses of free pairs contained in $\overline{\mathrm{St}}(v_{D_0})$, where here and in the following v_Z denotes the cell corresponding to a stratum Z.

We just mention how lc centers are coupled in free pairs. There exists a bijective correspondence between contracted lc centers $Z \subseteq D_0$ and contracted lc centers $W \nsubseteq D_0$, which assigns

$$Z = D_0 \cap \bigcap_{i \in J} D_i \quad \longrightarrow \quad Z^+ := \text{irr. comp. in } \bigcap_{i \in J} D_i \text{ containing } Z,$$

$$W^- := W \cap D_0 \quad \longleftarrow \quad W.$$

Hence, the corresponding cells come in pairs

$$(v_{D_0 \cap W} = v_{D_0} * v_W, v_W),$$

where $W \subset X$ is a contracted lc center. Because of the bijective correspondence $Z \mapsto Z^+$ and $W \mapsto W^-$, a maximal dimensional pair of the type described is a free pair (cf. [10, Theorem 19]). The result is achieved by collapsing iteratively free pairs of maximal dimension. $\square$

The existence of a divisor D_0, which is positive along R in Theorem 7.3(3), is essential. For instance, in [10, Remark 20] the authors have shown that if $(D_i \cdot R) < 0$ for every $D_i \subset \Delta^{=1}$ and Z is a contracted lc center, then the exceptional locus of f lies in the intersection of all the divisors D_i and coincides with Z. Therefore, $\mathcal{D}(\Delta_Y^{=1})$ is obtained from $\mathcal{D}(\Delta^{=1})$ by removing the cell of maximal dimension v_Z. The two complexes do not have the same homotopy type, since their Euler characteristics differ from a unit.

The existence of the prime divisor in Theorem 7.3(3) is satisfied if one of the following criteria is satisfied. For the proof of the criteria, we refer the interested reader to the main reference [**10**].

CRITERION 7.4 ([**10**, Lemma 21]). *Assume the following*:

(1) (X, Δ) *is a $\mathbb{Q}$-factorial dlt pair, endowed with a morphism g : $X \to Z$;*

(2) $f : X \dashrightarrow Y$ *is a birational transformation of a $(K_X + \Delta)$-MMP over Z;*

(3) *there exists a numerically g-trivial effective divisor fully supported on $\Delta^{=1}$.*

Then Theorem 7.3(3) holds, unless $\mathrm{Ex}(f)$ *does not contain any lc centers.*

CRITERION 7.5 ([**10**, Lemma 23]). *Assume that:*

(1) (X, Δ) *is a $\mathbb{Q}$-factorial lc pair, endowed with a projective birational morphism $g : X \to Z$ onto a normal $\mathbb{Q}$-factorial variety Z;*

(2) $f : X \dashrightarrow Y$ *is a birational transformation of a $(K_X + \Delta)$-MMP over Z;*

(3) f *does not contract any divisor E with the property that $Y \to Z$ is an isomorphism at the generic point of $f(E)$.*

Then, there exists a prime divisor $E_R \subseteq \mathrm{Ex}(g)$ such that $E_R \cdot R > 0$.

REMARK 7.6. The $\mathbb{Q}$-factoriality in Criterion 7.5 is essential to guarantee that $\mathrm{Ex}(g)$ is divisorial, and to ensure the existence of exceptional divisors with the prescribed positivity.

8. Proof of Main Theorem (2)

Main Theorem (2) asserts that the dual complex of a dlt modification of an isolated singularity $\mathcal{DMR}(x \in X)$ is a distinguished homeomorphism class inside the homotopy class of $\mathcal{DR}(x \in X)$. We present two proofs, both of which appear in [**10**]. The first one works under the stronger assumption that X is lc and $\mathbb{Q}$-factorial, while the second is completely general. The proofs are slightly different in nature, since they rely respectively on Criterion 7.5 and 7.4. We find it instructive to describe both.

PROOF OF MAIN THEOREM (2) UNDER THE HYPOTHESIS THAT X IS LC AND $\mathbb{Q}$-FACTORIAL.
Let $f : (Y, E := \mathrm{Ex}(f)) \to X$ be any resolution of X. In particular, (Y, E) is log smooth. We run a $(K_Y + E)$-MMP over X to get

$$(Y, E) = (Y_0, E_0) \xrightarrow{\varphi_1} (Y_1, E_1) \xrightarrow{\varphi_2} \cdots \xrightarrow{\varphi_l} (Y_l, E_l) \xrightarrow{f_l} X.$$

The fact that X is lc assures that the MMP terminates and contracts all the f-exceptional divisors with discrepancy $a(D_i, X, 0) > -1$ (cf. [11; 21, §1.35]). A single step of the MMP is here represented in the following diagram:

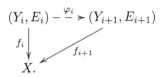

Since the morphisms f_i are isomorphisms away from the isolated singularity $x \in X$ and the MMP is relative to the base X, the divisors E_i, as well as any divisor contracted by the MMP, are f_i-exceptional and dominate x.

Further, at each step of the MMP, the hypotheses of Criterion 7.5 hold. Indeed, if φ_i contracts a divisor D_i and f_{i+1} is an isomorphism at the generic point of $\varphi_i(D_i)$, then Y_{i+1} is isomorphic to X and the MMP terminates. Hence, we can suppose Criterion 7.5 holds and by Theorem 7.3, this implies that $\mathcal{D}(E)$ collapses to $\mathcal{D}(E_l)$.

There are two options: either X is klt or not. In the latter case, the MMP terminates with (Y_l, E_l), a dlt modification of X. Hence, $\mathcal{D}(E_l) \simeq \mathcal{DMR}(x \in X)$ by definition and the proof ends. In the former case, the MMP terminates with X itself so that f_l is a birational transformation of the $(K_Y + E)$-MMP. Hence, since X is $\mathbb{Q}$-factorial and the relative Picard number $\rho(Y_l/X)$ is one, f_l contracts a single divisor, thus $\mathcal{D}(E_l)$ is a point. This proves Main Theorem (2) and a weaker version of Main Theorem (3).

REMARK 8.1. In the previous proof the $\mathbb{Q}$-factoriality of X is needed to invoke Criterion 7.5 and to guarantee that f_l is a divisorial contraction.

GENERAL PROOF OF MAIN THEOREM (2). We proceed in several steps.

Step 1. There exists a dlt modification $q : (X^{\mathrm{dlt}}, \Delta^{\mathrm{dlt}}) \to X$ with the property that $q^{-1}(0)$ has pure codimension 1 (cf. [10, Lemma 29]).
    In particular, $(\Delta^{\mathrm{dlt}})^{=1} = \Delta^{\mathrm{dlt}} = \mathrm{Supp}(q^{-1}(0))$.

Step 2. Let $f : (Y, E := \mathrm{Ex}(f)) \to X$ be any resolution of X. After a composition of admissible blow-ups π, we can resolve the inde-terminacy of the birational map $Y \dashrightarrow X^{\mathrm{dlt}}$, i.e., there exists a smooth variety Y' fitting inside the following diagram of birational morphisms and maps:

Since π is a composition of blow-ups, $\mathcal{D}(\mathrm{Supp}(f \circ \pi)^{-1}(0))$ has the homotopy type of $\mathcal{D}(E)$ by Stepanov's lemma (cf. Lemma 6.2).

Step 3. We run the $(K_{Y'}+E' := K'_Y+\mathrm{Supp}(f \circ \pi)^{-1}(0))$-MMP relative to X^{dlt}. Since (Y', E') is a log smooth pair and $K_{Y'}+E'-g^*(K_{X^{\mathrm{dlt}}}+\Delta^{\mathrm{dlt}})$ is effective, the MMP terminates due to [**11**]. We obtain

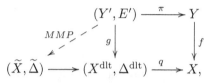

$$
\begin{array}{ccc}
(Y', E') & \xrightarrow{\ \pi\ } & Y \\
\scriptstyle{MMP} \nearrow \quad {\scriptstyle g}\downarrow & & \downarrow{\scriptstyle f} \\
(\widetilde{X}, \widetilde{\Delta}) \longrightarrow (X^{\mathrm{dlt}}, \Delta^{\mathrm{dlt}}) & \xrightarrow{\ q\ } & X,
\end{array}
$$

where $(\widetilde{X}, \widetilde{\Delta})$ is a dlt modification of $(X^{\mathrm{dlt}}, \Delta^{\mathrm{dlt}})$. In particular, $(\widetilde{X}, \widetilde{\Delta})$ is a dlt modification of X and by definition $\mathcal{D}(\widetilde{X}, \widetilde{\Delta}) \simeq \mathcal{DMR}(x \in X)$.

Since $E' := \mathrm{Supp}(f \circ \pi)^{-1}(0) = \mathrm{Supp}\, g^{-1}(\Delta^{\mathrm{dlt}})$, the divisor $g^*(\Delta^{\mathrm{dlt}})$ is numerically g-trivial and supported on E'. By Criterion 7.4 and Theorem 7.3, $\mathcal{D}(E')$ collapses to $\mathcal{D}(\widetilde{X}, \widetilde{\Delta})$.

To summarize, we have shown that $\mathcal{D}(E)$ has the homotopy type of $\mathcal{D}(E')$, which collapses to $\mathcal{D}(\widetilde{X}, \widetilde{\Delta}) \simeq \mathcal{DMR}(x \in X)$.   □

REMARK 8.2 (Logarithmic setting). If (X, Δ) is klt away from x and $K_X + \Delta$ is $\mathbb{Q}$-Cartier, the previous proof continues to hold in the logarithmic setting (cf. [**10**, Theorem 28(2)]). This means that the dual complex $\mathcal{DMR}(x \in (X, \Delta))$ is proved to have the same homotopy type as $\mathcal{D}(E + f_*^{-1}\Delta^{=1})$, where E is the divisorial part of the exceptional locus of a resolution f. In particular, in the klt case, $\mathcal{DMR}(x \in (X, \Delta))$ has the same homotopy type as $\mathcal{D}(E)$, so that it is independent of the choice of a klt boundary. In Section 9, we exploit the freedom of choosing a boundary Δ to construct a special dlt modification.

9. Proof of Main Theorem (3)

In this last section, we prove that the dual complex of the resolution of an isolated klt singularity $(x \in X)$ is contractible. A proof due to [**10**] has already been presented in Section 8 under the additional hypothesis that X is $\mathbb{Q}$-factorial.

PROOF OF MAIN THEOREM (3). We extract a Kollár component E of X (cf. [**43**, Lemma 1; **12**, Lemma 2.5]). This means that there exists a $\mathbb{Q}$-divisor H such that (X, H) is lc, and klt away from x, with a unique irreducible divisor E with discrepancy -1 over x. Moreover, one can construct a dlt modification of the pair (X, H) whose exceptional locus coincides with the support of E. The upshot is that $\mathcal{DMR}(x \in (X, H)) \simeq \mathcal{D}(E)$ is a point.

As explained in Remark 8.2, one can invoke an analogous version of Main Theorem (2) for lc pairs to conclude that $\mathcal{DR}(x \in X)$ has the homotopy type of $\mathcal{DMR}(x \in (X, H))$.   □

REMARK 9.1 (Potentially dlt). Replacing the hypothesis of X $\mathbb{Q}$-Gorenstein klt with potentially dlt, namely that there exists a boundary Δ such that (X, Δ) is dlt, the Main Theorem (3) continues to hold in this logarithmic setting (cf. [**10**, Theorem 2]).

Acknowledgments

I would like to thank the organizers of the conference *Arc schemes and singularity theory* held in Rennes in November 2016 for their hospitality. Especially, I thank Johannes Nicaise who motivated me to write this chapter and carefully read an early version, and my advisor Paolo Cascini for his advice and support. This work was supported by Imperial College London and the Engineering and Physical Sciences Research Council [EP/L015234/1]. The EPSRC Centre for Doctoral Training in Geometry and Number Theory (The London School of Geometry and Number Theory), University College London.

Bibliography

[1] D. Abramovich, K. Karu, K. Matsuki and J. Włodarczyk, Torification and factorization of birational maps. *J. Amer. Math. Soc.*, 15(3):531–572, 2002.

[2] D. Arapura, P. Bakhtary and J. Włodarczyk, Weights on cohomology, invariants of singularities, and dual complexes. *Math. Ann.*, 357(2):513–550, 2013.

[3] A. Beauville, Variétés Kähleriennes dont la première classe de Chern est nulle. *J. Differential Geom.*, 18(4):755–782, 1983.

[4] V. G. Berkovich, *Spectral Theory and Analytic Geometry Over Non-Archimedean Field*, Mathematical Surveys and Monographs, Vol. 33, American Mathematical Society, 1990.

[5] M. Brown and E. Mazzon, The essential skeleton of a product of degenerations. To appear in *Compos. Math.*, 2019.

[6] M. Brown and T. Foster, Rational connectivity and analytic contractibility, *J. Reine Angew. Math.*, 747:45–62, 2019.

[7] D. Cartwright, Excluded homeomorphism types for dual complexes of surfaces. In *Nonarchimedean and Tropical Geometry*, Simons Symposia, Springer, 2016, pp. 133–144.

[8] V. I. Danilov, Polyhedra of schemes and algebraic varieties, *Mat. Sb. (N.S.)*, 139(1):146–158, 160, 1975.

[9] M. A. A. de Cataldo, T. Hausel and L. Migliorini, Topology of Hitchin systems and Hodge theory of character varieties: the case A_1, *Ann. of Math.* (2), 175(3): 1329–1407, 2012.

[10] T. de Fernex, J. Kollár and C. Xu, The dual complex of singularities. In *Higher Dimensional Algebraic Geometry in Honour of Professor Yujiro Kawamata's Sixtieth Birthday*, Adv. Stud. Pure Math., Vol. 74, 2017, pp. 103–130.

[11] O. Fujino, Semi-stable minimal model program for varieties with trivial canonical divisor. *Proc. Japan Acad. Ser. A Math. Sci.*, 87(3):25–30, 2011.

[12] C. D. Hacon and J. McKernan, Boundedness of pluricanonical maps of varieties of general type. *Invent. Math.*, 166(1):1–25, 2006.

[13] L. H. Halle and J. Nicaise, Motivic zeta functions of degenerating Calabi-Yau varieties. *Math. Ann.*, 370(3-4):1277–1320, 2018.

[14] A. Hatcher. *Algebraic Topology*. Cambridge University Press, Cambridge, 2002.

[15] A. Hogadi and C. Xu, Degenerations of rationally connected varieties. *Trans. Amer. Math. Soc.*, 361(7):3931–3949, 2009.

[16] M. Kapovich and J. Kollár, Fundamental groups of links of isolated singularities. *J. Amer. Math. Soc.*, 27(4):929–952, 2014.

[17] L. Katzarkov, A. Noll, P. Pandit and C. Simpson, Harmonic maps to buildings and singular perturbation theory. *Comm. Math. Phys.*, 336(2):853–903, 2015.

[18] J. Kollár, R. Laza, G. Saccà and C. Voisin, Remarks on degenerations of hyper-Kähler manifolds. *Ann. Inst. Fourier*, 2019.

[19] J. Kollár, Shafarevich maps and plurigenera of algebraic varieties. *Invent. Math.*, 113(1):177–215, 1993.

[20] J. Kollár, New examples of terminal and log canonical singularities. Preprint, 2011.

[21] J. Kollár, Links of complex analytic singularities. In *Surveys in Differential Geometry. Geometry and Topology*, Surv. Differ. Geom., Vol. 18, International Press, Somerville, MA, 2013, pp. 157–193.

[22] J. Kollár, *Singularities of the Minimal Model Program*, Cambridge Tracts in Mathematics, Vol. 200, Cambridge University Press, Cambridge, 2013.

[23] J. Kollár, Simple normal crossing varieties with prescribed dual complex. *Algebr. Geom.*, 1(1):57–68, 2014.

[24] J. Kollár and S. Mori, *Birational Geometry of Algebraic Varietie*, Cambridge Tracts in Mathematics, Vol. 134, Cambridge University Press, Cambridge, 1998.

[25] J. Kollár and C. Xu, The dual complex of Calabi–Yau pairs. *Invent. Math.*, 205(3):527–557, 2016.

[26] M. Kontsevich and Y. Soibelman, Homological mirror symmetry and torus fibrations. In *Symplectic Geometry and Mirror Symmetry* (Seoul, 2000), World Sci. Publ., 2001, pp. 203–263.

[27] M. Kontsevich and Y. Soibelman, Affine structures and non-Archimedean analytic spaces. In *The Unity of Mathematics. In Honor of the Ninetieth Birthday of I. M. Gelfand*, Progress in Mathematics, Vol. 244, Birkhäuser Boston, 2006, pp. 321–385.

[28] V. S. Kulikov, Degenerations of $K3$ surfaces and enriques surfaces. *Izv. Akad. Nauk SSSR Ser. Mat.*, 41(5):1008–1042, 1977.

[29] D. R. Morrison, Semistable degenerations of Enriques' and hyperelliptic surfaces. *Duke Math. J.*, 48(1):197–249, 1981.

[30] M. Mustaţă and J. Nicaise, Weight functions on non-Archimedean analytic spaces and the Kontsevich–Soibelman skeleton. *Algebr. Geom.*, 2(3):365–404, 2015.

[31] J. Nicaise, C. Xu and T. Y. Yu, The non-archimedean SYZ fibration. Preprint, 2018.

[32] J. Nicaise and C. Xu, The essential skeleton of a degeneration of algebraic varieties. *Amer. J. Math.*, 138(6):1645–1667, 2016.

[33] S. Payne, Boundary complexes and weight filtrations. *Michigan Math. J.*, 62(2): 293–322, 2013.

[34] U. Persson, On degenerations of algebraic surfaces. *Mem. Amer. Math. Soc.*, 11(189):1–144, 1977.

[35] U. Persson and H. Pinkham, Degeneration of surfaces with trivial canonical bundle. *Ann. of Math.* (2), 113(1):45–66, 1981.

[36] C. Simpson, The dual boundary complex of the SL_2 character variety of a punctured sphere. *Ann. Fac. Sci. Toulouse Math.* (6), 25(2-3):317–361, 2016.

[37] D. A. Stepanov, A remark on the dual complex of a resolution of singularities. *Uspekhi Mat. Nauk*, 61(1(367)):185–186, 2006.

[38] D. A. Stepanov, Dual complex of a resolution of terminal singularities. In *Proceedings of the Fifth Kolmogorov Lectures*, Yaroslav. Gos. Ped. Univ. im. K. D. Ushinskogo, Yaroslavl', 2007, pp. 91–99 (in Russian).

[39] A. Strominger, S.-T. Yau and E. Zaslow, Mirror symmetry is T-duality. In *Winter School on Mirror Symmetry, Vector Bundles and Lagrangian Submanifolds* (Cambridge, MA, 1999), AMS/IP Stud. Adv. Math., Vol. 23, Amer. Math. Soc., 2001, pp. 333–347.

[40] S. Takayama, Local simple connectedness of resolutions of log-terminal singularities. *Internat. J. Math.*, 14(8):825–836, 2003.

[41] A. Thuillier, Géométrie toroïdale et géométrie analytique non archimédienne. Application au type d'homotopie de certains schémas formels. *Manuscripta Math.*, 123(4):381–451, 2007.

[42] C. Xu, Interaction between singularity theory and the minimal model program. Preprint, 2017.

[43] C. Xu, Finiteness of algebraic fundamental groups. *Compos. Math.*, 150(3):409–414, 2014.

15

Log-Regular Models for Products
of Degenerations

Morgan V. Brown[*,‡] **and Enrica Mazzon**[†,§]

Department of Mathematics, University of Miami
Coral Gables, FL 33146, USA
†*Department of Mathematics, Imperial College London*
180 Queen's Gate, London SW72AZ, UK
‡*mvbrown@math.miami.edu*
§*e.mazzon15@imperial.ac.uk*

This is a survey of the authors' recent paper [**3**]. We show how to use tools from log geometry to produce a skeleton for the product of snc degenerations. As an application, we are able to describe the homeomorphism type of the skeletons of some degenerations of hyper-Kähler varieties. We illustrate our results with many examples that are not in the original paper [**3**].

1. Introduction

(1.1) Let K be a field with a discrete valuation, R the associated valuation ring and k the residue field. We can think of a variety X over K as the generic member of a 1-parameter family of varieties. To compute the limit of such a family, we find an extension of X to R, which we denote by $\mathscr{X}$. However, even if X is smooth, it may not be possible to find a model $\mathscr{X}$, which is smooth over R. We need the weaker notion that $\mathscr{X}$ is snc, i.e., the special fiber $\mathscr{X}_k$ forms a strict normal crossings divisor inside $\mathscr{X}$.

So long as log resolution of singularities holds over R, we can always find an snc model for a smooth projective variety X. Such models are not unique because we can blow up smooth centers in the special fiber. Our object of study is the dual complex of the special fiber $\mathscr{X}_k$, which we denote $\mathcal{D}(\mathscr{X}_k)$.

This is a cell complex whose vertices are the components of the special fiber and whose cells record the irreducible components of intersections of those components.

We can compare the dual complexes of different models via an analytic construction called the Berkovich space X^{an} associated to X. Developed by Berkovich in [2], X^{an} is a K-analytic space, such that each point of X^{an} corresponds to a scheme theoretic point ξ_x of X, along with a valuation on the residue field $\kappa(\xi_x)$ extending the valuation on K. For any snc R-model $\mathscr{X}$, the dual complex $\mathcal{D}(\mathscr{X}_k)$ topologically embeds into X^{an} and its image is called the Berkovich skeleton $\mathrm{Sk}(\mathscr{X})$ of $\mathscr{X}$. Moreover, this skeleton captures the fundamental topological features of the Berkovich space, as $\mathrm{Sk}(\mathscr{X})$ is a strong deformation retract of X^{an} [2, 26].

(1.2) As an example, consider $\mathscr{X} = V(\pi(x^3 + y^3 + z^3) + xyz) \subset \mathbb{P}^2_R$, where π is a generator of the maximal ideal of R. The special fiber is the union of the coordinate hyperplanes, and so the dual complex is a triangle, where each vertex corresponds to a coordinate hyperplane. While blowing up an intersection point changes the combinatorial presentation of the dual complex, it preserves the skeleton inside the Berkovich space. In contrast, blowing up a smooth point enlarges the skeleton.

Birational geometry explains the difference between the pictures of Figure 1. Consider a pair $(\mathscr{X}, \Delta_{\mathscr{X}})$ over the germ of a curve. We can make birational modifications to this pair by running the minimal model program over the base curve. These operations remove rational curves which have negative intersection with the canonical class $K_{\mathscr{X}}$, with the eventual goal of producing a model whose canonical class is nef. In [4], de Fernex, Kollár, and Xu show that these operations induce homotopy equivalences on the dual complex. On the one hand, blowing up the intersection point extracts a divisor of log discrepancy 0, and this induces a PL homeomorphism on the dual complexes. On the other hand, when we blow up a different point, we get a retraction map.

This connection with birational geometry suggests the existence of a skeleton which does not depend on the choice of R-model. Mustaţă and Nicaise [21] define the essential skeleton by generalizing a construction of Kontsevich and Soilbeman [18] from Calabi–Yau varieties to varieties with non-negative Kodaira dimension. The key idea is to use pluricanonical forms to define weight functions on the Berkovich space. For a regular

Figure 1. Birational modifications of $V(\pi(x^3 + y^3 + z^3) + xyz)$.

form, the locus where the weight function is minimal is a cell complex; the essential skeleton is the union of these loci for all regular pluricanonical forms. It is expected that the essential skeleton $\mathrm{Sk}(X)$ is identified with the dual complex of a dlt minimal model, but there are technical obstacles to proving this. Nicaise and Xu demonstrate this in the case where X is smooth projective and K_X is semi-ample [**22**].

(**1.3**) The present work is an exposition of the authors' recent paper [**3**], where we investigate the behavior of dual complexes under products. The main obstacle to understanding the product is that the product of snc models is not itself snc, but rather toroidal. In Section 2, we use the theory of log structures and Kato fans to give a definition of Berkovich skeleton for log-regular models, extending the one for snc models.

THEOREM 1.3.1 ([**3**, Theorem 1.6.1]). *Assume that the residue field k is algebraically closed. Let (X, Δ_X) and (Y, Δ_Y) be snc pairs. Suppose that both pairs have non-negative Kodaira–Iitaka dimension and admit semistable log-regular models $\mathscr{X}^+$ and $\mathscr{Y}^+$ over S^+. Then, there exists a PL homeomorphism of essential skeletons*

$$\mathrm{Sk}(Z, \Delta_Z) \xrightarrow{\sim} \mathrm{Sk}(X, \Delta_X) \times \mathrm{Sk}(Y, \Delta_Y),$$

where Z and Δ_Z are the respective products.

We say that a degeneration is semi-stable if every component of the scheme theoretic special fiber has multiplicity 1. The condition of semi-stability is necessary, see Example 2.2.

In Section 3, we introduce the weight function and the essential skeleton. In birational geometry, it is natural to work with pairs instead of varieties, so we extend the notion of essential skeleton to pairs. As a result, we are able to extend many of the results relating Berkovich and birational geometry.

Finally, in Section 4, we give our main application to the degenerations of hyper-Kähler varieties. We are able to determine the homeomorphism type of the dual complex of two classes of hyper-Kähler varieties: the Hilbert scheme of a degenerating K3 surface and the Kummer construction applied to a degenerating abelian surface.

1.4. Notation

(**1.5**) Let R be a complete discrete valuation ring with maximal ideal $\mathfrak{m}$, residue field $k = R/\mathfrak{m}$ and quotient field K. We assume that the valuation v_K is normalized, namely $v_K(\pi) = 1$ for any uniformizer π of R. We define by $|\cdot|_K = \exp(-v_K(\cdot))$ the absolute value on K corresponding to v_K.

We write $S = \mathrm{Spec}\, R$. Given an R-scheme $\mathscr{X}$ of finite type, we denote by $\mathscr{X}_k$ the special fiber of $\mathscr{X}$ and by $\mathscr{X}_K$ the generic fiber.

(**1.6**) Let X be a proper K-scheme. A model for X over R is a flat separated R-scheme $\mathscr{X}$ of finite type endowed with an isomorphism of K-schemes

$\mathscr{X}_K \to X$. If X is smooth over K, we say that $\mathscr{X}$ is an snc model for X if it is proper, regular over R, and the special fiber $\mathscr{X}_k$ is a strict normal crossings divisor on $\mathscr{X}$. In equicharacteristic 0, such a model always exists, by Hironaka's resolution of singularities.

We say that a model $\mathscr{X}$ over R (not necessarily regular) is semi-stable if the special fiber $\mathscr{X}_k$ is reduced.

(1.7) We denote by $(\cdot)^{\mathrm{an}}$ the analytification functor from the category of K-schemes of finite type to Berkovich's category of K-analytic spaces. For every K-scheme of finite type X, as a set, X^{an} consists of the pairs $x = (\xi_x, |\cdot|_x)$, where ξ_x is a point of X and $|\cdot|_x$ is an absolute value on the residue field $\kappa(\xi_x)$ of X at ξ_x extending the absolute value $|\cdot|_K$ on K. We endow X^{an} with the Berkovich topology, i.e., the weakest one, such that

(i) the forgetful map $\phi : X^{\mathrm{an}} \to X$, defined as $(\xi_x, |\cdot|_x) \mapsto \xi_x$, is continuous,

(ii) for any Zariski open subset U of X and any regular function f on U, the map $|f| : \phi^{-1}(U) \to \mathbb{R}$ defined by $|f|(\xi_x, |\cdot|_x) = |f(\xi_x)|_x$ is continuous.

When X is irreducible, the set $\mathrm{Bir}(X)$ of birational points of X^{an} is defined as the inverse image under ϕ of the generic point of X. By definition, it is a birational invariant of X.

2. The Skeleton of a Log-Regular Model

2.1. Notions of log geometry

(2.2) Logarithmic geometry is a geometric theory which was founded by Fontaine and Illusie, then developed by Kato [13]. It generalizes the theory of schemes by introducing the new notion of *log structure*. A log structure is an additional structure on a scheme that can encode geometric information of compactifications or degenerations of the scheme. In particular, it provides a suitable framework to incorporate the theories of toroidal embeddings and of differential forms with logarithmic poles [15]. For a more extended dissertation of the topic, we refer to [13, 6]. We will briefly review the basic definitions and focus on log-regular log schemes [14].

(2.3) The algebra of monoids plays a central role in the theory of log schemes. We recall that a monoid is a commutative semi-group with a neutral element. A morphism of monoids is a map that respects internal laws and neutral elements. Given a monoid M, we denote by M^{gp} the groupification of M, where

$$M^{\mathrm{gp}} = M \times M/\sim, \quad (a,b) \sim (c,d) \Leftrightarrow \exists m \in M, m + a + d = m + b + c.$$

For example, the groupification of the monoid $\mathbb{N}$ is $\mathbb{Z}$.

A monoid M is called *integral* if the morphism $M \to M^{\mathrm{gp}}$ is injective; it is called *fine* if M is integral and finitely generated; it is called *saturated* if it is integral and, whenever $m \in M^{\mathrm{gp}}$ and $nm \in M$ for some positive integer n, then $m \in M$. We use the abbreviation *fs* for fine and saturated monoids.

(2.4) Given a scheme $\mathscr{X}$, a *log structure* on $\mathscr{X}$ consists of a sheaf of monoids $\mathcal{M}_{\mathscr{X}}$ with respect to the Zariski topology, together with a morphism of sheaves of monoids $\alpha : \mathcal{M}_{\mathscr{X}} \to (\mathcal{O}_{\mathscr{X}}, \times)$ such that $\alpha^{-1}(\mathcal{O}_{\mathscr{X}}^{\times}) \simeq \mathcal{O}_{\mathscr{X}}^{\times}$. We call and denote by

$\mathscr{X}^{+} = (\mathscr{X}, \mathcal{M}_{\mathscr{X}})$ the *logarithmic scheme* defined by the pair $(\mathscr{X}, \mathcal{M}_{\mathscr{X}})$,

$\mathcal{C}_{\mathscr{X},x} = \mathcal{M}_{\mathscr{X},x}/\mathcal{O}_{\mathscr{X},x}^{\times}$ the *characteristic monoid* of $\mathscr{X}^{+}$ at the point $x \in \mathscr{X}$,

$\mathcal{I}_{\mathscr{X},x}$ the ideal in $\mathcal{O}_{\mathscr{X},x}$ generated by $\mathcal{M}_{\mathscr{X},x} \setminus \mathcal{O}_{\mathscr{X},x}^{\times}$ at the point $x \in \mathscr{X}$.

In particular, any scheme $\mathscr{X}$ is a log scheme with trivial log structure $\mathcal{O}_{\mathscr{X}}^{\times}$. We say that a log scheme $\mathscr{X}^{+}$ is *integral/fine/saturated/fs* if at each point $x \in \mathscr{X}$, the monoid $\mathcal{M}_{\mathscr{X},x}$ is, respectively, integral/fine/saturated/fs.

A *morphism* of log schemes $(\mathscr{X}, \mathcal{M}_{\mathscr{X}}) \to (\mathscr{Y}, \mathcal{M}_{\mathscr{Y}})$ is a morphism $f : \mathscr{X} \to \mathscr{Y}$ of schemes together with a morphism of sheaves of monoids $f^{-1}\mathcal{M}_{\mathscr{Y}} \to \mathcal{M}_{\mathscr{X}}$ which is compatible with the structure morphisms $\alpha_{\mathscr{X}}$ and $\alpha_{\mathscr{Y}}$, i.e., the following diagram is commutative:

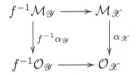

(2.5) Let $\mathscr{X}$ be a scheme with an effective divisor D. Then, the sheaf

$$\mathcal{M}_D(U) = \{f \in \mathcal{O}_{\mathscr{X}}(U) \mid f|_{\mathscr{X} \setminus D} \text{ is invertible}\}$$

together with the inclusion morphism in $\mathcal{O}_{\mathscr{X}}$ defines a log structure on $\mathscr{X}$. The log scheme consisting of $\mathscr{X}$ and $\mathcal{M}_D$ is denoted by $(\mathscr{X}, D)$ and is called the *divisorial log scheme* associated to the divisor D. We remark that the definition only depends on the support of the divisor D.

EXAMPLE 2.5.1. Let $\mathscr{X}$ be an snc model over R of a smooth and proper variety X over K. We denote by $\mathscr{X}^{+} = (\mathscr{X}, \mathscr{X}_k)$ the divisorial log scheme induced by the special fiber $\mathscr{X}_k$. Given a point $x \in \mathscr{X}$, we consider the irreducible components $E_1, \ldots, E_r$ of $\mathscr{X}_k$ passing through x and a local equation z_i for each E_i. Then,

$$\mathcal{M}_{\mathscr{X},x} = \left\{ u \cdot \prod_{i=1}^{r} z_i^{a_i} \,\middle|\, u \in \mathcal{O}_{\mathscr{X},x}^{\times}, a_i \in \mathbb{N} \right\} \quad \text{and} \quad \mathcal{C}_{\mathscr{X},x} \simeq \mathbb{N}^r.$$

In the case of a curve over K, the characteristic sheaf is locally isomorphic to

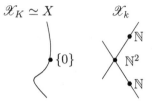

2.6. Log-regularity

(2.7) In [**14**], Kato introduces and studies the notion of log-regularity. A locally noetherian fs log scheme $\mathscr{X}^+$ is called log-regular at a point x if the following two conditions are satisfied: $\mathcal{O}_{\mathscr{X},x}/\mathcal{I}_{\mathscr{X},x}$ is a regular local ring, and $\dim \mathcal{O}_{\mathscr{X},x} = \dim \mathcal{O}_{\mathscr{X},x}/\mathcal{I}_{\mathscr{X},x} + \operatorname{rank} \mathcal{C}^{\mathrm{gp}}_{\mathscr{X},x}$.

Examples of log-regular log schemes are regular schemes with trivial log structure, divisorial log schemes with log structure coming from an snc divisor and toric varieties with divisorial log structure induced by the toric boundary.

(2.8) If $\mathscr{X}^+$ is a log-regular log scheme, then the locus where the log structure is non-trivial, i.e., non-isomorphic to $\mathcal{O}^{\times}_{\mathscr{X},x}$, is a divisor that we will denote by $D_{\mathscr{X}}$ and call the boundary divisor of $\mathscr{X}^+$. By [**14**, Theorem 11.6], the log structure on $\mathscr{X}^+$ is the divisorial log structure induced by $D_{\mathscr{X}}$.

Working over a perfect field, the divisor $D_{\mathscr{X}}$ is toroidal, and the inclusion of $\mathscr{X} \setminus D_{\mathscr{X}}$ in $\mathscr{X}$ defines a toroidal embeddings. We recall that, according to [**15**], a morphism is a toroidal embedding if étale locally it is the embedding of a dense open algebraic torus inside a toric variety. Since we consider log structures with respect to the Zariski topology, in our setting, only toroidal embeddings without self-intersections give rise to log-regular log schemes.

(2.9) We recall that in general a monoidal space is a topological space, endowed with a sheaf of fine and saturated monoids. Given a log-regular log scheme $\mathscr{X}^+$, we consider the subspace F of $\mathscr{X}$ consisting of generic points of intersections of irreducible components of $D_{\mathscr{X}}$ and the sheaf $\mathcal{M}_F$ given by the inverse image of the characteristic sheaf $\mathcal{C}_{\mathscr{X}}$ via the inclusion of F in $\mathscr{X}$. The monoidal space determined by F and $\mathcal{M}_F$ is called the Kato fan associated to $\mathscr{X}^+$ and is usually simply denoted by F.

By convention, when $\mathscr{X}$ is irreducible, the generic point of the empty intersection of irreducible components is the generic point of $\mathscr{X}^+$. By definition, this point is also included in the Kato fan F.

EXAMPLE 2.9.1. The Kato fan associated to $S^+ = (S, (\pi))$ consists of two points: the generic point of S that corresponds to the empty intersection and the closed point (π) corresponding to the unique irreducible component of the boundary divisor. The sheaf $\mathcal{M}_F$ is isomorphic to the skyscraper sheaf $\mathbb{N}$ at (π).

2.10. Skeleton of a log-regular log scheme

(2.11) Let $\mathscr{X}^+$ be a log-regular log scheme over S^+. The Kato fan F associated to $\mathscr{X}$ inherits a partial order from $\mathscr{X}^+$ as follows: given $x, y \in F$, $x \preceq y$ if x lies in the closure of $\{y\}$. We denote by $F(x)$ the subspace of such y. For any pair (x, y) with $y \in F(x)$, there is a cospecialization morphism

$$\tau_{x,y} : \mathcal{C}_{\mathscr{X},x} \to \mathcal{C}_{\mathscr{X},y}.$$

EXAMPLE 2.11.1. Proceeding as in the example of the curve in Example 2.5.1, we consider the log-regular log scheme $\mathscr{X}^+$ induced by the special fiber. We denote by η. the generic point of an intersection of irreducible components of $D_{\mathscr{X}}$. We obtain the following poset for F and the corresponding cospecializations:

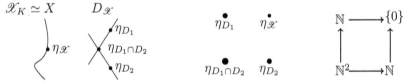

EXAMPLE 2.11.2 In the case of toric varieties, the sheaf $\mathcal{M}_F$ corresponds to the integral points of the toric fan. The following pictures illustrate the case of $\mathbb{P}^2$:

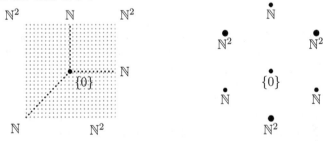

(2.12) For each point x in F, we denote by σ_x the set of morphisms of monoids

$$\alpha : \mathcal{C}_{\mathscr{X},x} \to (\mathbb{R}_{\geq 0}, +),$$

such that $\alpha(\pi) = 1$ for every uniformizer π in R. We endow σ_x with the topology of pointwise convergence, where $\mathbb{R}_{\geq 0}$ carries the usual Euclidean topology. Note that σ_x is a polyhedron, possibly unbounded, in the real affine space

$$\{\alpha : \mathcal{C}^{\mathrm{gp}}_{\mathscr{X},x} \to (\mathbb{R}, +) \,|\, \alpha(\pi) = 1 \text{ for every uniformizer } \pi \text{ in } R\}.$$

If y is a point of $F(x)$, then the morphism $\tau_{x,y}$ induces a topological embedding $\sigma_y \to \sigma_x$ that identifies σ_y with a face of σ_x.

(2.13) We denote by T the disjoint union of the topological spaces σ_x with x in F. On the topological space T, we consider the equivalence relation $\sim$ generated by $\alpha \sim \alpha \circ \tau_{x,y}$ where x and y are points in F, such that x lies in the closure of $\{y\}$ and α is a point of σ_y.

The skeleton of $\mathscr{X}^+$ is defined as the quotient of the topological space T by the equivalence relation $\sim$. We denote this skeleton by $\mathrm{Sk}(\mathscr{X}^+)$. It is clear that $\mathrm{Sk}(\mathscr{X}^+)$ has the structure of a polyhedral complex with cells $\{\sigma_x, \ x \in F\}$, so it comes equipped with a piecewise linear (PL) structure, and that the faces of a cell σ_x are precisely the cells σ_y with y in $F(x)$.

EXAMPLE 2.13.1. Following Examples 2.5.1 and 2.11.1, we consider local equations z_1 and z_2 for the irreducible components of $D_{\mathscr{X}}$ at their point of intersection. By a slight abuse of notation, we denote also by z_i their classes in the characteristic monoids. At each generic point x in F, we determine the conditions on the morphisms $\alpha : \mathcal{C}_{\mathscr{X},x} \to (\mathbb{R}_{\geqslant 0}, +)$ to define a point in the skeleton:

$D_{\mathscr{X}}$

$\alpha : \mathcal{C}_{\mathscr{X},\eta_{D_i}} \simeq \mathbb{N} \to (\mathbb{R}_{\geqslant 0}, +)$
$$\begin{cases} \alpha(\pi) = \alpha(z_i^{a_i}) = a_i \alpha(z_i) = 1 \\ \alpha(1) = \alpha(z_i) \end{cases}$$

$\alpha : \mathcal{C}_{\mathscr{X},\eta_{D_1 \cap D_2}} \simeq \mathbb{N}^2 \to (\mathbb{R}_{\geqslant 0}, +)$
$$\begin{cases} \alpha(\pi) = \alpha(z_1^{a_1} z_2^{a_2}) \\ \quad = a_1 \alpha(z_1) + a_2 \alpha(z_2) = 1 \\ \alpha((1,0)) = \alpha(z_1) \\ \alpha((0,1)) = \alpha(z_2) \end{cases}$$

$\alpha : \mathcal{C}_{\mathscr{X},\eta_{\mathscr{X}}} \simeq \{0\} \to (\mathbb{R}_{\geqslant 0}, +)$ but $\alpha(\pi) = \alpha(0) = 0 \neq 1$ for a
morphism of monoids.

We obtain that the faces of the skeleton $\mathrm{Sk}(\mathscr{X}^+)$ are $\sigma_{\eta_{D_i}} \simeq \{pt\}$, $\sigma_{\eta_{D_1 \cap D_2}} \simeq [0,1]$ and $\sigma_{\eta_{\mathscr{X}}} = \emptyset$. Thus, the skeleton is isomorphic to the standard 1-simplex

$$\sigma_{\eta_{D_1}} \qquad \sigma_{\eta_{D_1 \cap D_2}} \qquad \sigma_{\eta_{D_2}}$$

In particular, it is homeomorphic to the dual complex of the special fiber, $\mathscr{X}_k$. More generally, given an snc scheme $\mathscr{X}$, the skeleton associated to the log scheme $(\mathscr{X}, \mathscr{X}_k)$ is identified with the dual complex $\mathcal{D}(\mathscr{X}_k)$ (see [21, Proposition 3.1.4]).

REMARK 2.13.2. We note that σ_x is empty for any point $x \in F$ that does not lie in the special fiber $\mathscr{X}_k$: indeed, outside the special fiber, any uniformizer is an invertible element, so it is trivial in $\mathcal{C}_{\mathscr{X},x}$ and is mapped to 0 by any morphism of monoids. Therefore, the construction of the skeleton associated to $\mathscr{X}^+$ only concerns the points in the Kato fan F that lie in the special fiber.

2.14. Embedding the skeleton in the Berkovich Space

(2.15) Let (X, Δ_X) be an snc pair where X is a connected, smooth and proper K-variety and Δ_X is an effective $\mathbb{Q}$-divisor, such that $\Delta_X = \sum a_i \Delta_{X,i}$ with $0 < a_i \leqslant 1$. A log-regular log scheme $\mathscr{X}^+$ over S^+ is a model for $(X, \lceil \Delta_X \rceil)$ over S^+ if $\mathscr{X}$ is a proper model of X over R and $D_{\mathscr{X}} = \lceil \Delta_X \rceil + \mathscr{X}_{k,\mathrm{red}}$.

Let $\mathscr{X}^+$ be such a model. We denote by F its associated Kato fan.

PROPOSITION 2.15.1 ([**3**, Proposition 3.2.10]). *Let x be a point of F and let $\alpha : \mathcal{C}_{\mathscr{X},x} \to (\mathbb{R}_{\geqslant 0}, +)$ be an element of σ_x. Then, there exists a unique minimal real valuation*

$$v : \mathcal{O}_{\mathscr{X},x} \setminus \{0\} \to \mathbb{R}_{\geqslant 0},$$

such that $v(m) = \alpha(\overline{m})$ for each element m of $\mathcal{M}_{\mathscr{X},x}$, where $\overline{m}$ is the class of m in $\mathcal{C}_{\mathscr{X},x}$.

(2.16) Proposition 2.15.1 can be proved along the lines of [**21**, Proposition 2.4.4]. Indeed, every element $f \in \mathcal{O}_{\mathscr{X},x}$ can be written as a formal power series

$$f = \sum_{\gamma \in \mathcal{C}_{\mathscr{X},x}} c_\gamma \gamma$$

in $\widehat{\mathcal{O}}_{\mathscr{X},x}$, where each coefficient c_γ is either zero or a unit in $\mathcal{O}_{\mathscr{X},x}$. While such a formal expansion of f is not unique, the map $f \mapsto \min\{\alpha(\gamma) \,|\, c_\gamma \neq 0\}$ is well defined and satisfies all the requirements in the statement of the proposition.

(2.17) For any $x \in F$ and $\alpha \in \sigma_x$, the valuation obtained in the proposition induces a real valuation on the function field of X that extends the discrete valuation v_K on K. Thus, it defines a point of the K-analytic space X^{an}, which we will denote by $v_{x,\alpha}$.

The characterization of $v_{x,\alpha}$ in Proposition 2.15.1 implies that $v_{y,\alpha'} = v_{x,\alpha' \circ \tau_{x,y}}$ for every y in $F(x)$ and every α' in σ_y. Hence, we obtain a well-defined map

$$\iota : \mathrm{Sk}(\mathscr{X}^+) \to X^{\mathrm{an}}$$

by sending α to $v_{x,\alpha}$ for every point x of F and every $\alpha \in \sigma_x$.

PROPOSITION 2.17.1 ([**3**, Proposition 3.2.15]). *The map $\iota : \mathrm{Sk}(\mathscr{X}^+) \to X^{\mathrm{an}}$ is a topological embedding.*

(2.18) We denote by $D_{\mathscr{X},\mathrm{hor}}$ the sum with multiplicities of the components of $D_{\mathscr{X}}$ not contained in the special fiber $\mathscr{X}_k$. The inclusion $\iota : \mathrm{Sk}(\mathscr{X}^+) \to X^{\mathrm{an}}$ is actually an inclusion in $(X \setminus D_{\mathscr{X},\mathrm{hor}})^{\mathrm{an}}$ and it admits a continuous retraction

$$\rho_{\mathscr{X}} : (X \setminus D_{\mathscr{X},\mathrm{hor}})^{\mathrm{an}} \to \mathrm{Sk}(\mathscr{X}^+)$$

constructed as follows. Consider the reduction map [**21**, §2.2.2]

$$\mathrm{red}_{\mathscr{X}} : (X \setminus D_{\mathscr{X},\mathrm{hor}})^{\mathrm{an}} \to \mathscr{X}_k.$$

Let $x \in (X \setminus D_{\mathscr{X},\mathrm{hor}})^{\mathrm{an}}$ and let $E_1, \dots, E_r$ be the irreducible components of $D_{\mathscr{X}}$ passing through the point $\mathrm{red}_{\mathscr{X}}(x)$. The generic point η of the connected component of $E_1 \cap \cdots \cap E_r$ that contains $\mathrm{red}_{\mathscr{X}}(x)$ is a point in the associated Kato fan F. We set α to be the morphism of monoids $\alpha : \mathcal{C}_{\mathscr{X},\eta} \to \mathbb{R}_{\geqslant 0}$, such that $\alpha(\overline{m}) = v_x(m)$ for any element m of $\mathcal{M}_{\mathscr{X},\eta}$. Then, $\rho_{\mathscr{X}}(x)$ is the point of $\mathrm{Sk}(\mathscr{X}^+)$ corresponding to the couple (η, α). By construction, $\rho_{\mathscr{X}}$ is continuous and right inverse to the inclusion ι.

REMARK 2.18.1. The skeleton associated to a log-regular scheme $\mathscr{X}^+$, where $D_{\mathscr{X}}$ allows horizontal components, generalizes Berkovich's skeletons by considering strata corresponding to the intersection of components of the special and horizontal components. The contribution of such intersections yields unbounded faces in the direction of the horizontal components. It also generalizes the construction due to Gubler, Rabinoff and Werner in [**7**] of a skeleton associated to a strictly semi-stable snc pair.

EXAMPLE 2.18.2. Let $X = \mathbb{P}_K^1$ with homogeneous coordinates $[x : y]$ and $\Delta_X = (0 : 1) + (1 : 0)$. Consider $\mathscr{X}^+ = (\mathbb{P}_R^1, D_{\mathscr{X}})$, where $D_{\mathscr{X}} = \overline{\Delta_X} + \mathbb{P}_k^1$: it is a log-regular model of (X, Δ_X). Considering the local equations of $D_{\mathscr{X}}$ at its generic points, we compute the skeleton of $\mathscr{X}^+$:

$$
\begin{array}{c|l|l}
D_{\mathscr{X}} & & \\
\hline
x & \alpha : \mathcal{C}_{\mathscr{X},\eta_{\{\pi=0\}\cap\{x=0\}}} \simeq \mathbb{N}^2 \to (\mathbb{R}_{\geqslant 0}, +) & \left\{ \begin{array}{l} \alpha((1,0)) = \alpha(\pi) = 1 \\ \alpha((0,1)) = \alpha(x) \in \mathbb{R}_{\geqslant 0} \end{array} \right. \\
\hline
y \quad \pi & \alpha : \mathcal{C}_{\mathscr{X},\eta_{\{\pi=0\}\cap\{y=0\}}} \simeq \mathbb{N}^2 \to (\mathbb{R}_{\geqslant 0}, +) & \left\{ \begin{array}{l} \alpha((1,0)) = \alpha(\pi) = 1 \\ \alpha((0,1)) = \alpha(y) \in \mathbb{R}_{\geqslant 0} \end{array} \right. \\
\hline
& \alpha : \mathcal{C}_{\mathscr{X},\eta_{\{\pi=0\}}} \simeq \mathbb{N} \to (\mathbb{R}_{\geqslant 0}, +) & \alpha(1) = \alpha(\pi) = 1.
\end{array}
$$

Therefore, the faces of $\mathrm{Sk}(\mathscr{X}^+)$ are respectively two closed half lines and a point. By the identification of the equivalence relation $\sim$, we obtain that $\mathrm{Sk}(\mathscr{X}^+)$ is isomorphic to the real line

$$\underset{\cdots\cdots}{\overline{\quad\underset{\sigma\eta_{\{\pi=0\}\cap\{x=0\}}}{\qquad\qquad} \underset{\sigma\eta_{\{\pi=0\}}}{\bullet} \underset{\sigma\eta_{\{\pi=0\}\cap\{y=0\}}}{\qquad\qquad}\quad}}_{\cdots\cdots}$$

Moreover, the embedding of $\mathrm{Sk}(\mathscr{X}^+)$ in $(\mathbb{P}_K^1)^{\mathrm{an}}$ coincides with the canonical skeleton of the torus $\mathbb{G}_{m,K}^1$.

2.19. Product of skeletons

(2.20) Besides taking into consideration horizontal components, log-regular model has a second advantage: they behave well under products. Given morphisms of fs log schemes $s_{\mathscr{X}} : \mathscr{X}^+ \to S^+$ and $s_{\mathscr{Y}} : \mathscr{Y}^+ \to S^+$, their fibered product exists in the category of fs log schemes and is denoted by $\mathscr{X}^+ \times_{S^+}^{\mathrm{fs}} \mathscr{Y}^+$. If the morphisms $s_{\mathscr{X}}$ and $s_{\mathscr{Y}}$ are log-smooth, then the fibered product is log-regular as well [**14**, Theorem 8.2]. It is not true in general that the schematic fibered product of snc models still has an snc special fiber.

EXAMPLE 2.20.1. Let

$$\mathscr{X}^+ = \left(\operatorname{Spec} \frac{R[x,y]}{(xy = \pi)}, \mathscr{X}_k = \{xy = 0\}\right),$$

$$\mathscr{Y}^+ = \left(\operatorname{Spec} \frac{R[z,w]}{(zw = \pi)}, \mathscr{Y}_k = \{zw = 0\}\right).$$

Their special fibers have two irreducible components intersecting in a point, so the associated skeletons are both isomorphic to the closed unit interval. The underlying scheme of fs fibered product $\mathscr{X}^+ \times^{\text{fs}}_{S^+} \mathscr{Y}^+$ is isomorphic to the schematic fibered product, and the log structure of $\mathscr{X}^+ \times^{\text{fs}}_{S^+} \mathscr{Y}^+$ is induced by the special fiber. This consists of four irreducible components and is not an snc divisor: indeed, the four components intersect in a codimension 3 stratum. Thus, the dual complex is not well defined.

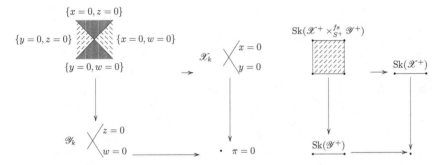

(2.21) Let (X, Δ_X) and (Y, Δ_Y) be snc pairs where Δ_X and Δ_Y are effective $\mathbb{Q}$-divisors with coefficients in the interval $[0,1]$. Let $\mathscr{X}^+$ and $\mathscr{Y}^+$ be log-smooth models over S^+ of $X^+ = (X, \lceil \Delta_X \rceil)$ and $Y^+ = (Y, \lceil \Delta_Y \rceil)$, respectively. Then, the fs fibered product $\mathscr{Z}^+ = \mathscr{X}^+ \times^{\text{fs}}_{S^+} \mathscr{Y}^+$ is a log-regular model of $Z^+ := X^+ \times^{\text{fs}}_{K} Y^+$. The continuous map of skeletons

$$\left(\operatorname{pr}_{\operatorname{Sk}(\mathscr{X})}, \operatorname{pr}_{\operatorname{Sk}(\mathscr{Y})}\right) : \operatorname{Sk}(\mathscr{Z}^+) \to \operatorname{Sk}(\mathscr{X}^+) \times \operatorname{Sk}(\mathscr{Y}^+)$$

is associated by functoriality to the projections $\operatorname{pr}_{\mathscr{X}} : \mathscr{Z}^+ \to \mathscr{X}^+$ and $\operatorname{pr}_{\mathscr{Y}} : \mathscr{Z}^+ \to \mathscr{Y}^+$.

PROPOSITION 2.21.1 ([**3**, Proposition 3.4.3]). *Assume that the residue field k is algebraically closed. If $\mathscr{X}^+$ is semi-stable, then the map $\left(\operatorname{pr}_{\operatorname{Sk}(\mathscr{X})}, \operatorname{pr}_{\operatorname{Sk}(\mathscr{Y})}\right)$ is a PL homeomorphism.*

(2.22) We recall that a log-regular log scheme $\mathscr{X}^+$ is said to be semi-stable if the special fiber is reduced. Proposition 2.21.1 can be proved in two steps. First, we show that the algebraic closedness of k and the semi-stability of $\mathscr{X}^+$ are sufficient conditions to conclude that the map of Kato fans $F_{\mathscr{X}} \to F_{\mathscr{X}} \times_{F_S} F_{\mathscr{Y}}$ is an isomorphism. Second, for any $z \in F_{\mathscr{X}}$, we build a bijective correspondence between the face σ_z of $\mathrm{Sk}(\mathscr{X}^+)$ and the product $\sigma_x \times \sigma_y$ of the faces associated to the unique corresponding points x and y in the Kato fans $F_{\mathscr{X}}$ and $F_{\mathscr{Y}}$.

REMARK 2.22.1. The assumption of semi-stability is crucial in the result of Proposition 2.21.1. The semi-stability hypothesis guarantees that the fibered product of $\mathscr{X}^+$ and $\mathscr{Y}^+$ in the category of log schemes is automatically saturated. Geometrically saturation corresponds to normalization. Therefore, if we do not require $\mathscr{X}^+$ to be semi-stable, we may end up with a logarithmic fibered product that is not normal and whose strata in the special fiber of the normalized fibered product are not in bijection with pairs of strata in the special fibers of two factors. We will illustrate this with an example.

EXAMPLE 2.22.1. Let q be the equation a generic quartic curve in $\mathbb{P}^2_{\mathbb{C}((t))}$. Then, $\mathscr{X} : tq + x^2y^2 = 0$ gives the equation of a family of genus 3 curves, degenerating to two double lines. The family $\mathscr{X}$ has four singularities of type A_1 in each component of the special fiber, corresponding to the base points of the family. Endowing $\mathscr{X}$ with the divisorial log structures induced by its special fiber, the skeleton $\mathrm{Sk}(\mathscr{X}^+)$ is a line segment. In this case, taking a semi-stable model of $\mathscr{X}_{\mathbb{C}((t))}$ requires a normalized base change R' of $R = \mathbb{C}[[t]]$ of order 2, which induces coverings branched at each of these singular points (see [11, p. 133] for details). Let $\mathscr{Y}$ be such a semi-stable reduction. Thus, the special fiber of $\mathscr{Y}$ consists of two elliptic curves, call them E_1 and E_2, which intersect in two points, p_A and p_B, which are the preimages of the point $(0 : 0 : 1)$. The skeleton $\mathrm{Sk}(\mathscr{Y}^+)$ is isomorphic to S^1, where $\mathscr{Y}^+$ has divisorial log structures induced by its special fiber.

$$\mathscr{X}_{\mathbb{C}} \qquad \mathcal{D}(\mathscr{X}_{\mathbb{C}}) \qquad\qquad \mathscr{Y}_{\mathbb{C}} \qquad \mathcal{D}(\mathscr{Y}_{\mathbb{C}})$$

We will compare the skeleton of fs fibered product $\mathscr{X}^+ \times_R^{\mathrm{fs}} \mathscr{X}^+$ with that of $\mathrm{Sk}(\mathscr{Y}^+ \times_{R'}^{\mathrm{fs}} \mathscr{Y}^+)$. For the product with a semi-stable model, the skeleton is the product of the skeletons, and $\mathrm{Sk}(\mathscr{Y}^+ \times_{R'}^{\mathrm{fs}} \mathscr{Y}^+)$ is therefore a real 2-torus $S^1 \times S^1$.

On the other hand, the skeleton of the product $\mathscr{X}^+ \times_R^{\mathrm{fs}} \mathscr{X}^+$ is given by a quotient of $S^1 \times S^1$ by the action of $\mathbb{Z}/2\mathbb{Z}$. $\mathrm{Sk}(\mathscr{Y}^+ \times_{R'}^{\mathrm{fs}} \mathscr{Y}^+)$ has a stratification, whose strata correspond to ordered pairs of strata in $\mathrm{Sk}(\mathscr{Y}^+)$, so zero-dimensional strata: $(E_1, E_2), (E_1, E_1), (E_2, E_1), (E_2, E_2)$

one-dimensional strata: $(E_1, p_A), (E_1, p_B), (E_2, p_A), (E_2, p_B)$

$$(p_A, E_1), (p_B, E_1), (p_A, E_2), (p_B, E_2)$$

two-dimensional strata: $(p_A, p_A), (p_A, p_B), (p_B, p_A), (p_B, p_B)$.

The action of $\mathbb{Z}/2\mathbb{Z}$ fixes E_1 and E_2, while switching p_A and p_B. Therefore, it fixes exactly the zero-dimensional strata while acting freely on the other points. The quotient, the skeleton $\mathscr{X}^+ \times_R^{\text{fs}} \mathscr{X}^+$, is piecewise-linearly homeomorphic to the sphere S^2. In particular, it is not isomorphic to the product of two line segments.

3. The Essential Skeleton of a Product

Inspired by the work of Kontsevich and Soibelman [18], in [21], Mustaţă and Nicaise define the weight function and the Kontsevich–Soibelman skeleton for a smooth variety X over a discretely valued field K and a pluricanonical form ω. They prove that a non-zero rational pluricanonical form ω on X induces a function wt_ω on X^{an}, such that the restriction to Berkovich skeletons of proper snc models $\mathscr{X}$ of X is piecewise affine. The function wt_ω is called the weight function associated to ω.

Moreover, if ω is a regular form, wt_ω is strictly decreasing under the retraction of X^{an} to $\text{Sk}(\mathscr{X})$. It follows that the locus of points with minimal weight is contained in any proper snc model of X; Mustaţă and Nicaise call this locus the Kontsevich–Soibelman skeleton associated to ω, denoted by $\text{Sk}(X, \omega)$. In the following, we will extend some of their results to the more general setting of pairs.

3.1. Kontsevich–Soibelman skeletons for a pair

(3.2) Let (X, Δ_X) be an snc pair where X is a connected, smooth and proper K-variety, and $\Delta_X = \sum a_i \Delta_{X,i}$ with $0 < a_i \leqslant 1$. We say that ω is a Δ_X-logarithmic m-pluricanonical form if it is a rational m-pluricanonical form on X^+ with poles of order at most ma_i along $\Delta_{X,i}$ for some m, such that $ma_i \in \mathbb{N}$ for any i; in other words, it is a regular section of $m(K_X + \Delta_X)$. Similar to [21], for such a form ω, the Kontsevich–Soibelman skeleton $\text{Sk}(X, \Delta_X, \omega)$ is the closure in $\text{Bir}(X)$ of the set of divisorial points of X^{an} where the weight function wt_ω reaches its minimal weight. *A priori* the weight function associated to a rational pluricanonical form may have minimal weight $-\infty$, hence the corresponding Kontsevich–Soibelman skeleton would be empty. We prove that this does not occur for Δ_X-logarithmic pluricanonical forms.

PROPOSITION 3.2.1 ([**3**, Proposition 4.1.6]). *Given a Δ_X-logarithmic m-pluricanonical form ω, for any log-regular model $\mathscr{X}^+$ of X^+, the inclusion $\text{Sk}(X, \Delta_X, \omega) \subseteq \text{Sk}(\mathscr{X}^+)$ holds.*

(3.3) The proof of Proposition 3.2.1 consists in showing the following property: the weight function associated to ω is strictly decreasing along the retraction $\rho_{\mathscr{X}}$ to the skeleton $\mathrm{Sk}(\mathscr{X}^+)$, i.e.,

$$\mathrm{wt}_\omega(\cdot) \geqslant \mathrm{wt}_\omega(\rho_{\mathscr{X}}(\cdot))$$

and that the equality holds only on $\mathrm{Sk}(\mathscr{X}^+)$. This computation is done by means of a reformulation of the weight function in terms of logarithmic differentials. More precisely, for every point x of $\mathrm{Sk}(\mathscr{X}^+)$, we have the equality

$$\mathrm{wt}_\omega(x) = v_x(\mathrm{div}_{\mathscr{X}^+}(\omega)) + m$$

(see [**3**, Lemma 4.1.4]) with the following notation. For any point $x = (\xi_x, |\cdot|_x) \in X^{\mathrm{an}}$ and for any $\mathbb{Q}$-Cartier divisor D on $\mathscr{X}^+$ whose support does not contain ξ_x, we set $v_x(D) = -\frac{1}{m}\ln|f(x)|$ where m is an integer such that mD is Cartier and f is any element of $K(X)^\times$, such that $mD = \mathrm{div}(f)$ locally at $\mathrm{red}_{\mathscr{X}}(x)$. The form ω defines a rational section of the relative logarithmic m-pluricanonical line bundle $\omega_{\mathscr{X}^+/S^+}^{\otimes m}$, hence it induces a divisor on $\mathscr{X}^+$ denoted by $\mathrm{div}_{\mathscr{X}^+}(\omega)$.

(3.4) The introduction of the boundary Δ_X allows us to construct non-empty Kontsevich–Soibelman skeletons even for varieties with Kodaira dimension $-\infty$, as in the following examples.

EXAMPLE 3.4.1. Let X be the projective line $\mathbb{P}^1_K$ with affine coordinates x and y and $\Delta_X = (0:1) + (1:0)$. Then, $a_i = 1$ for any i and there exist Δ_X-logarithmic canonical forms. For example, we consider

$$\omega = \frac{dx}{x} = -\frac{dy}{y}.$$

Let $\mathscr{X} = \mathbb{P}^1_R$ and $D_{\mathscr{X}} = (0:1) + (1:0) + \mathbb{P}^1_k$. The log scheme $\mathscr{X}^+ = (\mathscr{X}, D_{\mathscr{X}})$ is a log-regular model of $X^+ = (X, \lceil\Delta_X\rceil)$ and the associated skeleton $\mathrm{Sk}(\mathscr{X}^+)$ looks as follows:

Since $\mathrm{div}_{\mathscr{X}^+}(\omega) = 0$, the weight associated to ω is minimal at any point of the skeleton $\mathrm{Sk}(\mathscr{X}^+)$. Thus, $\mathrm{Sk}(X, \Delta_X, \omega) = \mathrm{Sk}(\mathscr{X}^+) \simeq \mathbb{R}$.

EXAMPLE 3.4.2. Let $X = \mathbb{P}^1_K$ and $\Delta_X = \frac{2}{3}(0:1) + \frac{2}{3}(1:0) + \frac{2}{3}(1:1)$. So, $a_i = \frac{2}{3}$ for any i and there exist Δ_X-logarithmic 3-pluricanonical forms. We set

$$\omega = \frac{1}{(x-1)^2} \cdot \frac{1}{x^2}(dx)^3 = -\frac{1}{(1-y)^2} \cdot \frac{1}{y^2}(dy)^3.$$

We consider $\mathscr{X} = \mathbb{P}^1_R$ and $D_{\mathscr{X}} = (0:1) + (1:0) + (1:1) + \mathbb{P}^1_k$, then $\mathscr{X}^+ = (\mathscr{X}, D_{\mathscr{X}})$ is a log-regular model of $X^+ = (X, \lceil \Delta_X \rceil)$ and $\mathrm{Sk}(\mathscr{X}^+)$ is

Since $\mathrm{div}_{\mathscr{X}^+}(\omega) = (0:1) + (1:0) + (1:1)$, the weight associated to ω is minimal at the divisorial point $v_{\mathbb{P}^1_k}$ corresponding to $\mathbb{P}^1_k$ and is strictly increasing with slope 1 along the unbounded edges, when we move away from the point $v_{\mathbb{P}^1_k}$. Therefore, $\mathrm{Sk}(X, \Delta_X, \omega) = \{v_{\mathbb{P}^1_k}\}$.

(3.5) These Kontsevich–Soibelman skeletons also behave nicely under birational morphisms. Let $f \colon X' \to X$ be a log resolution. Then, there is a $\mathbb{Q}$-divisor Γ' with snc support, and no coefficient exceeding 1, such that $K_{X'} + \Gamma' = f^*(K_X + \Delta_X)$. Take $\Delta_{X'}$ to be the positive part of Γ' and write $K_{X'} + \Delta_{X'} = f^*(K_X + \Delta_X) + N$. For any m, pullback along with multiplication by the divisor of discrepancies N induces an isomorphism of vector spaces

$$(3.5.1) \qquad H^0(X, mK_X + m\Delta_X) \cong H^0(X', mK_{X'} + m\Delta_{X'}).$$

Let ω and ω' be corresponding forms via this isomorphism.

PROPOSITION 3.5.2 ([**3**, Proposition 5.1.5]). *Under the identification of the birational points of X with those of X', $\mathrm{Sk}(X, \Delta_X, \omega)$ is identified with $\mathrm{Sk}(X', \Delta_{X'}, \omega')$.*

(3.6) Therefore, we extend the notion of Kontsevich–Soibelman skeleton to a dlt pair (X, Δ_X): we define $\mathrm{Sk}(X, \Delta_X, \omega)$ as the Kontsevich–Soibelman skeleton $\mathrm{Sk}(X', \Delta_{X'}, \omega')$, where $(X', \Delta_{X'})$ is any log-resolution of (X, Δ_X), and ω' is the form corresponding to ω under the isomorphism (3.5.1): Proposition 3.5.2 guarantees that this is well defined.

3.7. Essential skeleton of a pair

(3.8) We define the essential skeleton $\mathrm{Sk}(X, \Delta_X)$ of a dlt pair (X, Δ_X) as the union of all Kontsevich–Soibelman skeletons $\mathrm{Sk}(X, \Delta_X, \omega)$, where ω ranges over all Δ_X-logarithmic pluricanonical forms. In the case of an empty boundary and a smooth variety X, this recovers the notion introduced in [**21**].

(3.9) As the weight function is closely related to the log discrepancy from birational geometry, it is natural to expect that the essential skeleton in some way encodes some of the minimal model theory of X. Nicaise and Xu [**22**] show, when X is a smooth projective variety with K_X semi-ample, and $\mathscr{X}$ is a good dlt minimal model, that the dual complex of $\mathscr{X}_k$ can be identified with the essential skeleton of X. Hence, good minimal models represent a second equivalent approach to the construction of a canonical skeleton in X^{an} and provide a birational interpretation of the essential skeleton. This equivalence can be generalized to the context of pairs.

The dual intersection complex for the coefficient 1 part of a dlt pair is well defined and denoted by $\mathcal{D}^{=1}$. We denote by $\mathcal{D}_0^{=1}$ the open subset of $\mathcal{D}^{=1}$ corresponding to the strata supported on the special fiber.

PROPOSITION 3.9.1 ([**3**, Proposition 5.1.7]). *Let (X, Δ_X) be a dlt pair with $K_X + \Delta_X$ semi-ample and let $(\mathscr{X}, \Delta_{\mathscr{X}})$ be a good dlt minimal model of (X, Δ_X) over R. Then, the embedding of $\mathcal{D}_0^{=1}(\mathscr{X}, \Delta_{\mathscr{X}})$ into the birational points of X identifies $\mathcal{D}_0^{=1}(\mathscr{X}, \Delta_{\mathscr{X}})$ with $\mathrm{Sk}(X, \Delta_X)$.*

3.10. Essential skeleton of a product

THEOREM 3.10.1 ([**3**, Proposition 5.2.2]). *Assume that the residue field k is algebraically closed. Let (X, Δ_X) and (Y, Δ_Y) be snc pairs. Suppose that both pairs have non-negative Kodaira–Iitaka dimension and admit semi-stable log-regular models $\mathscr{X}^+$ and $\mathscr{Y}^+$ over S^+. Then, the PL homeomorphism of skeletons*

$$\mathrm{Sk}(\mathscr{Z}^+) \xrightarrow{\sim} \mathrm{Sk}(\mathscr{X}^+) \times \mathrm{Sk}(\mathscr{Y}^+),$$

given in Proposition 2.15.1, restricts to a PL homeomorphism of essential skeletons

$$\mathrm{Sk}(Z, \Delta_Z) \xrightarrow{\sim} \mathrm{Sk}(X, \Delta_X) \times \mathrm{Sk}(Y, \Delta_Y),$$

where $\mathscr{Z}^+ = \mathscr{X}^+ \times_{S^+}^{fs} \mathscr{Y}^+$, $Z = X \times_K Y$ and $\Delta_Z = X \times_K \Delta_Y + \Delta_X \times_K Y$ are the respective products.

(3.11) We recall that a pair (X, Δ_X) has non-negative Kodaira–Iitaka dimension if some multiple of the line bundle $K_X + \Delta_X$ has a non-zero regular section. Theorem 3.10.1 establishes the behavior of essential skeletons under products.

On the one hand, if two points $x \in \mathscr{X}^+$ and $y \in \mathscr{Y}^+$ are respectively in the essential skeleton of (X, Δ_X) and (Y, Δ_Y), there exist two Δ_X-logarithmic and Δ_Y-logarithmic m-pluricanonical forms ω_{X^+} and ω_{Y^+}, such that $x \in \mathrm{Sk}(X, \Delta_X, \omega_{X^+})$ and $y \in \mathrm{Sk}(Y, \Delta_Y, \omega_{Y^+})$. Let z be the unique point in $\mathrm{Sk}(\mathscr{Z}^+)$ corresponding to the pair (x, y) via the isomorphism of Proposition 2.15.1. We prove that z lies in the Kontsevich–Soibelman skeleton associated to the form $\varpi = \mathrm{pr}_X^* \omega_{X^+} \otimes \mathrm{pr}_Y^* \omega_{Y^+}$ (see [**3**, Proposition 4.2.9]). The weights of x, y and z satisfy the following equation:

$$\mathrm{wt}_\varpi(z) = \mathrm{wt}_{\omega_{X^+}}(x) + \mathrm{wt}_{\omega_{Y^+}}(y) - m.$$

Thus, $\mathrm{Sk}(X, \Delta_X) \times \mathrm{Sk}(Y, \Delta_Y) \subseteq \mathrm{Sk}(Z, \Delta_Z)$.

On the other hand, given a point $z \in \mathrm{Sk}(\mathscr{Z}^+)$ in the essential skeleton and ω, such that $\mathrm{wt}_\omega(z)$ is minimal, we consider its projection $x \in \mathrm{Sk}(\mathscr{X}^+)$ via the isomorphism of Proposition 2.15.1. We construct a suitable embedding of $\mathscr{X}^+$ in $\mathscr{Z}^+$, such that ω induces a form on X whose weight at x is minimal [3, Proposition 5.2.2]. Repeating the same construction for the other factor, we obtain that $\mathrm{Sk}(Z, \Delta_Z)$ maps to $\mathrm{Sk}(X, \Delta_X) \times \mathrm{Sk}(Y, \Delta_Y)$, concluding the proof.

(3.12) A similar version of Theorem 3.10.1 holds for projective varieties over the germ of a punctured curve, using techniques from the birational geometry.

THEOREM 3.12.1 ([3, Proposition 5.3.5]). *Let (X, Δ_X) and (Y, Δ_Y) be dlt pairs over the germ of a punctured curve C. We denote by (Z, Δ_Z) their product, where $\Delta_Z = X \times_C \Delta_Y + \Delta_X \times_C Y$. Let $(\mathscr{X}, \Delta_{\mathscr{X}})$ and $(\mathscr{Y}, \Delta_{\mathscr{Y}})$ be semi-stable good projective dlt minimal models over the pointed curve. Then, the product (Z, Δ_Z) has a semi-stable good projective dlt minimal model $(\mathscr{Z}', \Delta_{\mathscr{Z}'})$ and $\mathcal{D}_0^{=1}(\Delta_{\mathscr{Z}'}) \simeq \mathcal{D}_0^{=1}(\Delta_{\mathscr{X}}) \times \mathcal{D}_0^{=1}(\Delta_{\mathscr{Y}})$.*

REMARK 3.12.2. We can compare Theorem 3.10.1 with Theorem 3.12.1. Indeed, the results look analogous, but it is only under the additional assumption of semi-ampleness that good minimal dlt models exist and that the essential skeleton coincides with the dual complex of a good minimal dlt model. However, it is convenient to have the two versions since semi-stability is not well behaved under birational transformations: in fact, there exist degenerations that admit a semi-stable good dlt minimal model but no semi-stable log-regular model. For instance, see [10, Example 5.3.5] for the explicit example of a degeneration of $K3$ surfaces that has no semi-stable log-regular model.

4. Applications

4.1. Hyper-Kähler varieties

(4.2) A smooth projective variety is said to be K-trivial if its canonical class is trivial. These varieties are very important in the birational classification of algebraic varieties. Beauville–Bogomolov's decomposition theorem says that any K-trivial variety can, up to a finite cover, be decomposed into a product of irreducible factors of three types: abelian varieties, Calabi–Yau varieties, and hyper-Kähler varieties [1].

(4.3) A smooth projective variety is said to be hyper-Kähler if it admits a global non-degenerate 2-form. Such a variety is necessarily K-trivial and has even dimension. While there is a rich theory of hyper-Kähler varieties, up to deformation, few examples are known. For any K3 surface S, the Hilbert scheme $\mathrm{Hilb}^n(S)$ of n points on S is hyper-Kähler. Another family arises from generalizing the Kummer construction: let A be an abelian surface, then the Hilbert scheme $\mathrm{Hilb}^{n+1}(A)$ admits a multiplication map to A and the fiber $K_n(A)$ over the identity is a hyper-Kähler variety. Together, with two examples due to O'Grady in dimensions 6 and 10 [**24**, **23**], these four constructions give the only known deformation classes of hyper-Kähler varieties.

(**4.4**) Semi-stable degenerations of hyper-Kähler varieties can be classified according to the *type*, which is the index of nilpotency of the log of the monodromy operator. The possible types are I, II, and III, and the higher types correspond to higher dimensional dual complexes.

Kollár, Laza, Saccà and Voisin investigate the dual complexes of hyper-Kähler varieties [**16**]: they find that in type II, the $\mathbb{Q}$-cohomology is always trivial, and in type III, the $\mathbb{Q}$-cohomology is that of complex projective space. Gulbrandsen, Halle, Hulek, and Zhang are able to produce a model for the Hilbert scheme of some type-II degenerations of K3 surfaces using GIT [**8**, **9**]. In this case, they find that the dual complex of a semi-stable minimal model of the resulting Hilbert scheme is a simplex.

4.5. Skeletons of Hyper-Kähler degenerations

THEOREM 4.5.1 ([**3**, Theorem 5.2.2, Proposition 6.1.9]). *Let Y be a smooth variety over K of dimension at least 2 admitting a semi-stable log-regular model or a semi-stable good dlt minimal model. Let $Y \times_K \cdots \times_K Y$ be the product of n copies of Y, let G be a finite group acting as permutations on the factors, and let X be the quotient. Then as topological spaces $\mathrm{Sk}(X) \simeq (\mathrm{Sk}(Y) \times \cdots \times \mathrm{Sk}(Y))/G$.*

REMARK 4.5.2. Theorem 4.5.1 is not true for curves because there is a ramification divisor for the action of the symmetric group on the product. Thus, pluricanonical forms on the quotient do not pull back to pluricanonical forms on the product. In fact, for any curve C of genus g, the symmetric power $\mathrm{Sym}^{g+1}(C)$ is uniruled because every linear series of degree $g + 1$ has positive dimension as a projective space. The essential skeleton of $\mathrm{Sym}^{g+1}(C)$ is empty, even if C has positive genus and $\mathrm{Sk}(C)$ is non-empty.

We can apply Theorem 4.5.1 in both the case of the Hilbert scheme of a smooth $K3$ surface and the case of the Kummer construction.

COROLLARY 4.5.3 ([**3**, Corollary 6.2.3]). *Let S be a smooth K3 surface over K admitting a semi-stable log-regular model or a semi-stable good dlt minimal model. Then, $\mathrm{Sk}(\mathrm{Hilb}^n(S)) \simeq \mathrm{Sym}^n(\mathrm{Sk}(S))$.*

For the case of Hilbert schemes of surfaces, we need only Theorem 4.5.1 and that $\mathrm{Hilb}^n(S)$ is a crepant resolution of the symmetric power. For the Kummer construction, we will also need to use some facts about the group structure on the skeleton of an abelian variety, see [**2**, **25**, **10**].

COROLLARY 4.5.4 ([**3**, Proposition 6.3.3]). *Let A be an abelian surface over K admitting a semi-stable log-regular model or a semi-stable good dlt minimal model. Let G be the symmetric group on $n + 1$ elements. Then,*

$$\mathrm{Sk}(K_n(A)) \simeq \Big(\mathrm{Sk}(A) \otimes_{\mathbb{Z}} \mathbb{Z}^{n+1} / \big\langle \textstyle\sum e_i \big\rangle\Big) \Big/ G,$$

where e_i represent the standard basis elements of the lattice $\mathbb{Z}^{n+1}$ and G acts by permuting the e_i.

4.6. Homeomorphism types of skeletons of hyper-Kähler degenerations

(**4.7**) The degeneration of hyper-Kähler varieties X constructed via the Hilbert scheme or as a Kummer variety always has same type as the original surface [**1**, Lemma 2, p. 767]. Our goal is to compute the topological space $\mathrm{Sk}(X)$ for each type of degeneration of such hyper-Kähler variety.

(**4.8**) The special fiber of a minimal model of a type-I degeneration of either K3 surfaces S or abelian surfaces A consists of a single component, so the skeleton is a single point. Likewise, in this case the skeleton of the Hilbert scheme $\mathrm{Hilb}^n(S)$ and of the Kummer variety $K_n(A)$ are both a single point.

For the other types, the topology of the skeleton is more interesting. In type II, the skeleton of a K3 surface is homeomorphic to an interval, and the skeleton of an abelian surface is homeomorphic to a circle S^1. In type III, the skeleton of a K3 surface is homeomorphic to S^2, and the skeleton of an abelian surface is a torus $S^1 \times S^1$. See an overview of these results in [**5**].

THEOREM 4.8.1 ([**3**, Proposition 6.2.4, Proposition 6.3.4]). *Let X be a type-II dimension $2n$ hyper-Kähler variety arising as either the Hilbert scheme of a K3 surface or the Kummer construction applied to an abelian surface. Then, $\mathrm{Sk}(X) \cong D^n$, the closed n-disk.*

In both cases, we show that $\mathrm{Sk}(X)$ is homeomorphic to the n-simplex by familiar constructions from combinatorics.

PROOF. For the Hilbert scheme case, we view the skeleton of the K3 surface as the interval $[0, 1]$. By Corollary 4.5.3, $\mathrm{Sk}(X) \cong \mathrm{Sym}^n[0, 1]$. A point in this space is given by an n-tuple $(x_1, x_2, \ldots, x_n)$, satisfying $0 \leq x_1 \leq x_2 \leq \cdots \leq x_n \leq 1$. After making the substitution $y_1 = x_1$, $y_i = x_i - x_{i-1}$ for $i > 1$, y_i satisfy the familiar inequalities $y_i \geq 0, \sum y_i \leq 1$.

For the Kummer variety case, consider the real torus $T = S^1 \otimes_{\mathbb{Z}} \mathbb{Z}^{n+1}/\langle \sum e_i \rangle$. Write $T = V/\Gamma$, where V is the universal cover and Γ is a lattice. The vector space V inherits the action of G as the Weyl group $W(A_n)$ acting by reflections. The quotient $\mathrm{Sk}(X)$ is identified with a fundamental domain of V for the action of the semi-direct product of

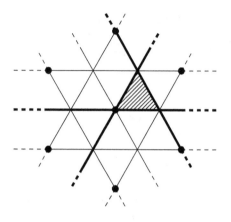

Figure 2. The fundamental domain for $\widetilde{A_3}$ acting by reflections on $\mathbb{R}^2$. The bold lines are the generating reflections, and the bold points belong to the lattice of translations.

$W(A_n)$ with Γ. But this semi-direct product is the affine Weyl group $\widetilde{A_{n+1}}$, so the quotient is an n-simplex [**20**]. Figure 2 illustrates the case $n = 2$.   □

THEOREM 4.8.2 ([**3**, Proposition 6.2.4, Proposition 6.3.4]). *Let X be a type-III dimension $2n$ hyper-Kähler variety arising as either the Hilbert scheme of a K3 surface or the Kummer construction applied to an abelian surface. Then, $\mathrm{Sk}(X) \cong \mathbb{CP}^n$, complex projective space.*

Remarkably, the proof for the type-III case requires algebraic geometry, even though the statement is purely about group actions on cell complexes.

PROOF. In each case, we give the skeleton of the surface the structure of a complex algebraic curve. In the K3 case, we are asked to compute $\mathrm{Sym}^n(\mathbb{CP}^1)$. This is the space of effective divisors of degree n on $\mathbb{CP}^1$, which is $\mathbb{CP}^n$.

In the Kummer case, we choose a point and a complex structure to give the skeleton the structure of an elliptic curve (C, q). The skeleton is the zero section of $\mathrm{Sym}^{n+1}(C)$. The addition map sends a cycle $p_1 + \cdots + p_{n+1}$ to the unique point whose divisor is linearly equivalent to $p_1 + \cdots + p_{n+1} - nq$. The fiber over q is a complete linear series on C of degree $n + 1$, so it is a projective space of dimension n (see also [**19**] for a generalization).   □

In both cases, the topological type of the essential skeleton of a type-III degeneration is that of $\mathbb{CP}^n$. This is consistent both with previous results [**16**] as well as expectations from mirror symmetry [**12, 18, 17**].

Acknowledgments

The authors would like to thank Johannes Nicaise for recommending this project and Drew Armstrong, Lorenzo Fantini, Phillip Griffiths, Lars Halvard Halle, Sean Keel, Janos Kollár, Mirko Mauri, Johannes Nicaise, Chenyang Xu, and David Zureick-Brown for the helpful conversations, comments, and suggestions in writing the paper [3]. Brown was supported by the Simons Foundation Collaboration Grant 524003. Mazzon was partially supported by the ERC Starting Grant MOTZETA (project 306610) of the European Research Council (PI: Johannes Nicaise) and by the Engineering and Physical Sciences Research Council [EP/L015234/1], The EPSRC Centre for Doctoral Training in Geometry and Number Theory (The London School of Geometry and Number Theory), University College London.

Bibliography

[1] A. Beauville, Variétés Kähleriennes dont la première classe de Chern est nulle. *J. Differential Geom.*, 18(4):755–782, 1983.

[2] V.G. Berkovich, *Spectral Theory and Analytic Geometry over Non-Archimedean Fields*, Mathematical Surveys and Monographs, Vol. 33. American Mathematical Society, Providence, RI, 1990.

[3] M. Brown and E. Mazzon, The essential Skeleton of a product of degenerations. Preprint, December 2017.

[4] T. de Fernex, J. Kollár and C. Xu, The dual complex of singularities. In *Higher Dimensional Algebraic Geometry, in Honour of Professor Yujiro Kawamatas 60th birthday*, Vol. 74. Adv. Stud. Pure Math., pp. 103–130. December 2017.

[5] R. Friedman and D. R. Morrison, The birational geometry of degenerations: an overview. In *The birational Geometry of Degenerations* (Cambridge, Mass., 1981), Progress in Mathematics, Vol. 20, Birkhäuser, Boston, MA, 1983, pp. 1–32.

[6] O. Gabber and L. Ramero, Foundations for almost ring theory — release 6.95. Preprint, September 2004.

[7] W. Gubler, J. Rabinoff and A. Werner, Skeletons and tropicalizations. *Adv. Math.*, 294(Supplement C):150–215, 2016.

[8] M.G. Gulbrandsen, L. H. Halle and K. Hulek, A git construction of degenerations of Hilbert schemes of points. Preprint, April 2016.

[9] M.G. Gulbrandsen, L. H. Halle, K. Hulek and Z. Zhang, The geometry of degenerations of Hilbert schemes of points. Preprint, February 2018.

[10] L.H. Halle and J Nicaise, Motivic zeta functions of degenerating Calabi-Yau varieties. *Math. Ann.*, 370:1277–1320, 2017.

[11] J. Harris and I. Morrison, *Moduli of Curves*, Graduate Texts in Mathematics, Vol. 187, Springer-Verlag, New York, 1998.

[12] J.-M. Hwang, Base manifolds for fibrations of projective irreducible symplectic manifolds. *Invent. Math.*, 174(3):625–644, 2008.

[13] K. Kato, Logarithmic structures of Fontaine-Illusie. In *Algebraic Analysis, Geometry, and Number Theory (Baltimore, MD, 1988)*, Johns Hopkins University Press, Baltimore, MD, 1989, pp. 191–224.

[14] K. Kato, Toric singularities. *Amer. J. Math.*, 116(5):1073–1099, 1994.

[15] G. Kempf, F. F. Knudsen, D. Mumford and B. Saint-Donat, *Toroidal Embeddings. I*. Lecture Notes in Mathematics, Vol. 339. Springer-Verlag, Berlin, 1973.

[16] J. Kollár, R. Laza, G. Saccà and C. Voisin, Remarks on degenerations of hyper-Kähler manifolds. Preprint, April 2017.

[17] M. Kontsevich and Y. Soibelman. Homological mirror symmetry and torus fibrations. In *Symplectic Geometry and Mirror Symmetry* (Seoul, 2000), World Sci. Publ., River Edge, NJ, 2001, pp. 203–263.

[18] M. Kontsevich and Y. Soibelman, *Affine Structures and Non-Archimedean Analytic Spaces*. Birkhäuser Boston, Boston, MA, 2006, pp. 321–385.

[19] E. Looijenga, Root systems and elliptic curves. *Invent. Math.*, 38(1):17–32, 1976/77.

[20] H. R. Morton, Symmetric products of the circle. *Proc. Cambridge Philos. Soc.*, 63:349–352, 1967.

[21] M. Mustaţă and J. Nicaise, Weight functions on non-Archimedean analytic spaces and the Kontsevich–Soibelman skeleton. *Algebr. Geom.*, 2(3):365–404, 2015.

[22] J. Nicaise and C. Xu, The essential skeleton of a degeneration of algebraic varieties. *Amer. J. Math.*, 138(6):1645–1667, 2016.

[23] K. G. O'Grady, Desingularized moduli spaces of sheaves on a *K*3. *J. Reine Angew. Math.*, 512:49–117, 1999.

[24] K. G. O'Grady, A new six-dimensional irreducible symplectic variety. *J. Algebraic Geom.*, 12(3):435–505, 2003.

[25] M. Temkin, Metrization of differential pluriforms on Berkovich analytic spaces. In M. Baker and S. Payne (eds.), *Nonarchimedean and Tropical Geometry*, Simons Symposia, 2016, pp. 195–285.

[26] A. Thuillier, Géométrie toroïdale et géométrie analytique non archimédienne. Application au type d'homotopie de certains schémas formels. *Manuscripta Math.*, 123(4):381–451, 2007.

16

Arc Scheme and Bernstein Operators

Michel Gros[*,‡], Luis Narváez Macarro[†,§]
and Julien Sebag[*,¶]

*Institut de recherche mathématique de Rennes
UMR 6625 du CNRS, Campus de Beaulieu
Université de Rennes, CNRS, F-35000 Rennes, France
†Departamento de Álgebra & Instituto de
Matemáticas (IMUS), Facultad de Matemáticas
Universidad de Sevilla, E-41012 Sevilla, Spain
‡michel.gros@univ-rennes1.fr
§narvaez@us.es
¶julien.sebag@univ-rennes1.fr

Let k be a field of characteristic zero. Let $f \in k[x, y]$ be an irreducible polynomial. In this chapter, we link the differential operators of $k[x, y]$ that appear in the Bernstein functional equation for f to the nilpotent elements of the k-algebra of the arc scheme associated with the affine plane curve defined by the datum of f.

1. Introduction

1.1. Let k be a field of characteristic zero. Let $\mathscr{C}$ be an integral affine plane k-curve. In this chapter, we explain an original relation between the nilradical of the k-algebra $\mathcal{O}(\mathscr{L}_\infty(\mathscr{C}))$ of the arc scheme $\mathscr{L}_\infty(\mathscr{C})$ associated with $\mathscr{C}$ and differential operators on $\mathbf{A} = k[x, y]$ and precisely those which appear in Bernstein's functional equation attached to the datum of f. We prove the following theorem (see Theorem 5.3 for a more complete and precise statement).

THEOREM 1.1. *Let k be a field of characteristic zero. Let $f \in \mathbf{A} = k[x, y]$ be a reduced polynomial, $\langle f \rangle$ the ideal generated by f and $\mathscr{C} = \mathrm{Spec}(\mathbf{A}/\langle f \rangle)$. If $P(s)$ is a differential operator of $\mathbf{A}$ of order $\alpha \geq 2$, such that $P(s)f^{s+1} = b(s)f^s$*

(*with* $b \in \mathbf{Q}[s]$), *then the coefficients of its symbol* $\sigma_\alpha(P(s))$, *considered as a polynomial in* s, *define nilpotent elements in* $\mathcal{O}(\mathscr{L}_\infty(\mathscr{C}))$.

Let us denote by $\mathscr{D}$ the set of differential operators on $\mathbf{A}$. This statement is a consequence of a more general result which describes a relation between a part of the graded ring $\mathrm{gr}(\mathscr{D})$ of the differential operators on $\mathbf{A}$ and nilpotent functions on the arc scheme. Indeed, we have (see Theorem 4.2 and Corollary 4.8) the following theorem.

THEOREM 1.2. *Let* k *be a field of characteristic 0. Let* $f \in \mathbf{A}$ *be a reduced polynomial with* $\mathscr{C} = \mathrm{Spec}(\mathbf{A}/\langle f \rangle)$. *Let* $P \in \mathbf{A}[x_1, y_1]$ *be a homogeneous polynomial with* $\deg_1(P) = d \geq 0$ *(where we denote by* $\deg_1(P)$ *the total degree with respect to* x_1, y_1*). The following assertions are equivalent:*

(1) *The element* P *induces a nilpotent element in the ring* $\mathcal{O}(\mathscr{L}_\infty(\mathscr{C}))$.

(2) *For every differential operator* $D \in \mathscr{D}^d$ *whose principal symbol* $\sigma(x, y, \xi, \eta)$ *in the polynomial ring* $\mathrm{gr}(\mathscr{D})$ *verifies* $\sigma(x, y, -y_1, x_1) = P$, *we have* $D(f^d) \in \langle f \rangle$.

1.2. The key technical ingredient of our differential interpretation of this class of nilpotent functions on $\mathcal{O}(\mathscr{L}_\infty(\mathscr{C}))$ is based on various results of differential algebra. The parts of the computations in Section 6 concerning arc scheme have been realized under SAGE following the algorithms introduced in [8] (see also [2]), and those concerning Bernstein's symbol f^s have been realized with [10]. We also refer to [8, 9, 12, 15, 17, 18] for related works. In the end, let us stress that this work is a first step in the direction of understanding the differential nature of arc scheme and that it seems very plausible to us, with respect to various computations, that it can be generalized in higher dimensions.

2. Recollection on Arc Scheme

2.1. To every k-variety V and every integer $m \in \mathbf{N}$, one attaches its *jet scheme* $\mathscr{L}_m(V)$ of level m by the bi-functorial property (in S, V):

$$(2.1) \qquad \mathrm{Hom}_{\mathbf{Sch}_k}(S, \mathscr{L}_m(V)) \cong \mathrm{Hom}_{\mathbf{Sch}_k}(S \otimes_k k[[T]]/\langle T^{m+1} \rangle, V)$$

for every k-scheme S. The *arc scheme* $\mathscr{L}_\infty(V)$ of V is then defined by the functorial formula

$$\mathrm{Hom}_{\mathbf{Sch}_k}(S, \mathscr{L}_\infty(V)) \cong \varinjlim_m \mathrm{Hom}_{\mathbf{Sch}_k}(S \otimes_k k[[T]]/\langle T^{m+1} \rangle, V).$$

For example, see [3, 4, 6, 14, 16] for details on arc schemes.

2.2. Because of formula (2.1) and the universal property of symmetric algebras, we easily conclude that $\mathscr{L}_1(V) \cong \mathrm{Spec}(\mathrm{Sym}(\Omega^1_{V/k}))$, i.e., it is the *tangent space* $\pi_V \colon T_{V/k} \to V$ of V. We set $\pi_1^\infty \colon \mathscr{L}_\infty(V) \to \mathscr{L}_1(V)$ for the canonical morphism and $\mathrm{Reg}(V)$ for the regular locus of V. We denote by $\mathscr{G}(V)$ the closed subscheme $\overline{\pi_1^\infty(\mathscr{L}_\infty(V))}$ of $\mathscr{L}_\infty(V)$ endowed with its reduced structure. It is not hard to check that

$$(2.2) \qquad \mathscr{G}(V) := \overline{\pi_1^\infty(\mathscr{L}_\infty(V))} = \overline{\pi_V^{-1}(\mathrm{Reg}(V))}.$$

Indeed, both closed subsets contain, as a dense open subset, the subset $\pi_V^{-1}(\mathrm{Reg}(V))$. If the k-variety V is assumed to be integral, then $\mathscr{G}(V)$ (endowed with its reduced structure) is an irreducible component of $T_{V/k}$.

3. Recollection of Differential Algebra

For simplicity, we will restrict ourselves in this presentation to the very special case of the polynomial ring $\mathbf{A} := \mathbf{A}_0 := k[x, y]$. The following definitions and constructions can be *verbatim* extended to higher dimensions. We set $\mathbf{A}_1 = k[x, y, x_1, y_1] = \mathbf{A}[x_1, y_1]$, $\mathbf{A}_n = \mathbf{A}[x_i, y_i, i \in \{1, \ldots, n\}]$ for every integer $n \geq 1$ and $\mathbf{A}_\infty := k[x_i, y_i; i \in \mathbf{N}]$. We denote by $\deg_1$ the total degree with respect to the variables x_1, y_1 in the polynomial ring A_1.

3.1. We endow the k-algebra $\mathbf{A}_\infty$ with the k-derivation Δ defined by $\Delta(\dagger_i) = \dagger_{i+1}$, for every integer $i \in \mathbf{N}$ and every symbol $\dagger \in \{x, y\}$. The injective morphism of k-algebras $\mathbf{A} \to \mathbf{A}_\infty$, defined by $\dagger \mapsto \dagger_0$, identifies the polynomial ring $\mathbf{A}$ in $\mathbf{A}_\infty$ with $k[x_0, y_0]$ and gives rise to a structure of $\mathbf{A}$-algebra on $\mathbf{A}_\infty$. We will freely identify the rings $\mathbf{A}, \mathbf{A}_1$ with their images in the ring $\mathbf{A}_\infty$, and the variables x, y respectively with x_0, y_0. For every polynomial $f \in \mathbf{A}$, we denote by $[f]$ the differential ideal generated by f in the differential ring $\mathbf{A}_\infty$ and by $\{f\}$ the radical of the ideal $[f] = \langle \Delta^i(f); i \in \mathbf{N} \rangle$. Let us note that

$$(3.1) \qquad \Delta(f) = \partial_x(f)x_1 + \partial_y(f)y_1.$$

We consider the ideal $\{f\} \cap \mathbf{A}_1$ of the ring $\mathbf{A}_1$. Obviously, this ideal only depends on the ideal $\langle f \rangle$ in $\mathbf{A}$; finally, if $\mathscr{C}$ is the affine plane k-curve defined by the datum of f, we set

$$(3.2) \qquad \mathcal{N}_1(\mathscr{C}) = \{f\} \cap \mathbf{A}_1.$$

The Kolchin irreducibility theorem (see [**7**, Chapter IV/S17/Proposition 10]) implies, if the polynomial f is assumed to be irreducible, that the ideal $\{f\}$ is prime (hence, the ideal $\mathcal{N}_1(\mathscr{C})$ is also prime in this case). The main purpose of this chapter is to study this ideal $\mathcal{N}_1(\mathscr{C})$.

3.2. Following formula (3.2), we also set $\mathcal{N}_n(\mathscr{C}) = \{f\} \cap \mathbf{A}_n$ for every integer $n \geq 1$.

3.3. Section 3.1 is the algebraic counterpart of Section 2. Precisely, if $\mathcal{O}(\mathscr{C}) = \mathbf{A}/\langle f \rangle$, the universal property which defines arc scheme implies that $\mathscr{L}_\infty(\mathscr{C}) \cong \mathrm{Spec}(\mathbf{A}_\infty/[f])$ and $\mathscr{L}_m(V) \cong \mathrm{Spec}(\mathbf{A}_m/\langle \Delta^i(f); i \in \{0, \ldots, m\} \rangle)$; hence, we have $\mathscr{G}(\mathscr{C}) = \mathrm{Spec}(\mathbf{A}_1/\mathcal{N}_1(\mathscr{C}))$.

3.4. Let $f \in \mathbf{A}$ be a reduced polynomial (not assumed to be irreducible). Let $f_1, \ldots, f_n$ be irreducible polynomials, such that $f = f_1 \cdots f_n$. It also follows from [**7**, Chapter IV/S17/Proposition 10] that $\{f\} = \{f_1\} \cap \cdots \cap \{f_n\}$, and thus

$$(3.3) \qquad \mathcal{N}_1(\mathscr{C}) = \mathcal{N}_1(\mathscr{C}_1) \cap \cdots \cap \mathcal{N}_1(\mathscr{C}_n),$$

where, for every integer $i \in \{1, \ldots, n\}$, we have $\mathscr{C}_i := \operatorname{Spec}(\mathbf{A}/\langle f_i \rangle)$.

3.5. Let $f \in \mathbf{A}$ be an irreducible polynomial. Let $\partial(f)$ be a non-zero partial derivative of f. For simplicity, we assume that $\partial(f) = \partial_x(f)$. We have the formula

$$(3.4) \qquad \{f\} = ([f] : \partial(f)^\infty) := \{P \in \mathbf{A}_\infty \mid \exists N \in \mathbf{N} \; \partial(f)^N P \in [f]\}.$$

Indeed, the factorization lemma implies the existence of a morphism of differential k-algebras

$$\mathbf{A}_\infty/([f] : \partial(f)^\infty) \to (\mathbf{A}_\infty/[f])_{\partial(f)},$$

which is injective ($(\mathbf{A}_\infty/[f])_{\partial(f)}$ denoting the localization of $\mathbf{A}_\infty/[f]$ at $\partial(f)$ as usual). Since $(\mathbf{A}_\infty/[f])_{\partial(f)} \cong (\mathbf{A}/\langle f \rangle)_{\partial(f)}[y_i; i \geq 1]$ and since the ideal $\langle f \rangle$ is prime, we conclude that the differential ideal $([f] : \partial(f)^\infty)$ is prime; hence radical. Thus, we have $\{f\} \subset ([f] : \partial(f)^\infty)$. Now, by [**7**, Chapter IV/S17/Proposition 10], we know that the ideal $\{f\}$ is prime. Let $P \in ([f] : \partial(f)^\infty)$. By the very definition, there exists an integer Nm such that $(\partial(f)P)^N \in [f]$; hence, $\partial(f)P \in \{f\}$. Since $\partial(f)^\infty \notin \langle f \rangle$, we conclude by a degree argument that $P \in \{f\}$ which concludes the proof of formula (3.4).

REMARK 3.2. This kind of results are classical statements of differential algebra and can be generalized to arbitrary ideals. They geometrically correspond to density property in arc scheme for open subsets avoiding singularities and are very useful in particular for our current purpose. Stronger versions which works for algebraic *and* differential equations can be directly deduced from works of Lazard, Rosenfeld, etc. (e.g., see [**7**, Chapter IV/§9/Lemma 2]).

We denote by $\langle f, \Delta(f) \rangle$ the ideal of the ring $\mathbf{A}_1$ generated by f and $\Delta(f)$ and define

$$(\langle f, \Delta(f) \rangle : \partial(f)^\infty) = \{P \in \mathbf{A}_1 \mid \exists N \in \mathbf{N} \; \partial(f)^N P \in \langle f, \Delta(f) \rangle\}.$$

Then, we have the following formula:

$$(3.6) \qquad (\langle f, \Delta(f) \rangle : \partial(f)^\infty) = ([f] : \partial(f)^\infty) \cap \mathbf{A}_1 = \mathcal{N}_1(\mathscr{C}).$$

Thanks to formula (3.4), we only have to justify that the ideal $(\langle f, \Delta(f) \rangle : \partial(f)^\infty)$ contains $([f] : \partial(f)^\infty) \cap \mathbf{A}_1$. Let $P \in ([f] : \partial(f)^\infty) \cap \mathbf{A}_1$. The image of P in $\mathbf{A}_\infty/([f] : \partial(f)^\infty)$ is zero. Since we have the following commutative diagram of morphisms of k-algebras:

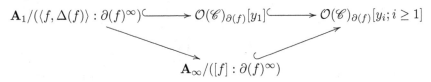

we conclude that the image of P is zero in the ring $\mathcal{O}(\mathscr{C})_{\partial(f)}[y_i; i \geq 1] \cong (\mathbf{A}_\infty/[f])_{\partial(f)}$, which means that $P \in (\langle f, \Delta(f) \rangle : \partial(f)^\infty)$. As a direct consequence of formula (3.6), we observe that $\mathcal{N}_1(\mathscr{C})$, seen as an ideal of the ring $\mathbf{A}[x_1, y_1]$ of polynomials in two variables x_1, y_1 over $\mathbf{A}$ is homogeneous (for $\deg_1$).

3.6. More generally, following the former ideas, we prove that

$$(\langle f, \Delta(f), \dots, \Delta^n(f) \rangle : \partial(f)^\infty) = ([f] : \partial(f)^\infty) \cap \mathbf{A}_n = \mathcal{N}_n(\mathscr{C})$$

for every integer $n \geq 1$ and every non-zero partial derivative $\partial(f)$ of f.

4. Nilpotent Functions on Arc Scheme and Differential Operators

4.1. Recall that a *differential operator* D of $\mathbf{A}$ (over k) of order $\leq d$ is a k-linear map $D : \mathbf{A} \to \mathbf{A}$, such that the bracket $[D, a]$ is a differential operator of order $\leq d - 1$ for every $a \in \mathbf{A}$, the differential operators of order 0 being the (multiplication by) elements of $\mathbf{A}$. Let $\mathscr{D} = \cup_{d \geq 0} \mathscr{D}^d$ be the filtered (by the order) ring of differential operators of $\mathbf{A}$ (see [5]). In coordinates, such a datum can be represented as a formal combination with coefficients in $\mathbf{A}$ of the form $D = \sum_{i+j \leq d} a_{i,j}(x, y) \partial_x^i \partial_y^j$ whenever $D \in \mathscr{D}^d$. The graded ring $\mathrm{gr}(\mathscr{D})$ is a commutative polynomial ring which can be identified with the graded ring $\mathbf{A}[\xi, \eta] = k[x, y][\xi, \eta]$, with $\deg(x) = \deg(y) = 0$ and $\deg(\xi) = \deg(\eta) = 1$. The variable ξ corresponds to the symbol $\sigma_1(\partial_x)$ and the variable η corresponds to the symbol $\sigma_1(\partial_y)$. The above identification is the coordinate-dependent version of the intrinsic isomorphism between the symmetric algebra of the $\mathbf{A}$-module of k-derivations of $\mathbf{A}$ and $\mathrm{gr}(\mathscr{D})$. If a differential operator D has order d, i.e., if $D \in \mathscr{D}^d \setminus \mathscr{D}^{d-1}$, then its d-symbol $\sigma_d(D)$ is called its *principal symbol*, which will be denoted by $\sigma(P)$.

4.2. One of the cornerstones of $\mathscr{D}$-module theory is the *Bernstein construction*. It consists in considering the polynomial ring $\mathscr{D}[s]$, where s is a commuting variable, and the symbol f^s, on which differential operators act formally (see [1]). In that way, the $\mathbf{A}$-module $\mathbf{A}[s, f^{-1}]f^s$ becomes a left $\mathscr{D}[s]$-module. The order of a differential operator $P(s) = \sum_i P_i s^i \in \mathscr{D}[s]$ is defined as $\mathrm{ord}(P(s)) = \max\{\mathrm{ord}(P_i)\}$ and the corresponding graded ring $\mathrm{gr}(\mathscr{D}[s])$ can be identified with the (commutative) graded ring $\mathbf{A}[s][\xi, \eta] = k[x, y, s][\xi, \eta]$, with $\deg(x) = \deg(y) = \deg(s) = 0$ and $\deg(\xi) = \deg(\eta) = 1$. If $\mathrm{ord}(P(s)) = d$, then its d-symbol $\sigma_d(P(s)) = \sum_{i \in J} \sigma_d(P_i)s^i$, with

$J = \{i \mid \mathrm{ord}(P_i) = d\}$, is a homogeneous polynomial of degree d in the variables ξ, η with coefficients in $\mathbf{A}[s]$, which will be called its *principal symbol* and denoted by $\sigma(P(s))$. Let us note that, for each $c \in k$, we have

$$\sigma_d(P(s))_{s=c} = \sum_{i \in J} \sigma_d(P_i)c^i = \sigma_d\left(\sum_{i \in J} c^i P_i\right) = \sigma_d(P(c)).$$

If $D \in \mathscr{D}$ has order $\leq d$, one has the following well-known identity:

$$(4.1) \qquad D(f^s) = d!\sigma_d(D)(\partial_x(f), \partial_y(f))\binom{s}{d}f^{s-d} + \sum_{m=0}^{d-1} T_m,$$

where, for every integer $m \in \{0, \dots, d-1\}$, the term T_m is $\mathbf{A}$-divisible by $\binom{s}{m}f^{s-m}$.

4.3. Each affine plane k-curve $\mathscr{C}$ gives rise to the *V-filtration* $V_\star^{\mathscr{C}}$ of $\mathscr{D}$ along $\mathscr{C}$ defined as follows: for every integer $r \in \mathbf{Z}$, one sets

$$(4.2) \qquad V_r^{\mathscr{C}} = \{D \in \mathscr{D} \mid \forall \ell \in \mathbf{Z}\ D(\langle f\rangle^\ell) \subset \langle f\rangle^{\ell-r}\},$$

where $f \in \mathbf{A}$ is a reduced equation of $\mathscr{C}$. In this formula, one adopts the convention that $\langle f\rangle^t = \mathbf{A}$ for every negative integer $t \in \mathbf{Z}$. It is clear from the inductive definition of differential operators that $\mathscr{D}^d \subset V_d^{\mathscr{C}}$ for each $d \geq 0$.

LEMMA 4.2. *Let k be a field of characteristic zero. Let $f \in \mathbf{A}$ be a reduced polynomial (not assumed to be irreducible). Let D be a differential operator on $\mathbf{A}$ of order $d \geq 0$. The following assertions are equivalent:*

  (1) *The operator D verifies $D(f^d) \in \langle f\rangle$.*
  (2) *There is an integer $e \geq d$ such that $D(f^e) \in \langle f^{e-d+1}\rangle$.*
  (3) *The operator D belongs to $V_{d-1}^{\mathscr{C}}$.*
  (4) *The principal symbol $\sigma_d(D)$ of the operator D satisfies*

$$\sigma_d(D)(\partial_x(f), \partial_y(f)) \in \langle f\rangle.$$

PROOF. The implications $(3) \Rightarrow (1) \Rightarrow (2)$ are obvious. $(1) \Rightarrow (3)$ For each $a \in \mathbf{A}$ and each integer $\ell \geq d$, we have $D(af^\ell) = [D, af^{\ell-d}](f^d) + af^{\ell-d}D(f^d) \in \langle f\rangle$, since $[D, af^{\ell-d}]$ has order $\leq d-1$. $(2) \Leftrightarrow (4)$ It follows from formula (4.1) with $s = e$ that $D(f^e) - d!\binom{e}{d}\sigma_d(D)(\partial_x(f), \partial_y(f))f^{e-d} \in \langle f^{e-d+1}\rangle$. The same observation with $s = d$ provides a proof of $(4) \Leftrightarrow (1)$. $\qquad\square$

4.4. We consider the set $\mathfrak{G}$ formed by the Hasse–Schmidt derivations (or higher derivations [11, §27]) H of $\mathbf{A}_1$ corresponding to the morphisms $H: \mathbf{A}_1 \to \mathbf{A}_1[[T]]$, such that $H_0 = \mathrm{Id}$ and $H_i(P)$ is zero or a homogeneous polynomial with $0 \leq \deg_1(H_i(P)) \leq d - i$ for every integer $i \geq 1$ and every homogeneous polynomial $P \in \mathbf{A}_1$ with $\deg_1(P) = d$. Let $f \in \mathbf{A}$ be a polynomial. We say that $H \in \mathfrak{G}$ is *f-good* if $H(\Delta(f)) \in \langle f, \Delta(f)\rangle \mathbf{A}_1[[T]]$. We denote by $\mathfrak{G}^f$ the set of the f-good Hasse–Schmidt derivations in $\mathfrak{G}$.

EXAMPLE 4.1. Let $D = a\partial_x + b\partial_y \in \mathrm{Der}_k(\mathbf{A})$. Then, the exponential E of the k-derivation $a\partial_{x_1} + b\partial_{y_1}$ of $\mathbf{A}_1$ belongs to $\mathfrak{G}^f$. Indeed, we observe that $(a\partial_{x_1} + b\partial_{y_1})(\Delta(f)) = a\partial_x(f) + b\partial_y(f)$.

For every k-derivation $D = a\partial_x + b\partial_y \in \mathrm{Der}_k(\mathbf{A})$, we set $D_1 = a\partial_{x_1} + b\partial_{y_1} \in \mathrm{Der}_k(\mathbf{A}_1)$.

4.5. Let us now state the main result of this section. We denote by $\sharp\colon \mathrm{gr}(\mathscr{D}) \to \mathbf{A}_1$ the morphism of $\mathbf{A}$-algebras defined by

(4.5) $$P(x, y, \xi, \eta) \mapsto P^\sharp := P(x, y, -y_1, x_1).$$

THEOREM 4.2. *Let k be a field of characteristic zero. Let $f \in \mathbf{A}_0$ be an irreducible polynomial with $\mathscr{C} = \mathrm{Spec}(\mathbf{A}/\langle f \rangle)$. Let $P \in \mathbf{A}_1$ be a homogeneous polynomial with $\deg_1(P) = d \geq 0$. The following assertions are equivalent:*

1. *The polynomial P belongs to $\mathcal{N}_1(\mathscr{C})$.*
2. *For every f-good Hasse–Schmidt derivation $H \in \mathfrak{G}^f$, we have $H_d(P) \in \langle f \rangle$.*
3. *For every f-logarithmic k-derivation $D \in \mathrm{Der}_k(\mathbf{A})$, we have $D_1^d(P) \in \langle f \rangle$.*
4. *If $\delta = \partial_x(f)\partial_y - \partial_y(f)\partial_x \in \mathrm{Der}_k(\mathbf{A})$, we have $\delta_1^d(P) \in \langle f \rangle$.*
5. *For every differential operator $D \in \mathscr{D}^d$, such that $\sigma_d(D)^\sharp = P$, we have $D(f^d) \in \langle f \rangle$.*

(5') *There exists a differential operator $D \in \mathscr{D}^d$, such that $\sigma_d(D)^\sharp = P$ with $D(f^d) \in \langle f \rangle$.*

6. *For every $Q \in \mathrm{gr}(\mathscr{D})$ with $Q^\sharp = P$, we have $Q(\partial_x(f), \partial_y(f)) \in \langle f \rangle$.*

PROOF. We may assume that $\partial(f) := \partial_x(f)$ is non-zero up to switching the variables. $(1) \Rightarrow (2)$ By the very definition, there exist an integer N and homogeneous polynomials $Q_1, Q_2 \in \mathbf{A}_1$, respectively, with $\deg_1(Q_1) = d$ and $\deg_1(Q_2) = d - 1$, such that

(4.7) $$\partial(f)^N P = fQ_1 + \Delta(f)Q_2.$$

Since the Hasse–Schmidt derivation H is f-good, we observe that $H(\Delta(f)) = \Delta(f) + H_1(\Delta(f))T$ with $H_1(\Delta(f)) = fh_1$ and $h_1 \in \mathbf{A}$. By applying the morphism H to equation (4.7), we deduce that

(4.8) $$\partial(f)^N H(P) = (H(Q_1) + Th_1H(Q_2))f + \Delta(f)H(Q_2).$$

Then, by identifying the T^d-coefficients in formula (4.7), we deduce that $H_d(P) \in \langle f, \Delta(f) \rangle$. Since $H_d(P) \in \mathbf{A}$, we conclude that f divides $H_d(P)$ thanks to the homogeneity of $\Delta(f)$. $(2) \Rightarrow (3)$ If D is f-logarithmic, the Hasse–Schmidt derivation of $\mathbf{A}_1$ defined by

$$\sum_{i \geq 0} T^i \frac{D_1^i}{i!}$$

belongs to $\mathfrak{G}^f$. Then, we apply (2). (3) $\Rightarrow$ (4) The proof follows from the fact that δ is f-logarithmic. (4) $\Rightarrow$ (5) A direct computation gives the following formula:

$$(4.9) \qquad \delta_1^d(P) = d!\,P(-\partial_y(f), \partial_x(f)).$$

The conclusion easily follows from (4) and Lemma 4.2. (5) $\Rightarrow$ (4) It is also a direct consequence of Lemma 4.2 by formula (4.9). (4) $\Rightarrow$ (1) We have

$$(4.10) \qquad (\partial(f))^d P \equiv \eta^d P(-\partial_y(f), \partial_x(f)) \pmod{\langle \Delta(f) \rangle}.$$

By the very definition of the ideal $\mathcal{N}_1(\mathscr{C})$ and formula (4.10), we observe that $P \in \mathcal{N}_1(\mathscr{C})$ if and only if $\eta^d P(-\partial_y(f), \partial_x(f)) \in \mathcal{N}_1(\mathscr{C})$. Then, by assumption and formula (4.9), we conclude the proof. By Lemma 4.2, we observe that $(5') \Leftrightarrow (5)$ holds true since a differential operator D, such that $\sigma_d(D)^\sharp = P$ always exists. The assertion (6) is a direct translation of assertion (4) by Lemma 4.2. $\qquad \square$

COROLLARY 4.7. *Let k be a field of characteristic zero, $f \in \mathbf{A}$ be an irreducible polynomial and $\mathscr{C} = \mathrm{Spec}(\mathbf{A}/\langle f \rangle)$. Let $P = ax_1 + by_1 \in \mathbf{A}_1$ be a homogeneous polynomial of degree 1. Then, the following assertions are equivalent:*

(1) *There exists a polynomial $\alpha \in \mathbf{A} \setminus \langle f \rangle$, such that the Kähler differential form $\omega = a\,dx + b\,dy \in \Omega^1_{\mathbf{A}/k}$ satisfies $\alpha\omega \in f\Omega^1_{\mathbf{A}/k} + \mathbf{A}df$.*

(2) *The k-derivation $D = b\partial_x - a\partial_y \in \mathrm{Der}_k(\mathbf{A})$ satisfies $D(f) \in \langle f \rangle$.*

(3) *The polynomial P belongs to $\mathcal{N}_1(\mathscr{C})$.*

This equivalence provides an isomorphism of $\mathbf{A}$-modules

$$\mathrm{Tors}(\Omega^1_{\mathcal{O}(\mathscr{C})/k}) \cong \mathcal{N}_1^1(\mathscr{C})/\langle fx_1, fy_1, \Delta(f) \rangle,$$

where we denote by $\mathrm{Tors}(\Omega^1_{\mathcal{O}(\mathscr{C})/k})$ the torsion submodule of the module $\Omega^1_{\mathcal{O}(\mathscr{C})/k}$ of the Kähler differential forms of the ring $\mathcal{O}(\mathscr{C})$ and by $\mathcal{N}_1^1(\mathscr{C})$ the homogeneous part of degree 1 of the homogeneous ideal $\mathcal{N}_1(\mathscr{C})$.

See [2] for related topics.

PROOF. Equivalence (1) $\Leftrightarrow$ (2) can be proved by a direct argument of linear algebra. Let us construct the isomorphism. We consider the $\mathbf{A}$-linear map $\mathcal{N}_1(\mathscr{C}) \to \Omega^1_{\mathbf{A}/k}$, which sends $\omega_1\xi + \omega_2\eta$ to $\omega_1 dx + \omega_2 dy$, and compose it by the surjective $\mathbf{A}$-linear map $\Omega^1_{\mathbf{A}/k} \to \Omega^1_{\mathcal{O}(\mathscr{C})/k}$. The obtained $\mathbf{A}$-linear map θ takes its values in $\mathrm{Tors}(\Omega^1_{\mathcal{O}(\mathscr{C})/k})$ by (3) $\Rightarrow$ (1). Its kernel coincides with $\langle f\xi, f\eta, \Delta(f) \rangle$ by the very definition of θ. The surjectivity directly follows from (1) $\Rightarrow$ (3). $\qquad \square$

By formula (3.3), we also deduce the following statement.

COROLLARY 4.8. *Let k be a field of characteristic zero, $f \in \mathbf{A}$ be a reduced polynomial (not assumed to be irreducible) and $\mathscr{C} = \mathrm{Spec}(\mathbf{A}/\langle f \rangle)$. Let $f_1, \ldots, f_n$ be the irreducible factors of f, with $\alpha_i = \deg(f_i)$. For every* *integer* $i \in \{1, \ldots, n\}$, *we denote by* $\mathscr{C}_i$ *the affine plane k-curve attached to the datum of* f_i. *Let* $P \in \mathbf{A}_1$ *be a homogeneous polynomial with* $\deg_1(P) = d \geq 0$. *Then, the following statements are equivalent:*

(1) *The polynomial* P *belongs to* $\mathcal{N}_1(\mathscr{C})$.

(2) *For every integer* $i \in \{1, \ldots, n\}$, *for every Hasse–Schmidt derivation* $H \in \mathfrak{G}^f$, *we have* $H_d(P) \in \langle f_i \rangle$.

(3) *For every Hasse–Schmidt derivation* $H \in \mathfrak{G}^f$, *we have* $H_d(P) \in \langle f \rangle$.

(4) *For every integer* $i \in \{1, \ldots, n\}$, *for every differential operator* $D \in \mathscr{D}^d$, *such that* $\sigma_d(D)^\sharp = P$, *we have* $D(f_i^d) \in \langle f_i \rangle$.

(5) *For every differential operator* $D \in \mathscr{D}^d$, *such that* $\sigma_d(D)^\sharp = P$, *we have* $D(f^d) \in \langle f \rangle$.

(6) *There are a differential operator* $D \in \mathscr{D}^d$, *and an integer* $e \geq d$, *such that* $\sigma_d(D)^\sharp = P$ *and* $D(f^e) \in \langle f^{e-d+1} \rangle$.

PROOF. Equivalence (1) $\Leftrightarrow$ (2) $\Leftrightarrow$ (4) follows from formula (3.3). Let us prove (2) $\Leftrightarrow$ (3). Since f_i are irreducible and mutually distinct, we know that f divides $H_d(P)$ if and only if each f_i divides $H_d(P)$; it proves the equivalence. Let us prove (4) $\Rightarrow$ (5). By the Taylor expansion, we have the following relation for every integer $\ell \in \{1, \ldots, n\}$:

$$
\begin{aligned}
(4.13) \quad P(-\partial_y(f), \partial_x(f)) &= P\left(-\sum_{i=1}^{n} \partial_y(f_i) \prod_{j \neq \ell} f_j, \sum_{i=1}^{n} \partial_x(f_i) \prod_{j \neq \ell} f_j\right) \\
&= P\left(-\partial_y(f_\ell) \prod_{j \neq \ell} f_j, \partial_x(f_\ell) \prod_{j \neq \ell} f_j\right) + G(x, y) \\
&= \left(\prod_{j \neq \ell} f_j\right) P(-\partial_y(f_\ell), \partial_x(f_\ell)) + G(x, y),
\end{aligned}
$$

where the polynomial G is divisible by f_ℓ. Since f_i are irreducible and mutually distinct, we deduce that f divides $P(-\partial_y(f), \partial_x(f))$. We conclude the proof from Lemma 4.2. (5) $\Rightarrow$ (4) Let $\ell \in \{1, \ldots, n\}$. By formula (4.13) and assertion (5), we deduce that f_ℓ divides $P(-\partial_y(f_\ell), \partial_x(f_\ell))$ since f_i are irreducible and mutually distinct. Once again, we conclude the proof by Lemma 4.2. Equivalence (5) $\Leftrightarrow$ (6) is a direct consequence of Lemma 4.2. $\qquad\square$

4.6. Let $n \geq 2$. Let us consider the *weight* of a monomial $x_0^{a_0} \ldots x_n^{a_n} y_0^{b_0} \ldots y_n^{a_n}$ as the integer $(\sum_{i=1}^{n} i a_i + \sum_{i=1}^{n} i b_i)$. We say that a polynomial of $\mathbf{A}_n$ is *isobaric* of weight w if each of its (non-zero) monomials is of weight w. Let $P \in \mathbf{A}_n$ be an isobaric polynomial of weight $w \geq 1$. We assume that $\partial(f) = \partial_x(f) \neq 0$. We observe that there exists an integer $N \in \mathbf{N}$, such that

$$
(4.14) \quad \partial(f)^N P \equiv \sum_{a \in \mathbf{N}^{n-1}} P_a y_2^{a_2} \ldots y_n^{a_n} \pmod{\Delta(f), \ldots, \Delta^n(f)},
$$

with $P_a \in \mathbf{A}_1$ for every tuple $a \in \mathbf{N}^{n-1}$. In formula (4.14), we deduce from the isobaricity of the $\Delta^i(f)$ that, for every tuple $a \in \mathbf{N}^{n-1}$, the polynomial P_a is homogeneous with $\deg_1(P_a) = w - \sum_{i=2}^n ia_i$. By [8], one knows that $P \in \mathcal{N}_n(\mathscr{C})$ if and only if $P_a(-\partial_y(f), \partial_x(f)) \in \langle f \rangle$ for every tuple $a \in \mathbf{N}^{n-1}$. So, Theorem 4.2 provides an interpretation in terms of differential operators of the elements in $\mathcal{N}_n(\mathscr{C})$.

5. Bernstein Operators

Let us recall Bernstein's theorem [1].

THEOREM 5.1 (Bernstein). *There exist a non-zero polynomial $b(s) \in k[s]$ and a differential operator $P(s) \in \mathscr{D}[s]$, such that the following functional equation holds:*

$$(5.2) \qquad P(s)\left(f^{s+1}\right) = b(s)f^s.$$

5.1. *The set of polynomials $b(s) \in k[s]$ for which a functional equation of the form (5.2) holds is easily seen to be an ideal of $k[s]$ and Bernstein's theorem tells us that this ideal is non-zero. Its monic generator is called the* Bernstein polynomial, *or the* b-function *of f, and it is denoted by $b_f(s)$. By definition, what we will call a* Bernstein operator *(or b-operator) for f will be any differential operator $P(s) \in \mathscr{D}[s]$, such that $P(s)\left(f^{s+1}\right) = b(s)f^s$ for certain $b(s) \in k[s]$, i.e., $P(s)f - b(s) \in \text{ann}_{\mathscr{D}[s]}(f^s)$. Let us note that $b_f(s)$ is the generator of the contraction of the left ideal $\mathscr{D}[s]f + \text{ann}_{\mathscr{D}[s]}(f^s) \subset \mathscr{D}[s]$ in $k[s]$.*

5.2. Let us prove our main application.

THEOREM 5.3. *Let k be a field of characteristic 0. Let $f \in \mathbf{A}$ be a reduced polynomial (not assumed to be irreducible). Let $P(s) = \sum_i P_i s^i \in \mathscr{D}[s]$ be a differential operator of order $\alpha \geq 0$, i.e., $\max_i\{\text{ord}(P_i)\} = \alpha$. Then, for every integer i, such that $\text{ord}(P_i) = \alpha$, we have $\sigma_\alpha(P_i)^\sharp \in \mathcal{N}_1(\mathscr{C})$ provided that one of the following hypotheses holds:*

(1) $P(s) \in \text{ann}_{\mathscr{D}[s]}(f^s)$.
(2) $\alpha \geq 2$ and $P(s)$ is a b-operator for f.

PROOF. Let J denote the set of integers i, such that $\text{ord}(P_i) = \alpha$, $P'(s) = \sum_{i \in J} P_i s^i$ and $P''(s) = P(s) - P'(s)$. We have $\sigma_\alpha(P(s)) = \sum_{i \in J} \sigma_\alpha(P_i)s^i = \sigma_\alpha(P'(s))$ and $\sigma_\alpha(P(c)) = \sigma_\alpha(P'(c)) = \sigma_\alpha(P(s))_{s=c}$ for each $c \in k$. (1) Since $\text{ord}(P''(c)) < \alpha$ for every $c \in k$, we have that, for each $\beta \geq \alpha$, $P''(\beta)(f^\beta) \in \langle f^{\beta-\alpha+1} \rangle$, and so, from

$$(5.4) \qquad 0 = P(\beta)(f^\beta) = P'(\beta)(f^\beta) + P''(\beta)(f^\beta),$$

we deduce that $P'(\beta)(f^\beta) \in \langle f^{\beta-\alpha+1} \rangle$ and we apply Corollary 4.8 to obtain that $\sigma_\alpha(P'(\beta))^\sharp = \sum_{i \in J} \sigma_\alpha(P_i)^\sharp \beta^i \in \mathcal{N}_1(\mathscr{C})$ for every $\beta \gg 0$. From there, we conclude that $\sigma_\alpha(P_i)^\sharp \in \mathcal{N}_1(\mathscr{C})$ for every integer $i \in J$. Indeed, let us consider the polynomial $R = \sum_{i \in J} \overline{\sigma_\alpha(P_i)^\sharp} T^i \in (\mathbf{A}_1/\mathcal{N}_1(\mathscr{C}))[T]$. We have

proved that this polynomial has infinitely many roots in $k \hookrightarrow \mathbf{A}_1/\mathcal{N}_1(\mathscr{C})$; hence, $R = 0$, which means that $\sigma_\alpha(P_i)^\sharp \in \mathcal{N}_1(\mathscr{C})$ for every integer $i \in J$.
(2) Assume that $P(s)(f^{s+1}) = b(s)f^s$. We proceed in a similar way to (1). For each $\beta \geq \alpha - 1$, we have

$$b(\beta - 1)f^{\beta-1} = P(\beta - 1)(f^\beta) = P'(\beta - 1)(f^\beta) + P''(\beta - 1)(f^\beta),$$

and so, $P'(\beta - 1)(f^\beta) \in \langle f^{\beta-1} \rangle + \langle f^{\beta-\alpha+1} \rangle = \langle f^{\beta-\alpha+1} \rangle$. We obtain again $\sigma_\alpha(P'(\beta - 1))^\sharp \in \mathcal{N}_1(\mathscr{C})$ for every $\beta \gg 0$, and we conclude that $\sigma_\alpha(P_i)^\sharp \in \mathcal{N}_1(\mathscr{C})$ for every integer $i \in J$. $\qquad\square$

6. Examples and Further Comments

6.1. Let (r, q) be a pair of coprime integers with $r > q \geq 2$. Let us consider $f = x^r - y^q \in \mathbf{A}$. For every integer $i \in \{0, \ldots, q\}$, we set $D_i = q^i y^{q-i} y_1^i - r^i x^{r-i} x_1^i$. The family of the polynomials in $\mathbf{A}_1$, formed by the polynomials D_i, for every integer $i \in \{0, \ldots, q\}$, and $E = qy_1 x - rx_1 y$, is denoted by $\mathfrak{B}$. By [9], we have the following result.

THEOREM 6.1. *Let k be a field of characteristic zero. Let (r, t) be a pair of coprime integers, with $r > q \geq 2$. We set $f = x^r - y^q$ and $\mathscr{C} = \mathrm{Spec}(\mathbf{A}/\langle f \rangle)$. Then, the family $\mathfrak{B}$ is a Gröbner basis of the ideal $\mathcal{N}_1(\mathscr{C})$ (for the monomial ordering $y_1 >_{\mathrm{lex}} x_1 >_{\mathrm{lex}} y_0 >_{\mathrm{lex}} x_0$ on the ring $\mathbf{A}_1$).*

In this case, since the partial derivatives of f form a regular sequence and f is quasi-homogeneous, it is well known that the annihilator of f^s over $\mathscr{D}[s]$ is generated by $\delta = \partial_y(f)\partial_x - \partial_x(f)\partial_y$ and $\chi = qx\partial_x + ry\partial_y - rqs$ (corresponding to the Euler operator). One observes that $\sigma_1(\delta)^\sharp$ coincides with E up to a scalar and that $\sigma_1(\chi)^\sharp$ coincides with D_1 up to a scalar.

• In the particular case $f = x^5 - y^3$, a b-operator for f is given by $P(s) = \sum_{i=0}^6 P_i s^i$ with

$$P_6 = -18509765625\partial_y^3,$$

$$P_5 = 569531250x\partial_x\partial_y^3 + 5695312500y\partial_y^4 - 101091796875\partial_y^3,$$

$$P_4 = 56953125x^2\partial_x^2\partial_y^3 + 759375000xy\partial_x\partial_y^4 + 12301875\partial_x^5$$
$$+ 4328437500x\partial_x\partial_y^3 + 30375000000y\partial_y^4 - 223414453125\partial_y^3,$$

$$P_3 = 75937500x^2y\partial_x^2\partial_y^4 + 16402500y\partial_x^5\partial_y + 341718750x^2\partial_x^2\partial_y^3$$
$$+ 3493125000xy\partial_x\partial_y^4 + 738112500\partial_x^5 + 12339843750x\partial_x\partial_y^3$$
$$+ 64757812500y\partial_y^4 - 252523828125\partial_y^3,$$

$$P_2 = 227812500x^2y\partial_x^2\partial_y^4 + 49207500y\partial_x^5\partial_y + 687234375x^2\partial_x^2\partial_y^3$$
$$+ \ 5906250000xy\partial_x\partial_y^4 + 148442625\partial_x^5 + 16741687500x\partial_x\partial_y^3$$
$$+ \ 68934375000y\partial_y^4 - 150418631250\partial_y^3,$$

$$P_1 = 226125000x^2y\partial_x^2\partial_y^4 + 48843000y\partial_x^5\partial_y + 574593750x^2\partial_x^2\partial_y^3$$
$$+ \ 4360500000xy\partial_x\partial_y^4 + 124112250\partial_x^5 + 10914120000x\partial_x\partial_y^3$$
$$+ \ 36611250000y\partial_y^4 - 42536025000\partial_y^3,$$

$$P_0 = 74250000x^2y\partial_x^2\partial_y^4 + 16038000y\partial_x^5\partial_y + 172062000x^2\partial_x^2\partial_y^3$$
$$+ \ 1188000000xy\partial_x\partial_y^4 + 37165392\partial_x^5 + 2752992000x\partial_x\partial_y^3$$
$$+ \ 7755000000y\partial_y^4 - 38090800000\partial_y^3.$$

In this case, we observe that the maximal order among the P_i is 6, and $\sigma_6(P) = \sigma_6(P_0) + \sigma_6(P_1)s + \sigma_6(P_2)s^2 + \sigma_6(P_3)s^3$. We check Theorem 5.3 by observing that

$$\begin{cases} \sigma_6(P_0)^\sharp = 594000yy_1^2x_1(125x^2x_1^3 - 27y_1^3) = -594000yy_1^2x_1D_3, \\ \sigma_6(P_1)^\sharp = 1809000yy_1^2x_1(125x^2x_1^3 - 27y_1^3) = -1809000yy_1^2x_1D_3, \\ \sigma_6(P_2)^\sharp = 1822500yy_1^2x_1(125x^2x_1^3 - 27y_1^3) = -1822500yy_1^2x_1D_3, \\ \sigma_6(P_3)^\sharp = 607500yy_1^2x_1(125x^2x_1^3 - 27y_1^3) = -607500yy_1^2x_1D_3. \end{cases}$$

Let us stress that the other operators $\sigma(P_4)^\sharp, \sigma(P_5)^\sharp, \sigma(P_6)^\sharp$ do not belong to $\mathcal{N}_1(\mathscr{C})$.

• From the above b-operator $P(s)$ and by using the Euler derivation $\chi = (1/5)x\partial_x + (1/3)y\partial_y$, we find another b-operator $Q = P(\chi - 1) = \sum_{i=3}^{9} Q_i$ with

$$Q_9 = -911250x^6\partial_x^6\partial_y^3 - 6075000x^5y\partial_x^5\partial_y^4 - 16453125x^4y^2\partial_x^4\partial_y^5$$
$$- \ 23625000x^3y^3\partial_x^3\partial_y^6 - 19687500x^2y^4\partial_x^2\partial_y^7 - 9375000xy^5\partial_x\partial_y^8$$
$$- \ 1953125y^6\partial_y^9 + 19683x^4\partial_x^9 + 262440x^3y\partial_x^8\partial_y + 984150x^2y^2\partial_x^7\partial_y^2$$
$$+ \ 1458000xy^3\partial_x^6\partial_y^3 + 759375y^4\partial_x^5\partial_y^4,$$

$$Q_8 = -36450000x^5\partial_x^5\partial_y^3 - 202500000x^4y\partial_x^4\partial_y^4 - 440437500x^3y^2\partial_x^3\partial_y^5$$
$$- \ 484312500x^2y^3\partial_x^2\partial_y^6 - 282187500xy^4\partial_x\partial_y^7 - 70312500y^5\partial_y^8$$
$$+ \ 708588x^3\partial_x^8 + 6692220x^2y\partial_x^7\partial_y + 16839900xy^2\partial_x^6\partial_y^2$$
$$+ \ 12757500y^3\partial_x^5\partial_y^3,$$

$$Q_7 = -521032500x^4\partial_x^4\partial_y^3 - 2317950000x^3y\partial_x^3\partial_y^4 - 3822750000x^2y^2\partial_x^2\partial_y^5$$

$$- 2911125000xy^3\partial_x\partial_y^6 - 896875000y^4\partial_y^7 + 7846956x^2\partial_x^7$$

$$+ 48668040xy\partial_x^6\partial_y + 62694000y^2\partial_x^5\partial_y^2,$$

$$Q_6 = -3318570000x^3\partial_x^3\partial_y^3 - 11114100000x^2y\partial_x^2\partial_y^4 - 12525750000xy^2\partial_x\partial_y^5$$

$$- 5053125000y^3\partial_y^6 + 31370328x\partial_x^6 + 99144000y\partial_x^5\partial_y,$$

$$Q_5 = -9573948000x^2\partial_x^2\partial_y^3 - 21607740000xy\partial_x\partial_y^4 - 12783650000y^2\partial_y^5$$

$$+ 37165392\partial_x^5,$$

$$Q_4 = -11223828000x\partial_x\partial_y^3 - 13029200000y\partial_y^4,$$

$$Q_3 = -38090800000\partial_y^3.$$

We check Theorem 5.3 by observing that $\sigma_9(Q_9)^\sharp = BE \in \mathcal{N}_1(\mathscr{C})$ with

$$B := 390625y^5x_1{}^8 - 1640625xy^4x_1{}^7y_1 + 2953125x^2y^3x_1{}^6y_1{}^2$$

$$- 2953125x^3y^2x_1{}^5y_1{}^3 + 1518750x^4yx_1{}^4y_1{}^4 - 303750x^5x_1{}^3y_1{}^5$$

$$+ 151875y^3x_1{}^3y_1{}^5 - 200475xy^2x_1{}^2y_1{}^6 + 76545x^2yx_1y_1{}^7 - 6561x^3y_1{}^8.$$

6.2. Let us consider the irreducible polynomial $f = x^4 + y^5 + y^4x \in \mathbf{A}$ with $\mathscr{C} = \text{Spec}(\mathbf{A}/\langle f \rangle)$. In this case, by using the algorithm in [8], we easily compute a Gröbner basis of the ideal $\mathcal{N}_1(\mathscr{C})$, whose first terms are given by the following polynomials f_i (let us stress that we only express here the elements of the Gröbner basis, which are used in our comparison):

$$f_1 = y^3y_1 + 1/4y^2y_1x - 3/16yy_1x^2 - 27/16yx^2x_1 - 20yxx_1 + 9/4y_1x^3$$

$$+ 25y_1x^2 + x^2x_1,$$

$$f_2 = y^2y_1^2 + 51/4yy_1xx_1 + 150yy_1x_1 + 27/4yxx_1^2 + 86yx_1^2 - 16y_1^2x^2$$

$$- 375/2y_1^2x - 9y_1x^2x_1 - 115y_1xx_1 + 1/2xx_1^2,$$

$$f_3 = y^2x_1 - 5/4yy_1x + 3/4yxx_1 - y_1x^2,$$

$$f_4 = yy_1^4 - 1726540272/103759765625yx^3x_1^4$$

$$- 56619636336/103759765625yx^2x_1^4$$

$$- 605419911792/103759765625yxx_1^4 - 670482101/33203125yx_1^4$$

$$+ 262888848/20751953125y_1^4x^4 + 20666534601/66406250000y_1^4x^3$$

$$+ 10327829/5312500y_1^4x^2 + 6/5y_1^4x + 28420416/6103515625y_1^3x^4x_1$$

$$+ \ 44571634617/415039062500 y_1^3 x^3 x_1 + 82868877/132812500 y_1^3 x^2 x_1$$

$$+ \ 13/250 y_1^3 x x_1 - 255783744/103759765625 y_1^2 x^4 x_1^2$$

$$- \ 47992863213/830078125000 y_1^2 x^3 x_1^2 - 45537923/132812500 y_1^2 x^2 x_1^2$$

$$- \ 191/6250 y_1^2 x x_1^2 + 2302053696/103759765625 y_1 x^4 x_1^3$$

$$+ \ 296865044361/415039062500 y_1 x^3 x_1^3$$

$$+ \ 622128693111/83007812500 y_1 x^2 x_1^3 + 335496263/13281250 y_1 x x_1^3$$

$$+ \ 2306050317/332031250000 x^3 x_1^4 + 403553493/2441406250 x^2 x_1^4$$

$$+ \ 330962023/332031250 x x_1^4 + 256/625 x_1^4,$$

$$f_5 = y y_1^2 x - 42 y y_1 x x_1 - 500 y y_1 x_1 - 27 y x x_1^2 - 340 y x_1^2 + 52 y_1^2 x^2$$

$$+ \ 625 y_1^2 x + 36 y_1 x^2 x_1 + 450 y_1 x x_1 + x x_1^2.$$

In this case, f is not quasi-homogeneous and the annihilator of f^s over $\mathscr{D}[s]$ cannot be generated by total order 1 operators (see [13, Theorem 4.7]). It is generated by

$$P^{(1)}(s) = (16y^2 + 12xy)\partial_y + (20xy + 16x^2)\partial_x - (80y + 64x)s,$$

$$P^{(2)}(s) = (12y^3 - 100y^2 + 5xy - 4x^2)\partial_y + (4y^3 + 16y^2 x - 125xy)\partial_x$$

$$- \ (64y^2 - 500y)s,$$

$$P^{(3)}(s) = P_2^{(3)} s^2 + P_1^{(3)} s + P_0^{(3)},$$

where $P_2^{(3)} = (1024y - 8000)$, $P_1^{(3)} = (-192y^2 + 1720y - 64x)\partial_y + (2000x - 256xy - 500y)\partial_x + 576y - 4280$, $P_0^{(3)} = (12x^2 - 24y^2 - 2xy)\partial_y^2 + (100y^2 - 16y^3 - 35xy)\partial_y \partial_x + (16y^3 + 125xy)\partial_x^2 + (827y - 108y^2 - 38x)\partial_y + (1070x - 36y^2 - 144xy + 125y)\partial_x$. We check Theorem 5.3 by observing that

$$\begin{cases} \sigma_1(P^{(1)})^\sharp = -(16y^2 + 12xy)x_1 + (20xy + 16x^2)y_1 = \dfrac{-1}{16} f_3, \\[2mm] \sigma_1(P^{(2)})^\sharp = -(12y^3 - 100y^2 + 5xy - 4x^2)x_1 \\[1mm] \qquad\qquad\quad + (4y^3 + 16y^2 x - 125xy)y_1 \\[1mm] \qquad = \qquad 4f_1 - 12yf_3 - 9xf_3 - 100f_3, \\[2mm] \sigma_2(P^{(3)})^\sharp = (12x^2 - 24y^2 - 2xy)x_1^2 - (100y^2 - 16y^3 - 35xy)x_1 y_1 \\[1mm] \qquad\qquad\quad + (16y^3 + 125xy)y_1^2 = (16x + 16y)f_2 \\[1mm] \qquad\qquad\quad + (-48xy_1 + 16yy_1 - 40x_1 - 500y_1)f_3 + (4x + 4y)f_5. \end{cases}$$

In this case, the computation of b-operators is a tricky question. In particular, the b-operator $P = \sum_{i=0}^{12} P_i s^i$ we have is very complicated and its writing too much large for this chapter. It has degree 12 in s and is of maximal order 6. Let us just mention that operators of order 6 appear in various degrees of s. For s^0, we have $P_0 = \sum_{i=0}^{6} P_{0,i}$. We set

$$
\begin{aligned}
Q := 10^{-11}P_{0,6} = (&-13789075251227814624023437500x^2 \\
&- 1723634406403476828002929 6875xy)\partial_x^5\partial_y \\
&+ (290325354442530881106445 31250x^2 \\
&+ 3630136227741031846728515 62500xy \\
&- 7281550520272593017578125 00y^2)\partial_x^4\partial_y^2 \\
&+ (-2383732328313721978170703 12500x^2 \\
&- 2923814301695528657431640 6250xy \\
&+ 2950262100035883767900390 62500y^2 \\
&- 4570873856947400932617187 50000x)\partial_x^3\partial_y^3 \\
&+ (4123201458954764144552343 7500x^2 \\
&- 1474058610373756564892109 37500xy \\
&- 1985959197835111378958203 12500y^2 \\
&+ 2011184497056856410351562 5000x \\
&- 3656699085557920746093750 0000y)\partial_x^2\partial_y^4 \\
&+ (-7462368319702775353875 000000x^2 \\
&+ 2503842790580865724811132 8125xy \\
&+ 2582068382118881354917968 7500y^2 \\
&- 3255258212031030596015625 0000x \\
&+ 1462679634223168298437500 00000y)\partial_x\partial_y^5 \\
&+ (-5596776239777081515406250 000xy \\
&- 4880585297245478836640625 000y^2 \\
&+ 2072880088806326487187500 0x \\
&- 2420274390479130122437500 0000y)\partial_y^6.
\end{aligned}
$$

We check Theorem 5.3 by observing that there exist polynomials C_1, C_2, C_3, C_4, such that $\sigma_6(Q)^\sharp = C_1 f_1 + C_2 f_2 + C_3 f_3 + C_4 f_4$ (hence, $\sigma_6(Q)^\sharp$ belongs to the ideal $\mathcal{N}_1(\mathscr{C})$). Let us stress that the expressions of C_i can be made explicit, but we prefer not to write them here because of their size. For s^7, we have $P_7 = \sum_{i=0}^{6} P_{7,i}$ and

$$R := 10^{-27} P_{7,6} = (-2729790765000x^2 - 3412238456250xy)dx^5 dy$$

$$+ (57873787110000x^2 + 72559144482750xy + 100526321250y^2)dx^4 dy^2$$

$$+ (-44994868478280x^2 - 2362298290110xy + 58779488577600y^2$$

$$- 94343902875000x)dx^3 dy^3 + (7143546701952x^2 - 28988982742014xy$$

$$- 36798270196500y^2 + 41511317265000x - 75475122300000y)dx^2 dy^4$$

$$+ (-1956315170016x^2 + 3618713208672xy + 3955477536000y^2$$

$$- 6732380909160x + 30190048920000y)dxdy^5 + (-1467236377512xy$$

$$- 1426479811470y^2 + 54342088056x - 4951168022880y)dy^6.$$

We check Theorem 5.3 by observing that there exist polynomials C_1', C_2', C_3', C_4', such that $\sigma_6(R)^\sharp = C_1' f_1 + C_2' f_2 + C_3' f_3 + C_4' f_4$ (hence, $\sigma_6(R)^\sharp$ belongs to the ideal $\mathcal{N}_1(\mathscr{C})$). Let us stress that, once again, the expressions of the C_i' can be made explicit, but we prefer not to write them here because of their size.

6.3. Let us finally consider the case of a reduced polynomial $f = xy(x+y)$ with associated affine plane curve $\mathscr{C}$. In this case, the annihilator of f^s over $\mathscr{D}[s]$ is generated by the differential operators $P^{(1)} = x\partial_x + y\partial_y - 3s$ and $P^{(2)} = x^2\partial_x + 2xy\partial_x - 2xy\partial_y - y^2\partial_y$. We check Theorem 5.3 by observing that

$$\begin{cases} \sigma_1(P^{(1)})^\sharp = xy_1 - yx_1 \\ \qquad = y(x_1 + y_1) - y_1(x + y), \\ \sigma_1(P^{(2)})^\sharp = x^2 y_1 + 2xy(y_1 + x_1) + y^2 x_1 \\ \qquad = y_1(x^2 - y^2) + 2xy(x_1 + y_1) + y^2(x_1 + y_1) \end{cases}$$

belong to $\mathcal{N}_1(\mathscr{C})$. In this case, we have $b_f(s) = (s+1)^2(3s+2)(3s+4)$ and a $b_f(s)$-operator is (up to scalar)

$$(6.1) \qquad P = 12y\partial_x^3\partial_y - 24y\partial_x^2\partial_y^2 + 24y\partial_x\partial_y^3 - 12y\partial_y^4 - 24\partial_x^3 + 36\partial_x^2\partial_y$$

$$(6.2) \qquad + 36\partial_x\partial_y^2 - 24\partial_y^3 + (12\partial_y^3 - 24\partial_x^3 + 54\partial_x^2\partial_y - 18\partial_x\partial_y^2)s.$$

We check Theorem 5.3 by observing that

$$\sigma_4(P)^\sharp = -12yx_1(y_1 + x_1)(y_1^2 - x_1y_1 + x_1^2) + 24yy_1x_1^2(y_1 + x_1)$$

clearly belong to $\mathcal{N}_1(\mathscr{C})$ which is, in this case, generated by $x_1 y_0 - y_1 x_0$ and $\langle x_0, x_1 \rangle \cdot \langle y_0, y_1 \rangle \cdot \langle x_0 + y_0, x_1 + y_1 \rangle$ (see [9]).

Acknowledgment

The second author is partially supported by MTM2016-75027-P, P12-FQM-2696 and FEDER.

Bibliography

[1] I. N. Bernšteĭn, Analytic continuation of generalized functions with respect to a parameter. *Funkcional. Anal. i Priložen.*, 6(4):26–40, 1972.

[2] D. Bourqui and J. Sebag, Arc schemes of affine algebraic plane curves and torsion Kähler differential forms. Chapter 6 in this volume.

[3] A. Chambert-Loir, J. Nicaise and J. Sebag, *Motivic Integration*. Progress in Mathematics, Vol. 325. Birhäuser, 2018.

[4] L. Ein and M. Mustaţă, Jet schemes and singularities. In *Algebraic Geometry* (Seattle, 2005. Part 2), Proc. Sympos. Pure Math., Vol. 80. Amer. Math. Soc., Providence, RI, 2009, pp. 505–546.

[5] A. Grothendieck (Rédigés avec la collaboration de J. Dieudonné), Éléments de géométrie algébrique. IV. Étude locale des schémas et des morphismes de schémas, Quatrième partie. *Inst. Hautes Études Sci. Publ. Math.*, (32):361, 1967.

[6] S. Ishii, Smoothness and jet schemes. In *Singularities* (Niigata–Toyama, 2007), Adv. Stud. Pure Math., Vol. 56 Math. Soc. Japan, Tokyo, 2009, pp. 187–199.

[7] E. R. Kolchin. *Differential Algebra and Algebraic Groups*. Pure and Applied Mathematics, Vol. 54, Academic Press, New York, 1973.

[8] K. Kpognon and J. Sebag, Nilpotence in arc schemes of plane curves. Preprint, submitted.

[9] K. Kpognon and J. Sebag, Nilpotency in tangent space of quasi-homogeneous plane curve singularities. Preprint, submitted.

[10] A. Leykin and H. Tsai, D-module package for Macaulay 2. Available at https://people.math.gatech.edu/~aleykin3/Dmodules.

[11] H. Matsumura, *Commutative Ring Theory*. Cambridge Studies in Advanced Mathematics, Vol. 8, Cambridge University Press, Cambridge, 1986.

[12] M. Mustaţă, Jet schemes of locally complete intersection canonical singularities. *Invent. Math.*, 145(3):397–424, 2001. With an appendix by David Eisenbud and Edward Frenkel.

[13] L. Narváez Macarro, A duality approach to the symmetry of Bernstein-Sato polynomials of free divisors. *Adv. Math.*, 281:1242–1273, 2015.

[14] J. Nicaise and J. Sebag, Greenberg approximation and the geometry of arc spaces. *Comm. Algebra*, 38(11):4077–4096, 2010.

[15] J. Sebag, On logarithmic differential operators and equations in the plane. *Illinois J. Math.*, 62(1–4):215–224, 2018.

[16] J. Sebag, Intégration motivique sur les schémas formels. *Bull. Soc. Math. France*, 132(1):1–54, 2004.

[17] J. Sebag, Arcs schemes, derivations and Lipman's theorem. *J. Algebra*, 347:173–183, 2011.

[18] J. Sebag, A remark on Berger's conjecture, Kolchin's theorem, and arc schemes. *Arch. Math. (Basel)*, 108(2):145–150, 2017.

Index

CPSIA information can be obtained
at www.ICGtesting.com
Printed in the USA
BVHW072233180320
575395BV00005B/13

9 781786 347190